T0295175

Water and Wastewater Treatment: Advanced Oxidation Processes

Water and Wastewater Treatment: Advanced Oxidation Processes

Editor: Leila Thomson

www.callistoreference.com

Callisto Reference,
118-35 Queens Blvd., Suite 400,
Forest Hills, NY 11375, USA

Visit us on the World Wide Web at:
www.callistoreference.com

© Callisto Reference, 2023

This book contains information obtained from authentic and highly regarded sources. Copyright for all individual chapters remain with the respective authors as indicated. All chapters are published with permission under the Creative Commons Attribution License or equivalent. A wide variety of references are listed. Permission and sources are indicated; for detailed attributions, please refer to the permissions page and list of contributors. Reasonable efforts have been made to publish reliable data and information, but the authors, editors and publisher cannot assume any responsibility for the validity of all materials or the consequences of their use.

ISBN: 978-1-64116-825-0 (Hardback)

Trademark Notice: Registered trademark of products or corporate names are used only for explanation and identification without intent to infringe.

Cataloging-in-Publication Data

Water and wastewater treatment : advanced oxidation processes / edited by Leila Thomson.
   p. cm.
Includes bibliographical references and index.
ISBN 978-1-64116-825-0
1. Water--Purification. 2. Sewage--Purification. 3. Water--Purificatio--Oxidation.
4. Sewage--Purificatio--Oxidation. I. Thomson, Leila.
TD430 .W38 2023
628.162--dc23

Table of Contents

Preface

Water and wastewater treatment involves the use of one or more physical, chemical and biological processes or their combination for removing solids and organic matter from the wastewater. Advanced oxidation processes (AOPs) are a set of chemical treatment procedures designed for effectively removing organic pollutants from water and wastewater by the process of oxidation. This emerging chemical technology is based on the in-situ generation of hydroxyl radicals, which are used as strong oxidants that can be used to oxidize a wide range of chemical compounds. In some AOPs, sulphate radicals, ozone, catalyst, or ultraviolet irradiation is also used to achieve more efficient treatment. AOP is beneficial for cleaning biologically toxic or non-degradable materials such as aromatics, pesticides, petroleum constituents, and volatile organic compounds dissolved in the wastewater. A major advantage of using AOPs for wastewater treatment is that these processes do not introduce any new hazardous substances into the water since the complete reduction product of hydroxyl radicals is water. One drawback of AOPs that limit its large-scale application and industrial usage relates to high costs. This book unravels the recent studies on water and wastewater treatment using the advanced oxidation process. Researchers and students in this field will be greatly assisted by it.

Various studies have approached the subject by analyzing it with a single perspective, but the present book provides diverse methodologies and techniques to address this field. This book contains theories and applications needed for understanding the subject from different perspectives. The aim is to keep the readers informed about the progresses in the field; therefore, the contributions were carefully examined to compile novel researches by specialists from across the globe.

Indeed, the job of the editor is the most crucial and challenging in compiling all chapters into a single book. In the end, I would extend my sincere thanks to the chapter authors for their profound work. I am also thankful for the support provided by my family and colleagues during the compilation of this book.

Editor

Degradation of Losartan in Fresh Urine by Sonochemical and Photochemical Advanced Oxidation Processes

John F. Guateque-Londoño [1,2], Efraím A. Serna-Galvis [1], Yenny Ávila-Torres [3,*]
and Ricardo A. Torres-Palma [1,*]

[1] Grupo de Investigación en Remediación Ambiental y Biocatálisis (GIRAB), Instituto de Química,
Facultad de Ciencias Exactas y Naturales, Universidad de Antioquia UdeA, Calle 70 No. 52-21, Calle Nueva,
050014 Medellín, Colombia; jfguateque@utp.edu.co (J.F.G.-L.); efrain.serna@udea.edu.co (E.A.S.-G.)
[2] Maestría en Ciencias Químicas, Facultad de Tecnología, Universidad Tecnológica de Pereira,
660001 Pereira, Colombia
[3] Grupo de Investigación QUIBIO, Facultad de Ciencias Básicas, Universidad Santiago de Cali,
Santiago de Cali, 760035 Pampalinda, Colombia
* Correspondence: yennytorres@usc.edu.co (Y.Á.-T.); ricardo.torres@udea.edu.co (R.A.T.-P.)

Abstract: In this work, the degradation of the pharmaceutical losartan, in simulated fresh urine (which was considered because urine is the main excretion route for this compound) by sonochemistry and UVC/H_2O_2 individually, was studied. Initially, special attention was paid to the degrading action of the processes. Then, theoretical analyses on Fukui function indices, to determine electron-rich regions on the pharmaceutical susceptible to attacks by the hydroxyl radical, were performed. Afterward, the ability of the processes to mineralize losartan and remove the phyto-toxicity was tested. It was found that in the sonochemical treatment, hydroxyl radicals played the main degrading role. In turn, in UVC/H_2O_2, both the light and hydroxyl radical eliminated the target contaminant. The sonochemical system showed the lowest interference for the elimination of losartan in the fresh urine. It was established that atoms in the imidazole of the contaminant were the moieties most prone to primary transformations by radicals. This was coincident with the initial degradation products coming from the processes action. Although both processes exhibited low mineralizing ability toward losartan, the sonochemical treatment converted losartan into nonphytotoxic products. This research presents relevant results on the elimination of a representative pharmaceutical in fresh urine by two advanced oxidation processes.

Keywords: advanced oxidation process; elimination routes; fresh urine; pharmaceutical degradation; processes selectivity; theoretical analysis

1. Introduction

Losartan was the first commercialized angiotensin II antagonist pharmaceutical. This is an antihypertensive consumed widely around the world [1]. Urine is the main route of excretion of losartan from the human body, ≈35% of the oral dose is expelled without alterations [2], reaching the wastewater systems. In fact, losartan has been determined in ranges of 0.0197–2.76 μg L^{-1} in wastewater treatment plants influent (WWTP) [3,4]. This indicates that losartan is not effectively removed by the conventional systems in WWTP.

In the aquatic environment, losartan can promote noxious effects on organisms, and it can be transformed into more toxic and persistent substances [5–7]. The recalcitrance to conventional treatment systems, negative environmental impact, and high excretion of losartan in urine lead to consider alternative options to eliminate this pharmaceutical from aqueous media. Particularly, the application of degradation processes should be focused on primary contamination sources, such as human fresh urine.

Advanced oxidation processes (AOPs, which are based on the production and utilization of radical species to attack pollutants) are interesting options for losartan elimination from urine, to avoid entering into the wastewater systems. Indeed, AOPs such as UVC/H_2O_2 and sonochemistry have been successfully applied for the elimination of different pharmaceuticals in diverse aqueous matrices [8].

In the UVC/H_2O_2 process, UVC light (e.g., photons of 254 nm) promotes the homolysis of hydrogen peroxide, generating hydroxyl radicals (Equation (1)) available to degrade organic contaminants (Equation (2)) [9].

$$H_2O_2 + hv_{(<290 \text{ nm})} \rightarrow 2HO^{\bullet} \tag{1}$$

$$HO^{\bullet} + \text{Pollutant} \rightarrow \text{degradation products} \tag{2}$$

Meanwhile, the sonochemical process, which uses high-frequency ultrasound waves ")))", produces hydroxyl radicals from the breaking of water molecules and dissolved oxygen (Equations (3)–(5)) [10].

$$H_2O +))) \rightarrow HO^{\bullet} + H^{\bullet} \tag{3}$$

$$O_2 +))) \rightarrow 2O^{\bullet} \tag{4}$$

$$H_2O + O^{\bullet}))) \rightarrow 2HO^{\bullet} \tag{5}$$

It should be mentioned that some previous works have evidenced the high potentiality of AOPs to eliminate pollutants in urine [11–20]. However, until now, the treatment of losartan in fresh urine, considering the intrinsic degradation abilities of UVC/H_2O_2 and sonochemistry has not been reported. Moreover, computational analyses about the reactivity of this pharmaceutical toward hydroxyl radical species or phytotoxicity tests of the treated water have not been considered. Thereby, the present research was focused on the losartan treatment in fresh urine by UVC/H_2O_2 and ultrasound individually. The selectivity of the processes toward the pollutant degradation in the urine matrix was established. Firstly, special attention was paid to the action routes of the processes involved in the elimination of losartan. Besides, computational analyses using DFT/Fukui functionals were performed to determine the most regions on losartan reactive to hydroxyl radicals, and these theoretical results were related to primary degradation products coming from the processes action. Additionally, considering the possible reuse of treated urine for water irrigation extra analyses such as mineralization and phytotoxicity were carried out.

2. Materials and Methods

2.1. Reagents

Losartan tablets (50 mg each) were purchased from La Santé S.A. Acetonitrile (HPLC grade), ammonium heptamolybdate (>99.3%), methanol (HPLC grade), potassium iodide (>99.5%), potassium perchlorate (>99.5%), sodium acetate (>99%), sodium chloride (99.9%), sodium dihydrogen phosphate (>99.0%), sodium hydroxide (>99.0%), sodium sulfate (>99.0%), sulfuric acid (95–97%), and urea (>99.0%) were provided by Merck. Ammonium chloride (>99.8%), calcium chloride dihydrate (>99.0%), ferrous sulfate heptahydrate (>99.0%), formic acid (99.0%), hydrogen peroxide (30% w/v), and magnesium chloride hexahydrate (>99.0%) were provided by PanReac. All the reagents were used as received.

The solutions were prepared using distilled water. In all cases, the initial losartan concentration was 43.38 μM (i.e., 20 mg L^{-1}, which is a plausible amount of the antihypertensive excreted in human urine [21]). The fresh urine used for the tests was prepared according to Table 1. The fresh urine was used immediately after its preparation and the pH was adjusted to 6.1.

Table 1. Composition of fresh urine [1].

Compound	Concentration (M)
Urea	0.2664
$NaCH_3COO$	0.1250
Na_2SO_4	0.01619
NH_4Cl	0.03365
NaH_2PO_4	0.02417
KCl	0.05634
$MgCl_2$	0.003886
$CaCl_2$	0.004595
NaOH	0.00300
pH: 6.1	

[1] Composition taken from Amstutz et al. [22].

2.2. Reaction Systems

For the UVC/H_2O_2 process, a homemade aluminum reflective reactor box equipped with UVC lamps (OSRAM HNS®, with the main emission peak at 254 nm, 60 W) was used (Figure 1a). Losartan solutions (50 mL) were placed in beakers under constant stirring. Meanwhile, the sonochemical treatments were performed in a Meinhardt cylindrical glass reactor containing 250 mL of losartan solution. Ultrasonic waves of 375 kHz and 106.3 W L^{-1} (actual ultrasound power density determined by the calorimetric method) were emitted from a transducer at the bottom of the reactor (Figure 1b). For both processes, the experimental conditions (i.e., reagents concentrations, ultrasonic frequency, light power) were selected based on previous works [23,24].

2.3. Analyses

2.3.1. Chromatographic Analyses

Losartan evolution was followed by using a UHPLC Thermo Scientific Dionex UltiMate 3000 instrument equipped with an Acclaim™ 120 RP C18 column (5 μm, 4.6 × 150 mm) and a diode array detector (operated at 230 and 254 nm). The mobile phase was methanol (10% v/v), acetonitrile (44% v/v), and formic acid (46% v/v, 10 mM, and pH 3.0) at a flow of 0.6 mL min^{-1}. Primary transformation products were elucidated by HPLC–MS analyses in our previous work [24]. For the chromatographic analyses, samples of 0.5 mL were periodically taken from the reaction systems (the total taken volume was always lower than 10% of the initial volume in each system). All experiments were performed at least in duplicate.

2.3.2. Oxidizing Species Accumulation

Accumulation of sonogenerated hydrogen peroxide was estimated by iodometry [25]. An aliquot of 600 μL from the reactors was added to a quartz cell containing 1350 μL of potassium iodide (0.1 M) and 50 μL of ammonium heptamolybdate (0.01 M). After 5 min, the absorbance at 350 nm was measured using a Mettler Toledo UV5 spectrophotometer.

2.3.3. Mineralization Determinations

Mineralization degree was established as removal of total organic carbon (TOC). TOC content of the samples was measured using a Shimadzu LCSH TOC analyzer (previously calibrated), according to Standard Methods 5310B (high-temperature combustion method), in which the water sample is homogenized and injected into a heated reaction chamber packed with an oxidative catalyst (platinum spheres). The water is vaporized, and the organic carbon is oxidized to CO_2. The CO_2 from oxidation is transported by a carrier gas stream and is then measured using an IR detector. The TOC analyzer performed the catalytic combustion at 680 °C using high-purity oxygen gas at a flow rate of 190 mL min^{-1}. The apparatus had a nondispersive infrared detector. For the TOC analyses, samples of

7.0 mL were taken from the reaction systems, and for the TOC analyses, the experiments were carried out independently from the initial tests of degradation (to avoid retire amounts higher than 10% of the initial volume in each system).

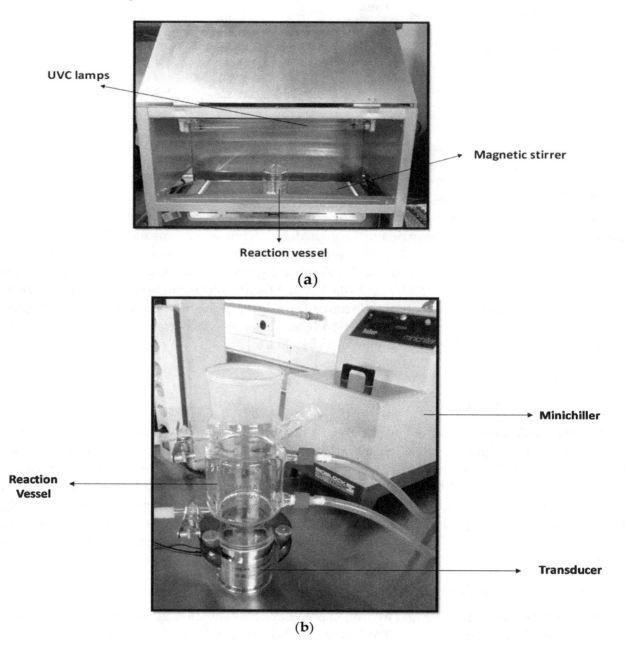

Figure 1. Reactors used in the degradation of losartan. (**a**) UVC/H$_2$O$_2$ process; (**b**) sonochemical treatment.

2.3.4. Phytotoxicity Tests

Toxicity against radish seeds (*Raphanus sativus*) was considered. For such purpose, the ratio of seeds germinated (RSG, Equation (6)) and the ratio of root length (RRG, Equation (7)) were determined. As a phytotoxicity parameter, the germination index (GI, Equation (8)) was assessed according to N.J. Hoekstra et al. [26]. For the phytotoxicity tests, samples of 5.0 mL were taken.

$$\text{RSG (\%)} = \frac{\text{Number of seeds germinated in sample}}{\text{Number of seeds germinated in control}} \times 100 \tag{6}$$

$$\text{RRG (\%)} = \frac{\text{mean root length in sample}}{\text{mean root length in control}} \times 100 \tag{7}$$

$$GI(\%) = \frac{RSG \times RRG}{100} \qquad (8)$$

2.3.5. Computational Analyses

For the determination of regions on losartan most susceptible to the attack of radical species and electrophilic oxidants, computational analyses were performed by applying the framework of functional density theory (DFT). The antihypertensive structure was optimized with the B3LYP hybrid functional density [27], 6-311++G(2d,2p) method [28] using the dielectric constant for water to simulate the aqueous environment. Thus, f^+ and f^- (i.e., nucleophilic and electrophilic Fukui function indices) values and the average between such values (f_{ave}) were calculated.

3. Results

3.1. Treatment of Fresh Urine Loaded with Losartan

The two processes were individually applied to degrade losartan in the simulated fresh urine (FU, whose composition is presented in Table 1). In addition to degradation in urine, losartan was also treated in distilled water (DW). The degradations followed pseudo-first-order kinetics, and their respective rate constants (k) in both matrices were established (see Figure S1 in Supplementary material). Then, the ratio between the degradation rate constants ($Rk:k_{FU}/k_{DW}$) was calculated. This Rk parameter is an indicator of both the selectivity of processes toward the antihypertensive degradation in the complex matrix and the inhibitory effect of losartan elimination caused by the fresh urine components. Table 2 contains the k and Rk values for each process.

Table 2. Kinetic constants (in min^{-1}) determined in the degradation of losartan in fresh urine (k_{FU}) and distilled water (k_{DW}) for each advanced oxidation processes [1].

AOP	k_{DW} (R^2)	k_{FU} (R^2)	$Rk = k_{FU}/k_{DW}$
Sonochemistry	0.0549 (0.9972)	0.0437 (0.9975)	0.796
UVC/H_2O_2	0.0532 (0.9987)	0.0245 (0.9981)	0.461

[1] Experimental conditions: [Pollutant] = 43.38 µM, pH: 6.1. Sonochemistry: 106.3 W L^{-1} (375 kHz). UVC/H_2O_2: [H_2O_2] = 500 µM, 60 W.

3.2. Degradation Routes of Losartan (LOS) in Different AOPs

To elucidate the routes of the processes action, some specific experiments and measures in distilled water were carried out and results were compared to those obtained in the urine to understand the effect of the matrix components. Results for each treatment in distilled water are detailed in the following subsections.

3.2.1. Action Routes of the UVC/H_2O_2 Process

The UVC/H_2O_2 process may include the action of light of 254 nm, hydrogen peroxide, and radicals. To identify the routes involved in the process, control tests for the individual effects of UVC and H_2O_2 were carried out. Figure 2 compares the degrading effect of individual components of the process in distilled water, plus the losartan elimination in both distilled water and fresh urine (FU) by UVC/H_2O_2.

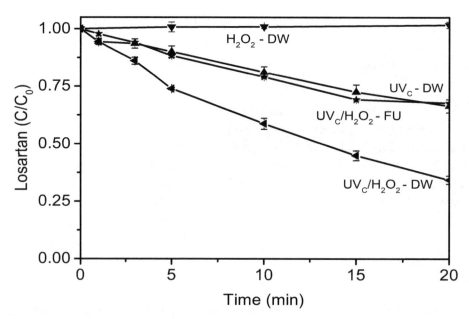

Figure 2. Determination of action routes of the UVC/H₂O₂ process on losartan degradation in distilled water (DW) and fresh urine (FU). Conditions: [LOS] = 43.38 μM, [H₂O₂] = 500 μM, UVC light: 60 W, pH: 6.1.

3.2.2. Degradation Routes Involved in the Sonochemical Treatment

Figure 3 depicts the degradation of LOS in both distilled water and fresh urine (FU) by ultrasound. To determine the degradation route in the sonochemical process, the accumulation of sonogenerated hydrogen peroxide in the presence and absence of the pollutant was also measured (results also presented in Figure 3).

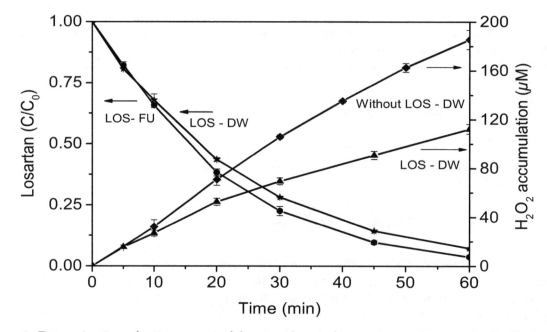

Figure 3. Determination of action routes of the sonochemical treatment on losartan in distilled water (DW) and fresh urine (FU). Conditions: [LOS] = 43.38 μM, 106.3 W L⁻¹ (375 kHz), pH = 6.1.

3.3. *Analysis of Losartan Susceptibility to Attacks by Radical Species*

To establish electron-rich regions on losartan susceptible to attacks of radicals, computational analyses were performed [29–31], and the results from the theoretical calculations were used to better understand the formation of the degradation intermediaries. Table 3 depicts the moieties on losartan having more electron density according to Fukui function indices. In addition to these indices, other related quantities such as local softness and global hardness were determined, the values of which were 17.513 and 0.0571 eV, respectively. Moreover, a donor–acceptor diagram (DAM), to show the donor capability of the pharmaceutical concerning hydroxyl radical (HO·), hydroperoxyl radical (HOO·), and superoxide anion radical ($\cdot O_2^-$), was elaborated (Figure S2).

Table 3. Results of computational analysis for losartan [1].

Structure and Numeration	Atoms	Fukui Function Indices		
		f^-	f^+	f ave
	1 C	0.045	0.054	0.049
	2 C	−0.027	0.005	−0.011
	3 C	0.066	−0.022	0.022
	4 C	0.006	0.004	0.005
	5 C	0.055	−0.015	0.020
	6 C	0.004	−0.007	−0.002
	7 C	−0.103	−0.012	−0.058
	8 C	−0.100	−0.031	−0.066
	10 C	−0.112	0.160	0.024
	11 C	−0.094	−0.088	−0.091
	12 C	0.160	−0.104	0.028
	13 C	−0.050	−0.045	−0.048
	14 C	−0.660	0.117	−0.272
	15 C	0.402	0.126	0.264
	16 C	0.007	0.009	0.008
	17 C	0.000	−0.044	−0.022
	18 C	1.841	1.119	1.480
	19 C	−0.661	−0.739	−0.700
	21 C	0.216	−0.116	0.050
	22 C	0.008	−0.141	−0.067
	1 N	0.053	−0.005	0.024
	2 N	−0.043	0.003	−0.020
	3 N	0.016	0.005	0.011
	4 N	0.022	0.000	0.011
	5 N	−0.258	0.044	−0.107
	6 N	−0.066	0.179	0.057
	Cl	0.061	0.058	0.060
	O	0.061	0.000	0.031

[1] Boxes in gray color contains atoms having high values for the Fukui function indices. It should be mentioned that the computational calculations were done for LOS in water.

3.4. Mineralization and Toxicity Evolution in Distilled Water

The ability of the two processes to mineralize losartan was analyzed. The experiments were carried out in distilled water to avoid interfering effects of matrix and understand the fundamental aspects of the mineralizing action of the processes. We can mention that if mineralization is carried out in the fresh urine matrix, the urea that has a higher concentration masks the contribution of losartan, making it difficult to evaluate the mineralization of the contaminant under the oxidation processes. The TOC removal, at different treatment times normalized concerning the time necessary to completely degrade losartan in distilled water, was evaluated. Two different treatment times were considered: T (when losartan is 100% degraded) and 2T (the double of time required to 100% remove the antihypertensive). Results for mineralization are presented in Figure 4A.

On the other hand, toxicity modifications exerted by treatment with ultrasound and UVC/H_2O_2 to the distilled water loaded with losartan were tested. Radish seeds (*Raphanus sativus*) were used as probe organisms. The growth index (GI) was used as the toxicity measure (phytotoxicity). Phytotoxicity was established at 2T of treatment for both processes (Figure 4B).

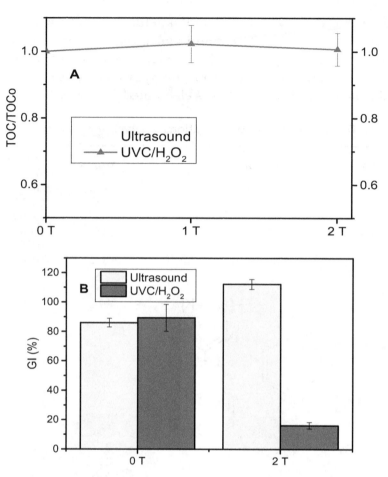

Figure 4. Extent of advanced oxidation treatments in distilled water. (**A**) Mineralization of losartan during the application of different processes; (**B**) evolution of the toxicity of losartan treated solutions against radish seeds. Note: the time was normalized concerning the time necessary to completely degrade losartan. Then, T is the time when losartan is 100% degraded, and 2T means the double of time required to 100% remove the antihypertensive. Experimental conditions as described in Figures 2 and 3.

4. Discussion

4.1. Treatment of Fresh Urine Loaded with Losartan

The Rk values for the ultrasound and UVC/H_2O_2 were 0.79 and 0.46, respectively (Table 2). It can be noted that ultrasound had the highest value for Rk; indicating that the losartan degradation through such process is affected at a low extent (21%) by urine matrix components. Meanwhile, for UVC/H_2O_2, the urine matrix presented a moderate inhibition (54%) of the antihypertensive elimination. These results suggest that the matrix components decreased the efficiency of the processes, which can be related to modifications of degradation routes. The explanations are presented in detail in the next subsections.

4.2. Degradation Routes of Losartan (LOS) in the Different AOPs

4.2.1. UVC/H_2O_2 Process

After the application of the individual components of the UVC/H_2O_2 system to LOS, it was found that hydrogen peroxide (even at 500 µM) did not induce significant removal of losartan (less than 5% elimination after 20 min of treatment). On the contrary, the treatment with the UVC light degraded ≈33.5% of the antihypertensive at 20 min of irradiation. The ultraviolet spectrum of losartan shows light absorption at 254 nm (Figure S3), which suggests that this molecule can be transformed by the UVC light. This is corroborated with the relative high photodegradation coefficient for losartan at UVC light (Cp, 123–190 L Einstein^{-1} cm^{-1} [32]). In fact, organic compounds having Cp values higher than 40 L Einstein^{-1} cm^{-1} can experience direct photolysis [33], which is currently related to the presence of aromatics rings, π-conjugated systems, and heteroatoms [34], as contained in the losartan structure (e.g., biphenyl, imidazole, and tetrazole). These aspects explain the losartan degradation by the UVC light. When losartan was treated by the complete UVC/H_2O_2 system, 65.7% of removal after 20 min was observed (Figure 2). The significant improvement of losartan elimination with the combination of hydrogen peroxide and UVC suggests the participation of radical species in the pollutant degradation. Indeed, as indicated earlier, the UVC/H_2O_2 process generates hydroxyl radical by homolytic rupture of peroxide by UVC light (Equation (1)). Hence, it can be indicated that in this process, the main action routes are the UVC photolysis and the attacks of hydroxyl radicals.

4.2.2. Ultrasound Process

The sonochemical system has three reaction zones: the inner part of cavitation bubbles, where volatile molecules are pyrolyzed by high temperatures and pressures [35–37]; the interfacial region, where hydrophobic substances can react with the sonogenerated hydroxyl radical [38]; the solution bulk, where a small number of hydroxyl radical can react with hydrophilic compounds [39].

When losartan in distilled water was treated by high-frequency ultrasound (375 kHz and 106.3 W L^{-1}), this process led to 97% of pollutant concentration reduction after 60 min of treatment (Figure 3). Since losartan is a nonvolatile compound, degradation by pyrolysis is negligible. Thus, the antihypertensive elimination would be associated with the attack of hydroxyl radicals. On the other hand, it is well-known that during the sonochemical process, hydrogen peroxide is formed by the combination of hydroxyl radicals (Equation (9)). Indeed, H_2O_2 production is an indicator of pollutant interaction with sonochemically formed HO$^\bullet$ [40]. Thereby, to prove the participation of sonogenerated HO$^\bullet$ in the pollutant degradation, the accumulation of H_2O_2 was determined. Figure 3 shows that the accumulations of H_2O_2 after 60 min of sonication in the absence and the presence of losartan were ≈180 and ≈110 μM, respectively.

$$HO^\bullet + HO^\bullet \rightarrow H_2O_2 \tag{9}$$

The oxidation of losartan by the accumulated hydrogen peroxide was discarded because pollutant removal by H_2O_2 even at 500 μM was not observed (Figure 2). Then, the lower accumulation of hydrogen peroxide in the presence of losartan is an indicator of the reaction between the HO$^\bullet$ and losartan. Moreover, due to the hydrophobic character of losartan (denoted by its high Log K_{OW} value, which is >4.0) [24]), its degradation is expected to occur in the interfacial zone of the system [41] by the sonogenerated hydroxyl radicals.

4.2.3. Understanding the Interference of Urine Matrix

Based on the degradation routes previously established, the interference of the urine matrix on the pharmaceutical degradation by the considered processes can be rationalized. During the application of UVC/H_2O_2 (which has both radical and photolytic routes), the antihypertensive removal was inhibited (by ≈55%) by the urine matrix components (see Rk value in Table 2). This was related to two aspects: the shielding of UVC light and scavenging of hydroxyl radicals. The shielding effect of the urine matrix was demonstrated through the evaluation of the only action of UVC light on losartan in both matrices (i.e., urine and distilled water), which showed a Rk value of 0.8 (Figure S4).

In turn, it is recognized that the inorganic anions such as chloride or bicarbonate, and organic substances like urea and acetate, present in the fresh urine, have relatively high rate constants with hydroxyl radicals (see Table 4), and as a result, they also affect the losartan degradation. It can be remarked the significant contribution of UVC photolysis to the degradation of losartan, as well as to the relative low interference of urine components for the light absorption (see Rk value for photolysis in Figure S4). Considering these findings, a scheme of losartan degradation by UVC/H_2O_2 was proposed in Figure 5a. It can be mentioned that the action of the photogenerated hydroxyl radicals induces transformations to losartan (such topic discussed below in Section 4.3), which is also schematized in this figure.

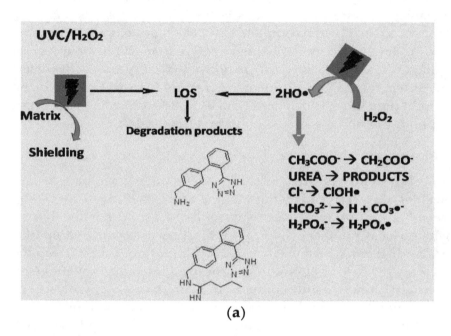

(a)

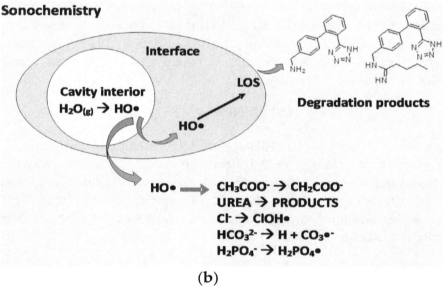

(b)

Figure 5. Scheme of degradation routes and interfering action of the urine components on the tested processes and generation of primary degradation products. (a) UVC/H_2O_2; (b) sonochemical treatment. Note: black arrows mean degradation routes and red arrows represent interfering action of the urine components.

Table 4. Rate constants of the reactions between hydroxyl radical and the diverse components of fresh urine.

Reaction	Second-Order Rate Constant (k^{2nd}, M^{-1} s^{-1})	References
$HO^{\bullet} + Cl^{-} \rightarrow ClOH^{\bullet-}$	4.3×10^9	[42]
$HO^{\bullet} + H_2PO_4^{-} \rightarrow HO^{-} + H_2PO_4^{\bullet}$	$\approx 2 \times 10^4$	[43]
$HO^{\bullet} + CH_3COO^{-} \rightarrow H_2O + CH_2COO^{\bullet-}$	7.0×10^7	[44]
$HO^{\bullet} + H_2NCONH_2 \rightarrow$ products	7.9×10^5	[43]
$HO^{\bullet} + HCO_3^{-} \rightarrow CO_3^{\bullet-} + H_2O$	8.5×10^6	[45]

In the case of the sonochemical process, for the rationalization of the low inhibitory effect of the urine matrix for the degradation of losartan (Figure 3), we must consider both the degradation route of losartan and the hydrophobic/hydrophilic nature of the substances in the matrix. The urine components are very hydrophilic, as evidenced by their Log Kow values (which are close to zero or negative, see Table S1). Thus, such components are mainly placed in the bulk of the solution and losartan is in the interfacial zone (where there is a high concentration of the sonogenerated HO$^\bullet$). Consequently, this pharmaceutical is slightly affected by the ions and/or organic compounds of the urine matrix (as schematized in Figure 5b). It should be indicated that the action of the sonogenerated hydroxyl radicals modifies the structure of losartan (such topic discussed below in Section 4.3), which is also schematized in this figure.

4.3. Analysis of Losartan Susceptibility to Attacks by Radical Species

The values of the hardness and softness for losartan indicate its high donor capacity. This is advantageous for attacks of the radicals to the pharmaceutical. Such behavior was also observed in the DAM (Figure S2), which shows that the losartan molecule has a better donor capacity concerning hydroxyl radical, hydroperoxyl radical, and superoxide anion radical. Besides, the computational analyses revealed that atoms on the imidazole moiety (15C, 18C, and 6N), aromatic rings (3C, 5C, 10C, and 12C), tetrazole (1C, 1N, 3N, and 4N), alcohol (O), and alkyl chain (21C) on losartan have the highest values for f_{ave} (this suggests that such regions on losartan are the most susceptible to transformations by radicals such as HO$^\bullet$). Indeed, we can mention that the atom with the highest Fukui function indices is more reactive to hydroxyl radical (the main degrading radical species in the tested AOPs). In the case of losartan, its C18 atom presents a f_{ave} of 1.480, the highest value concerning all the atoms in the entire molecule. This behavior can be associated with the stabilization by resonance among the imidazole ring for the radical generated (Figure S5). In contrast, the attack of hydroxyl radical on the C1 atom in the tetrazole ring for the hydroxyl radical does not lead to such stabilization (Figure S6). In fact, the Fukui function indices for the tetrazole system are smaller than for the imidazole ring. Additionally, in a previous work from our research team, it was reported that for losartan molecule, the HOMO is located in the imidazole ring, whereas LUMO is on the tetrazole ring [46].

The primary products of losartan degradation in distilled water present a good agreement to the computational analysis on reactive regions of losartan (see Table S2 and Table 3). In the sonochemical treatment, three transformation products coming from imidazole ring rupture (TP1, TP2, and TP3), several isomers of biphenyl hydroxylation (TP4a-f), and one product of alcohol moiety oxidation (TP5) have been observed. Additionally, products of hydroxylation/oxidation of the alkyl chain on the antihypertensive have been found (TP6 and TP7, Table S3). Furthermore, analogous primary transformations of losartan induced by UVC/H_2O_2 and photo-Fenton were recently reported. Kaur and Dulova also found the formation of TP2 TP3, TP4, TP5, TP6, and TP7, in addition to TP8 (product of hydroxylation at the imidazole ring) and TP9 (transformation coming from a chlorine removal of the imidazole structure, see Table S3) [4]. In this sense, the region attackable by the hydroxyl radical, indicated by theoretical results correlates with the reported primary transformation products. This highlights the usefulness of computational analysis as a tool to establish the regions on losartan susceptible to degradation by the radicals from the AOPs.

4.4. Mineralization and Toxicity Evolution

The ability of the two processes to mineralize losartan in distilled water was tested, showing that none of these processes transformed losartan into carbon dioxide, water, and inorganic ions even at longer treatment times (2T) (Figure 4A). These results can be understood based on the degradation routes involved in each process. In the case of ultrasound, the attack of sonogenerated radicals in the interfacial zone (main route above described) led to hydroxylations/oxidations and rupture of pollutant molecules (see Table S3), which typically generates products more hydrophilic than the parent compound [47]. Hence, due to the hydrophilic nature of losartan degradation products, they are placed far away from the cavitation bubble, and consequently far away from the sonogenerated $HO^{\bullet}$. Thereby, the mineralization of losartan by ultrasound is not observed.

In the case of losartan elimination by the UVC/H_2O_2 process, it was noted the high participation of light (Section 4.2.1). Although UVC has a strong degrading ability through isomerizations or carbon-heteroatoms bond cleavages, its mineralizing power is very low [48]. On the other hand, although the mineralizing ability of $HO^{\bullet}$ is widely recognized, under the tested conditions (moderate H_2O_2 concentration; i.e., 500 µM), the formed amount of such species seems to be not enough to reach some mineralization of losartan. Due to the nonmineralizing ability of ultrasound and UVC/H_2O_2 toward LOS, it was necessary to test the toxicity. To establish the potential reuse of the treated urine for irrigating crops; toxicity tests against radish seeds (*Raphanus sativus*) were performed (Figure 4B). It should be noted that the UVC/H_2O_2 process inhibits the germination of the seeds, this is associated with noxious substances generated in this system. In fact, recent research on losartan degradation by UVC/H_2O_2 process also evidenced that toxicity of solutions against *Daphnia magna* and *Desmodesmus subspicatus* augmented after the treatment [49].

Unlike UVC/H_2O_2, in the sonochemical process, the growth of the radish seeds increased with treatment (see 2T in Figure 4B). This suggests that the losartan by-products generated at large treatment periods of the sono-treatment are beneficial/less toxic for the indicator organism than the parent compound. Such results are coincident with several studies, which reported that the treatment of polluting substances using ultrasonic irradiation reduces the toxicity of solutions [50]. It must be indicated that although both UVC/H_2O_2 and sonochemistry can generate similar primary transformation products by hydroxyl radical attacks to losartan in distilled water (Section 4.3) at long treatment periods they may differ. Additionally, it must be considered that in the sonochemical process mainly acts hydroxyl radicals, whereas in the UVC/H_2O_2 both the radicals and UVC light are responsible for pollutant degradation (Section 4.2.2). Then, the observed differences in toxicity between both processes would be associated with their degradation mode. In the UVC/H_2O_2, the noxious substances could come from the action of UVC light on losartan or its primary degradation products (indeed, a previous work about the treatment of other emerging concern pollutants by UVC also reported the generation of toxic products for some organisms produced by this irradiation [32]).

5. Conclusions

It can be concluded that this research provides relevant information to understand the elimination of a representative pharmaceutical in fresh urine by two advanced oxidation processes having different nature (a photochemical treatment and other sonochemical system). The application of ultrasound and UVC/H_2O_2 individually, for the removal of the model pharmaceutical (antihypertensive losartan) in simulated fresh urine, showed that the sonochemical process was little affected by the urine matrix, exhibiting a high selectivity (Rk = 0.79) for the removal of losartan, which was related to degradation of the pharmaceutical at the interface of the cavitation bubble by the action of $HO^{\bullet}$. Meanwhile, the UVC/H_2O_2 process experienced moderate impacts of the matrix (Rk = 0.46) on the removal of losartan, because their degradation routes involved both photolysis and radical attacks. In turn, both ultrasound and UVC/H_2O_2 processes showed no mineralization of the pollutant in distilled water. Nevertheless, differently to UVC/H_2O_2, the sonochemical system transformed losartan into nonphytotoxic products (evidencing the potential reuse of sono-treated urine to irrigate crops). This illustrates the positive potentiality of ultrasound for the treatment of pharmaceuticals with hydrophobic characteristics in the simulated fresh urine. On the other hand, the computational analyses indicated that atoms on imidazole moiety on losartan were the most susceptible to transformations by the radical species. Such analysis was in good agreement with primary degradation products coming from UVC/H_2O_2 and sonochemical treatments, evidencing that theoretical methods are a useful tool to predict and rationalize the attacks of degrading species in the considered AOPs. Finally, it must be mentioned that losartan degradation was carried out at a pH value of 6.1; however, urine ranges from 4.5 to 8, and the modification of such parameter may change the results about the degradation of pharmaceuticals by the AOPs. Thus, the effect of the urine pH should be evaluated in more detail in future studies.

Supplementary Materials
Text S1: Determination of pseudo-first-order kinetic constants (k), Figure S1: Determination of the kinetic constants, Figure S2: Donor–acceptor diagram (DAM), Figure S3: Absorption spectra of losartan, Figure S4: Comparison of Rk for UVC/H_2O_2 and UVC alone, Figure S5: Resonance hybrid, Figure S6: Hydroxyl radical attack to the tetrazole ring, Table S1: Log K_{OW} of losartan and the components of urine, Table S2: Primary transformation products of losartan during sonochemical treatment, Table S3: Additional products of losartan transformation by UVC/H_2O_2 and photo-Fenton.

Author Contributions: Conceptualization: R.A.T.-P. and E.A.S.-G.; methodology: J.F.G.-L. and E.A.S.-G.; software: Y.Á.-T.; formal analysis: J.F.G.-L. and E.A.S.-G.; investigation: J.F.G.-L.; resources: R.A.T.-P. and Y.Á.-T.; writing—original draft preparation: J.F.G.-L.; writing—review and editing: E.A.S.-G., R.A.T.-P. and Y.Á.-T.; supervision: Y.Á.-T. and R.A.T.-P.; project administration: Y.Á.-T. and R.A.T.-P.; funding acquisition: Y.Á.-T. and R.A.T.-P. All authors have read and agreed to the published version of the manuscript.

Acknowledgments: Researchers from Grupo de Investigación en Remediación Ambiental y Biocatálisis (GIRAB) thanks Universidad de Antioquia UdeA for the support provided through "PROGRAMA DE SOSTENIBILIDAD"; E. A. Serna-Galvis thanks MINCIENCIAS COLOMBIA for his PhD fellowship during July 2015–June 2019 (Convocation 647 de 2014).

References

1. Gu, Q.; Burt, V.L.; Dillon, C.F.; Yoon, S. Trends in antihypertensive medication use and blood pressure control among United States adults with hypertension: The National Health And Nutrition Examination Survey, 2001 to 2010. *Circulation* **2012**, *126*, 2105–2114. [CrossRef]

2. Israili, Z.H. Clinical pharmacokinetics of angiotensin II (AT 1) receptor blockers in hypertension. *J. Hum. Hypertens.* **2000**, *14* (Suppl. 1), S73–S87. [CrossRef]

3. Gurke, R.; Rößler, M.; Marx, C.; Diamond, S.; Schubert, S.; Oertel, R.; Fauler, J. Science of the Total Environment Occurrence and removal of frequently prescribed pharmaceuticals and corresponding metabolites in wastewater of a sewage treatment plant. *Sci. Total Environ.* **2015**, *532*, 762–770. [CrossRef] [PubMed]

4. Kaur, B.; Dulova, N. UV-assisted chemical oxidation of antihypertensive losartan in water. *J. Environ. Manag.* **2020**, *261*, 110170. [CrossRef]

5. Osorio, V.; Larrañaga, A.; Aceña, J.; Pérez, S.; Barceló, D. Science of the Total Environment Concentration and risk of pharmaceuticals in freshwater systems are related to the population density and the livestock units in Iberian Rivers. *Sci. Total Environ.* **2016**, *540*, 267–277. [CrossRef]

6. Sanzi, F.; Souza, S.; Lopes, L.; Emanoel, J.; Hermes, F.; Alves, L.; Gonçalves, L.; Rodrigues, C.; Barbosa, B.; Moledo, D.; et al. Ecotoxicological effects of losartan on the brown mussel Perna perna and its occurrence in seawater from Santos Bay (Brazil). *Sci. Total Environ.* **2018**, *637–638*, 1363–1371. [CrossRef]

7. Bayer, A.; Asner, R.; Schüssler, W.; Kopf, W.; Weiß, K.; Sengl, M.; Letzel, M. Behavior of sartans (antihypertensive drugs) in wastewater treatment plants, their occurrence and risk for the aquatic environment. *Environ. Sci. Pollut. Res.* **2014**, *21*, 10830–10839. [CrossRef]

8. Miklos, D.B.; Remy, C.; Jekel, M.; Linden, K.G.; Hübner, U. Evaluation of advanced oxidation processes for water and wastewater treatment—A critical review. *Water Res.* **2018**, *139*, 118–131. [CrossRef]

9. Zhou, C.; Gao, N.; Deng, Y.; Chu, W.; Rong, W.; Zhou, S. Factors affecting ultraviolet irradiation/hydrogen peroxide (UV/H_2O_2) degradation of mixed N-nitrosamines in water. *J. Hazard. Mater.* **2012**, *231–232*, 43–48. [CrossRef]

10. Mahamuni, N.N.; Adewuyi, Y.G. Advanced oxidation processes (AOPs) involving ultrasound for waste water treatment: A review with emphasis on cost estimation. *Ultrason. Sonochem.* **2010**, *17*, 990–1003. [CrossRef]

11. Zhang, R.; Yang, Y.; Huang, C.-H.; Zhao, L.; Sun, P. Kinetics and modeling of sulfonamide antibiotic degradation in wastewater and human urine by UV/H_2O_2 and UV/PDS. *Water Res.* **2016**, *103*, 283–292. [CrossRef] [PubMed]

12. Giannakis, S.; Hendaoui, I.; Jovic, M.; Grandjean, D.; De Alencastro, L.F.; Girault, H.; Pulgarin, C. Solar photo-Fenton and UV/H_2O_2 processes against the antidepressant Venlafaxine in urban wastewaters and human urine. Intermediates formation and biodegradability assessment. *Chem. Eng. J.* **2017**, *308*, 492–504. [CrossRef]

13. Giannakis, S.; Jovic, M.; Gasilova, N.; Pastor Gelabert, M.; Schindelholz, S.; Furbringer, J.-M.M.; Girault, H.; Pulgarin, C. Iohexol degradation in wastewater and urine by UV-based Advanced Oxidation Processes (AOPs): Process modeling and by-products identification. *J. Environ. Manag.* **2017**, *195*, 174–185. [CrossRef]

14. Singla, J.; Verma, A.; Sangal, V.K. Parametric optimization for the treatment of human urine metabolite, creatinine using electro-oxidation. *J. Electroanal. Chem.* **2018**, *809*, 136–146. [CrossRef]

15. Giannakis, S.; Androulaki, B.; Comninellis, C.; Pulgarin, C. Wastewater and urine treatment by UVC-based advanced oxidation processes: Implications from the interactions of bacteria, viruses, and chemical contaminants. *Chem. Eng. J.* **2018**, *343*, 270–282. [CrossRef]

16. Yin, K.; He, Q.; Liu, C.; Deng, Y.; Wei, Y.; Chen, S.; Liu, T.; Luo, S. Prednisolone degradation by UV/chlorine process: Influence factors, transformation products and mechanism. *Chemosphere* **2018**, *212*, 56–66. [CrossRef]

17. Luo, J.; Liu, T.; Zhang, D.; Yin, K.; Wang, D.; Zhang, W.; Liu, C.; Yang, C.; Wei, Y.; Wang, L.; et al. The individual and Co-exposure degradation of benzophenone derivatives by UV/H_2O_2 and UV/PDS in different water matrices. *Water Res.* **2019**, *159*, 102–110. [CrossRef]

18. Montoya-Rodríguez, D.M.; Serna-Galvis, E.A.; Ferraro, F.; Torres-Palma, R.A. Degradation of the emerging concern pollutant ampicillin in aqueous media by sonochemical advanced oxidation processes—Parameters effect, removal of antimicrobial activity and pollutant treatment in hydrolyzed urine. *J. Environ. Manage.* **2020**, *261*, 110224. [CrossRef]

19. Lacasa, E.; Herraiz, M. The role of anode material in the selective oxidation of 2 penicillin G in urine. *Chemelectrochem* **2019**, *6*, 1376–1380. [CrossRef]

20. Cotillas, S.; Lacasa, E.; Sáez, C.; Cañizares, P.; Rodrigo, M.A. Removal of pharmaceuticals from the urine of polymedicated patients: A first approach. *Chem. Eng. J.* **2018**, *331*, 606–614. [CrossRef]

21. Schmidt, B.; Schieffer, B. Angiotensin II AT1 Receptor Antagonists. Clinical Implications of Active Metabolites. *J. Med. Chem.* **2003**, *46*, 2261–2270. [CrossRef]

22. Amstutz, V.; Katsaounis, A.; Kapalka, A.; Comninellis, C.; Udert, K.M. Effects of carbonate on the electrolytic removal of ammonia and urea from urine with thermally prepared IrO2 electrodes. *J. Appl. Electrochem.* **2012**, *42*, 787–795. [CrossRef]

23. Serna-Galvis, E.A.; Silva-Agredo, J.; Giraldo, A.L.; Flórez, O.A.; Torres-Palma, R.A. Comparison of route, mechanism and extent of treatment for the degradation of a β-lactam antibiotic by TiO_2 photocatalysis, sonochemistry, electrochemistry and the photo-Fenton system. *Chem. Eng. J.* **2016**, *284*, 953–962. [CrossRef]

24. Serna-Galvis, E.A.; Isaza-Pineda, L.; Moncayo-Lasso, A.; Hernández, F.; Ibáñez, M.; Torres-Palma, R.A. Comparative degradation of two highly consumed antihypertensives in water by sonochemical process. Determination of the reaction zone, primary degradation products and theoretical calculations on the oxidative process. *Ultrason. Sonochem.* **2019**, *58*, 104635. [CrossRef]

25. Serna-Galvis, E.A.; Silva-Agredo, J.; Giraldo-Aguirre, A.L.; Torres-Palma, R.A. Sonochemical degradation of the pharmaceutical fluoxetine: Effect of parameters, organic and inorganic additives and combination with a biological system. *Sci. Total Environ.* **2015**, *524–525*, 354–360. [CrossRef]

26. Hoekstra, N.J.; Bosker, T.; Lantinga, E.A. Effects of cattle dung from farms with different feeding strategies on germination and initial root growth of cress (Lepidium sativum L.). *Agric. Ecosyst. Environ.* **2002**, *93*, 189–196. [CrossRef]

27. Raghavachari, K. Perspective on "Density functional thermochemistry. III. The role of exact exchange". *Theor. Chem. Acc.* **2000**, *103*, 361–363. [CrossRef]

28. Tomasi, J.; Mennucci, B.; Cammi, R. Quantum Mechanical Continuum Solvation Models. *Chem. Rev.* **2005**, *105*, 2999–3094. [CrossRef]

29. An, T.; Yang, H.; Li, G.; Song, W.; Cooper, W.J.; Nie, X. Kinetics and mechanism of advanced oxidation processes (AOPs) in degradation of ciprofloxacin in water. *Appl. Catal. B Environ.* **2010**, *94*, 288–294. [CrossRef]

30. Gurkan, Y.Y.; Turkten, N.; Hatipoglu, A.; Cinar, Z. Photocatalytic degradation of cefazolin over N-doped TiO_2 under UV and sunlight irradiation: Prediction of the reaction paths via conceptual DFT. *Chem. Eng. J.* **2012**, *184*, 113–124. [CrossRef]

31. Li, L.; Wei, D.; Wei, G.; Du, Y. Transformation of cefazolin during chlorination process: Products, mechanism and genotoxicity assessment. *J. Hazard. Mater.* **2013**, *262*, 48–54. [CrossRef]

32. Clara, M.; Starling, V.M.; Souza, P.P.; Le, A.; Amorim, C.C.; Criquet, J. Intensification of UV-C treatment to remove emerging contaminants by $UV-C/H_2O_2$ and $UV-C/S_2O_8^{2-}$: Susceptibility to photolysis and investigation of acute toxicity UV-C. *Chem. Eng. J.* **2019**, *376*, 120856. [CrossRef]

33. Gerrity, D.; Lee, Y.; Von Gunten, U. Prediction of Trace Organic Contaminant Abatement with UV/H_2O_2: Development and Validation of Semi-Empirical Models for Municipal Wastewater Effluents. *Environ. Sci. Water Res. Technol.* **2016**, *2*, 460–473. [CrossRef]

34. Challis, J.K.; Hanson, M.L.; Friesen, K.J.; Wong, C.S. A critical assessment of the photodegradation of pharmaceuticals in aquatic environments: Defining our current understanding and identifying knowledge gaps. *Environ. Sci. Process. Impacts* **2014**, *16*, 672–696. [CrossRef]

35. Adewuyi, Y.G. Sonochemistry: Environmental Science and Engineering Applications. *Ind. Eng. Chem. Res.* **2001**, *40*, 4681–4715. [CrossRef]

36. Cheng, J.; Vecitis, C.D.; Park, H.; Mader, B.T.; Hoffmann, M.R. Sonochemical degradation of perfluorooctane sulfonate (PFOS) and perfluorooctanoate (PFOA) in landfill groundwater: Environmental matrix effects. *Environ. Sci. Technol. Ronmental.* **2008**, *42*, 8057–8063. [CrossRef]

37. Jiang, Y.; Pétrier, C.; David Waite, T. Kinetics and mechanisms of ultrasonic degradation of volatile chlorinated aromatics in aqueous solutions. *Ultrason. Sonochem.* **2002**, *9*, 317–323. [CrossRef]

38. Fernandez, N.A.; Rodriguez-freire, L.; Keswani, M.; Sierra-alvarez, R. Degradation of perfluoroalkyl and polyfluoroalkyl. *Environ. Sci.* **2016**, 975–983. [CrossRef]

39. Yasman, Y.; Bulatov, V.; Gridin, V.V.; Agur, S.; Galil, N.; Armon, R.; Schechter, I. A new sono-electrochemical method for enhanced detoxification of hydrophilic chloroorganic pollutants in water. *Ultrason. Sonochem.* **2004**, *11*, 365–372. [CrossRef]

40. Serna-Galvis, E.A.; Montoya-Rodríguez, D.M.; Isaza-Pineda, L.; Ibáñez, M.; Hernández, F.; Moncayo-Lasso, A.; Torres-Palma, R.A. Sonochemical degradation of antibiotics from representative classes-Considerations on structural effects, initial transformation products, antimicrobial activity and matrix. *Ultrason. Sonochem.* **2018**, *50*, 157–165. [CrossRef]

41. Tran, N.; Drogui, P.; Brar, S.K. Sonochemical techniques to degrade pharmaceutical organic pollutants. *Environ. Chem. Lett.* **2015**, *13*, 251–268. [CrossRef]

42. Lian, L.; Yao, B.; Hou, S.; Fang, J.; Yan, S.; Song, W. Kinetic Study of Hydroxyl and Sulfate Radical-Mediated Oxidation of Pharmaceuticals in Wastewater Effluents. *Environ. Sci. Technol.* **2017**, *51*, 2954–2962. [CrossRef] [PubMed]

43. Buxton, G.V.; Greenstock, C.L.; Helman, W.P.; Ross, A.B. Critical Review of rate constants for reactions of hydrated electron, hydrogen atoms and hydroxyl radicals ($\cdot OH/\cdot O$ in Aqueous Solution). *J. Phys. Chem. Ref. Data* **1988**, *513*. [CrossRef]

44. Minakata, D.; Song, W.; Crittenden, J. Reactivity of Aqueous Phase Hydroxyl Radical with Halogenated Carboxylate Anions: Experimental and Theoretical Studies. *Environ. Sci. Technol.* **2011**, *45*, 6057–6065. [CrossRef]

45. Toth, J.E.; Rickman, K.A.; Venter, A.R.; Kiddle, J.J.; Mezyk, S.P. Reaction kinetics and efficiencies for the hydroxyl and sulfate radical based oxidation of artificial sweeteners in water. *J. Phys. Chem. A.* **2012**, *116*, 9819–9824. [CrossRef]

46. Guateque-Londoño, J.F.; Serna-Galvis, E.A.; Silva-Agredo, J.; Ávila-Torres, Y.; Torres-Palma, R.A. Dataset on the degradation of losartan by TiO_2-photocatalysis and UVC/persulfate processes. *Data Br.* **2020**, *31*. [CrossRef]

47. Singla, R.; Grieser, F.; Ashokkumar, M. The mechanism of sonochemical degradation of a cationic surfactant in aqueous solution. *Ultrason. Sonochem.* **2011**, *18*, 484–488. [CrossRef]

48. Yang, L.; Yu, L.E.; Ray, M.B. Degradation of paracetamol in aqueous solutions by TiO_2 photocatalysis. *Water Res.* **2008**, *42*, 3480–3488. [CrossRef]

49. Adams, E. *Ecotoxicity and Genotoxicity Evaluation of Losartan Potassium after UVC Photolysis and UV/H_2O_2 Process*; Universidade Tecnológica Federal Do Paraná: Curitiba, Brazil, 2019.

50. Emery, R.J.; Papadaki, M.; Freitas dos Santos, L.M.; Mantzavinos, D. Extent of sonochemical degradation and change of toxicity of a pharmaceutical precursor (triphenylphosphine oxide) in water as a function of treatment conditions. *Environ. Int.* **2005**, *31*, 207–211. [CrossRef]

2

Oxidation of Selected Trace Organic Compounds through the Combination of Inline Electro-Chlorination with UV Radiation (UV/ECl$_2$) as Alternative AOP for Decentralized Drinking Water Treatment

Philipp Otter [1]⬤, **Katharina Mette** [2], **Robert Wesch** [2], **Tobias Gerhardt** [2], **Frank-Marc Krüger** [2], **Alexander Goldmaier** [1], **Florian Benz** [1], **Pradyut Malakar** [3] and **Thomas Grischek** [4],*⬤

[1] AUTARCON GmbH, D-34117 Kassel, Germany; otter@autarcon.com (P.O.); goldmaier@autarcon.com (A.G.); benz@autarcon.com (F.B.)

[2] GNF e.V. Volmerstr. 7 B, 12489 Berlin, Germany; k.mette@gnf-berlin.de (K.M.); r.wesch@gnf-berlin.de (R.W.); t.gerhardt@gnf-berlin.de (T.G.); f.krueger@gnf-berlin.de (F.-M.K.)

[3] International Centre for Ecological Engineering, University of Kalyani, Kalyani, West Bengal 741235, India; pradyutmalakar2@gmail.com

[4] Division of Water Sciences, University of Applied Sciences Dresden, Friedrich-List-Platz 1, 01069 Dresden, Germany

* Correspondence: thomas.grischek@htw-dresden.de

Abstract: A large variety of Advanced Oxidation Processes (AOPs) to degrade trace organic compounds during water treatment have been studied on a lab scale in the past. This paper presents the combination of inline electrolytic chlorine generation (ECl$_2$) with low pressure UV reactors (UV/ECl$_2$) in order to allow the operation of a chlorine-based AOP without the need for any chlorine dosing. Lab studies showed that from a Free Available Chlorine (FAC) concentration range between 1 and 18 mg/L produced by ECl$_2$ up to 84% can be photolyzed to form, among others, hydroxyl radicals (·OH) with an UV energy input of 0.48 kWh/m^3. This ratio could be increased to 97% by doubling the UV energy input to 0.96 kWh/m^3 and was constant throughout the tested FAC range. Also the achieved radical yield of 64% did not change along the given FAC concentration range and no dependence between pH 6 and pH 8 could be found, largely simplifying the operation of a pilot scale system in drinking water treatment. Whereas with ECl$_2$ alone only 5% of benzotriazoles could be degraded, the combination with UV improved the degradation to 89%. Similar results were achieved for 4-methylbenzotriazole, 5-methylbenzotriazole and iomeprol. Oxipurinol and gabapentin were readily degraded by ECl$_2$ alone. The trihalomethanes values were maintained below the Germany drinking water standard of 50 µg/L, provided residual chlorine concentrations are kept within the permissible limits. The here presented treatment approach is promising for decentralized treatment application but requires further optimization in order to reduce its energy requirements.

Keywords: trace organic compounds; emerging pollutants; rural regions; electrochlorination; UV; AOP; energy per order

1. Introduction

Annually on average 15 g of pharmaceuticals are consumed per capita, but human bodies are unable to fully metabolize pharmaceuticals, which are then excreted as parental components or metabolites [1,2]. They are discharged into wastewater together with Personal Care Products

(PCPs), sweeteners, illicit and non-controlled drugs, complexing agents, nanoparticles, perfluorinated compounds, pesticides, flame retardants, fuel additives and endocrine disrupting chemicals and detergents [3,4]. Conventional wastewater treatment such as activated sludge processes, exhibits limitations in the removal of Trace Organic Compounds (TOrCs) [5,6]. TOrCs are additionally released into the environment from irrigation with treated or untreated wastewater [7], disposal of animal waste on agricultural sites [8] and artificial groundwater recharge [9]. World-wide studies have confirmed the occurrence of pharmaceutical residues in the effluents of wastewater treatment plants (WWTPs) as well as in surface and groundwaters [10–14].

The problem of water contamination with TOrCs is even more prevalent in some developing areas. E.g., in India, the amount of ciprofloxacin, sulfamethoxazole, amoxicillin, norfloxacin, and ofloxacin in treated wastewater was up to 40 times higher compared to other countries in Europe, Australia, Asia, and North America [5,8]. Here even remote rural areas are affected.

Conventional drinking water treatment processes such as coagulation and flocculation are not designed to effectively remove TOrCs [8]. Powdered activated carbon (PAC), membrane filtration technologies [15–17] or advanced oxidation processes (AOP) have been studied in the past to evaluate their efficiency on the removal of selected TOrCs [18]. Most studied AOPs are based on ozonation, Fenton oxidation, or UV based AOPs such as UV/H_2O_2. For the application of AOPs in water treatment, the supply of chemicals such H_2O_2 in UV/H_2O_2 AOP has been identified as major challenge and cost factor [19].

A relatively new AOP is the combination of chlorination with UV radiation (UV/Cl_2) [20–24]. When aqueous chlorine solutions are exposed to UV, ·OH and ·Cl radicals are also formed (Figure 1). The reduction potential of ·OH radicals is with 2.8 V vs. Standard Hydrogen Electrode (SHE) substantially higher compared to the potential of ozone (2.07 V) or chlorine (1.37 V) [25]. UV/Cl_2 has proven to produce higher amounts of (·OH) radicals compared to UV/H_2O_2 mainly due to the low absorbance of UV light by H_2O_2 [20,26]. Whereas absorption coefficients (e) for UV/H_2O_2 of 19.6 M^{-1}·cm^{-1} have been identified for 254 nm [27] they have reached 59 M^{-1}·cm^{-1} and 66 M^{-1}·cm^{-1} for HOCl and OCl^- [28]. Higher quantum yields with regards to radical generation for UV/Cl_2 where reported by [26]. When radiated with UV light hypochlorous acid and its anion hypochlorite react in water not only to OH, but also to the reactive chlorine species (RCS) ·Cl and Cl_2 as well as to oxygen radicals. At higher pH, hydroxyl radicals might be consumed by chlorine itself also forming RCS [29].

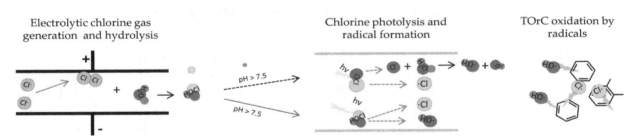

Figure 1. Pathway of chloride ions used for radical production (excerpt) in UV/ECl_2 process.

The RCS coexist with ·OH radicals and complement each other in degrading a wider variety of contaminants compared to e.g., UV/H_2O_2 AOP [29,30]. Depending on the chlorine species present two reaction pathways (Equations (1)–(3)) for the production of ·OH and ·Cl exist [26].

$$HOCl + h\nu \rightarrow \cdot OH + \cdot Cl, \tag{1}$$

$$OCl^- + h\nu \rightarrow \cdot O^- + \cdot Cl, \tag{2}$$

$$\cdot O^- + H_2O \rightarrow \cdot OH + OH^- \tag{3}$$

However, the need to supply and dose chlorine reagents also persists for UV/Cl_2-based AOPs. Here the production of chlorine through inline electrolysis (ECl_2) could offer an alternative to transportation, storage and handling of chlorine reagents. During ECl_2 elementary chlorine is produced at the anode of an electrolytic cell from the natural chloride content of the water itself (Equation (4)). The chlorine immediately hydrolyses to form hypochlorous acid (Equation (5)) or hypochlorite (Equation (6)). Anodic side reactions and cathodic reactions are given in Equations (7) and (8):

Anodic reaction chlorine:

$$2Cl^- \rightarrow Cl_2 + 2e^-, \tag{4}$$

Hydrolysis of chlorine gas:

$$Cl_2 + 2H_2O \leftrightarrow HClO + Cl^- + H_3O^+, \tag{5}$$

Equilibrium at pH = 7.5:

$$HClO \leftrightarrow OCl^- + H^+, \tag{6}$$

Anodic side reaction oxygen:

$$2H_2O \rightarrow O_2 + 4H^+ + 4e^-, \tag{7}$$

Cathodic reaction:

$$2H_3O^+ + 2e^- \rightarrow H_2 + 2H_2O \tag{8}$$

In past studies this approach has already proven feasible to meet the disinfection requirements for remote drinking water supply [31–33]. In the presented study ECl_2 was combined with UV radiation (Figure 1).

The novelty of this approach is the operation of a chlorine based AOP for the degradation of TOrCs that is completely independent from any external chemical supply. On a small scale, ECl_2 as well as UV could be easily operated by photovoltaic (PV) which makes the process also independent of any external energy supply.

Available, studies to evaluate the degradation potential of AOPs using chlorine have been generally carried out for single compounds investigating concentrations much higher than those found in real waters. Further, deionized water free of other organic substances was used as solvent and sophisticated equipment was applied for e.g., determining reaction constants (k′). Therefore, uncertainties exist with regards to real case scenarios [24].

In this study a UV/ECl_2 setting was tested under lab conditions in order to evaluate chlorine production and radical formation in dependence of chloride concentrations, pH, cell currents and UV energy input applied. Following the lab test a UV/ECl_2 pilot setting was tested for the first time under real case conditions treating Elbe river water. In two short term sampling campaigns the removal efficiency of selected TOrCs, the energy consumption and the formation of disinfection by products (DBPs) by analyzing trihalomethanes (THM) was evaluated. The pilot system was operated long term to observe technical challenges that may occur under real case scenarios. The hypothesis of the here conducted work is, that the combination of ECl_2 with UV poses a technically feasible alternative to reduce TOrCs without the need for any external chemicals and electricity supply, which can be applied in decentralized water treatment. The conducted trials hereby allow a first insight into the application of UV/ECl_2 as part of an actual drinking water treatment system.

2. Methodology

To produce chlorine by means of ECl_2 mixed oxide electrodes (MOX) (GNF, Berlin, Germany) coated with ruthenium (Ru) and iridium (Ir) oxides. Ru- and Ir-mixed oxide electrodes have been selected due to their low overpotential for the oxidation reaction of chloride to chlorine and therefore offer a higher current efficiency for chlorine evolution compared to e.g., platinum coated electrodes [34].

2.1. Lab Test Setting

In order to assure sufficient chlorine generated by inline electrolysis as precursors for the chlorine photolysis and disinfection, lab experiments were conducted. Those included the variation of chloride ion concentration, flow rate, current density, pH and electrical conductivity.

During lab tests water was pumped through an array of two MOX electrodes and up to two UV lamps (Figure 2). The distance between the cell plates was 5 mm. Currents of up to 8 A were applied on the electrodes. With a total surface area of 959 cm^2, the current density accounted for 16.7 mA/cm^2. These currents were chosen because previous (unpublished) studies showed that this is the maximum applicable current density for long term chlorine evolution without damaging the coating. The MOXs were powered using a BaseTech BT-305 power supply unit (Hirschau, Germany).

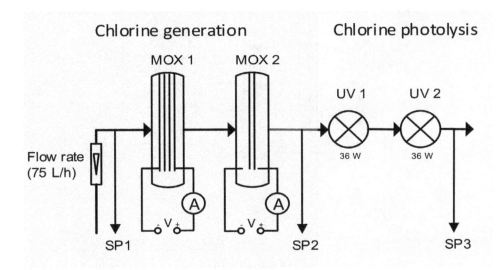

Figure 2. Lab test setting for performance evaluation of chlorine and radical formation.

Distilled water was pH adjusted with 10% HCl (Carl Roth, Karlsruhe, Germany) and 10% NaOH (Carl Roth). NaCl (Carl Roth) was added to the water to achieve concentrations of 25, 50 and 100 mg/L. The electric conductivity was adjusted to the desired value using NaHCO$_3$ (Carl Roth, Germany) and Na$_2$SO$_4$ (Carl Roth). Samples for chlorine measurement were taken directly behind the MOXs at Sampling Point (SP) 2 and SP 3 (Figure 2). The chlorine was then photolyzed by up to two PURION® 2500 36 W low pressure UV-C reactors with a volume of 0.75 L which were equipped with a calibrated silicon semiconductor-based UV-irradiance sensor (SUV-13A1Y2C, Purion, Zella-Mehlis, Germany) on each UV unit to assure constant irradiation of the UV lamps. The radical yield was determined at SP3.

The input of the electrical energy for the UV photolysis was 0.48 kWh/m^3 for one and 0.96 kWh/m^3 for two lamps, respectively. Samples to evaluate radical formation were taken prior (SP2) and after the UV reactor(s) (SP3).

The amount of HOCl and OCl$^-$ decomposed (Equations (1) and (2)) during FAC photodegradation is directly related to the amount of ·OH radicals [35] and the radical yield factor η can thus be quantified following Equation (9):

$$\eta = \frac{\Delta n_{\cdot OH} + \Delta n_{\cdot Cl}}{\Delta n_{FAC}} \tag{9}$$

The quantification of radical formation was carried out following [36] as described below. The yield factors were calculated for pH values of 6, 7 and 8 in order to determine its dependence of chlorine species present (Equation (6)).

2.2. Field Test Settings

The field test was carried out for 10 months and was conducted with Elbe river water, which was filtrated by a UF system (150 kDa, 0.01 μm, Pall, Port Washington, NY, USA) before the UV/ECl$_2$ AOP in order to provide turbidity free water.

As natural water matrices contain a variety of radical scavengers the here tested setting will generate only site-specific results. Better pre-treatment hereby is expected to generate higher degradation as shown in [37].

The water was pumped by the pilot system in a flow through setting through three MOX with a total area of 1918 cm^2 operated in series followed by up to two Purion® 2500 36 W low pressure UV-C lamps (Figure 3).

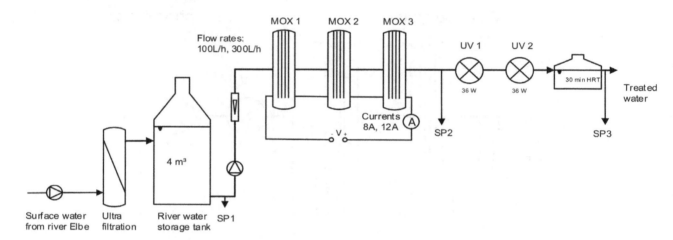

Figure 3. Pilot system tested with Elbe river water.

To release potentially formed calcareous deposits the polarity was inverted once every three hours. In own past studies, these comparable large intervals have proven to have a negligible effect on the lifetime of the electrolytic cells.

For the evaluation of TOrC degradation during drinking water treatment two short term sampling campaigns were conducted.

In order to evaluate the influence of the chlorine concentration and the UV irradiation on the degradation of TOrCs the ECl$_2$ current and resulting current densities, the number of UV lamps and the flow rate were varied during the field test. This tested pilot settings are documented in Table 1 including the specific energy demand for each of the tested settings.

A current of 8 A was hereby selected to meet a minimum total chlorine concentration of 2 mg$_{/L}$ after the electrolytic cell. Sampling was done before treatment (SP1), behind the ECl$_2$ (SP2) and after a hydraulic retention time (HRT) of 30 min behind the respective number of UV reactors. Samples for TOrCs, DOC, and THM analysis were quenched after 30 min using sulfite.

The MOX cell voltage was recorded in order to calculate energy demand and the "electric energy per order" (E$_{EO}$) of the ECl$_2$ (Equation (10)). As handling of contaminants at a waterworks site was prohibited, no spiking of the water with TOrCs was performed. Only selected substances, that were regularly present in the Elbe river water at sufficient concentrations (Table 2) were used for the evaluation of TOrC degradation. Due to the high costs of TOrC analysis the number of analyses and test settings was very limited in the here presented work and the results can only give a tendency of the degradation behavior.

Table 1. Test matrix and energy requirements for the different field test settings of UV/ECl$_2$.

Test Setting	Pump		ECl$_2$					UV		ECl$_2$/UV
	Flow Rate (L/h)	Energy Input (kWh/m^3)	Linear Flow Velocity (m/h)	Current (A)	Voltage (V)	Current Density (mA/cm^2)	Energy Input (kWh/m^3)	No. of UV Lamps	Energy Input (kWh/m^3)	Total Energy Input (kWh/m^3)
A	100	0.07	42	8	7.1	8.4	0.57	0	0.00	0.64
								1	0.36	0.99
								2	0.72	1.35
B	100	0.07	42	12	9.5	12.6	1.14	0	0.00	1.21
								2	0.72	1.93
C	300	0.20	126	12	11.7	12.6	0.47	0	0.00	0.67
								1	0.12	0.79
								2	0.24	0.91

Table 2. Analyzed TOrCs concentrations of UF Filtrate and relevant water quality parameters.

Parameter	Initial Concentrations/Values	Description
TOrC	1st and 2nd trial	
Benzotriazole (ng/L) CAS 95-14-7	410 and 440	Corrosion inhibitor
4-Methylbenzotriazole (ng/L) CAS 29878-31-7	190 and 170	Benzotriazole derivative used as a corrosion inhibitor
5-Methylbenzotriazole (ng/L) CAS 136-85-6	82 and 80	Antifreeze agent, corrosion inhibitor
Gabapentin (ng/L) CAS 60142-96-3	230 and 250	Anticonvulsant (Antiepileptic)
Iomeprol (ng/L) CAS 78649-41-9	380 and 520	X-ray contrast media
Oxipurinol (ng/L) CAS 2465-59-0	850 and 930	Active metabolite of Allopurinol (uricostatic)
Standard water quality parameters		
pH	8.1 and 7.9	
Temperature (°C)	18 and 25	
DOC (mg/L)	5.5 and 6.7	
Nitrate (mg/L NO_3^-) *	13 ± 2.7	
Nitrite (mg/L NO_2^-)*	0.043 ± 0.02	
Ammonium (mg/L NH_4) *	0.082 ± 0.07	
Total Hardness (mg/L $CaCO_3$)	67.5 ± 3.8	
Chloride (mg/L)	22.6 ± 3.7	

* Elbe river data derived from [38] and calculated by authors.

2.3. Energy Demand and Energy Efficiency

In order to estimate the energetic efficiency of the UV/ECl$_2$ AOP, the "electric energy per order" (E_{EO}), representing the electrical energy necessary for the degradation of one log unit of a contaminant, was calculated following Equation (10) [39]:

$$E_{EO} = \frac{P}{q \log\left(\frac{c_0}{c}\right)},$$ (10)

where P: Electrical power applied to run the process (W), q: Flow rate (m^3/h), c$_0$: Initial contaminant concentration (μg/L) and c: Final contaminant concentration (μg/L)

The E_{EO} of the here generated data allows a comparison with alternative AOPs and may serve as requisite data for the evaluation of their economic feasibility and sustainability. It should be considered that for the generation of E_{EO} values of UV based AOPs in literature often only the energy required to power the UV lamp is considered. The energy required to produce, transport and dose e.g., H$_2$O$_2$ or chlorine are often neglected. Therefore, the E_{EO} calculated for the here presented approach distinguishes between the energy required for the UV and UV/ECl$_2$. Further deviations are to be expected when varying process capacity and source water quality [40].

2.4. Water Analysis

The quantification of hydroxyl radicals was carried out following [36] by adding methanol (Carl Roth, Karlsruhe, Germany) in excess as radical scavenger. According to [41] methanol reacts with OH and Cl radicals to formaldehyde following the Equations (11)–(14):

$$CH_3OH + \cdot OH \rightarrow CH_2OH\cdot + H_2O,$$ (11)

$$CH_3OH + \cdot Cl \rightarrow CH_2OH\cdot + HCl,$$ (12)

$$CH_2OH\cdot + O_2 \rightarrow O_2CH_2OH\cdot,$$ (13)

$$O_2CH_2OH\cdot \rightarrow CH_2O + HO_2\cdot$$ (14)

Formaldehyde concentrations were analyzed following [42], were the extinction coefficient E_{DDL} caused by diacetyldihydrolutidin (DDL) formed during Hantzsch reaction was determined with a calibrated UV-VIS spectrometer (UV-1602, Shimadzu, Jena, Germany) using a 1 cm cuvette. Since the efficiency of ·OH radicals reacting with methanol is 93% [36], the OH radical production can be determined by the amount of DDL produced. As ·Cl is also an active species in this reaction [43], the calculated values only represent an upper limit of the ·OH production. Still, the obtained value is seen as useful to determine the capability of the formed radical species to react even with relatively inert C-H bonds like those of methanol.

Quantitative analysis of oxipurinol, gabapentin and iomeprol was done using HPLC-MS/MS following DIN 38407-47 using Agilent 1260 HPLC-System (Agilent Technologies, Waldbronn, Germany) und API 6500 MS/MS-System (AB Sciex, Darmstadt, Germany). Benzotriazoles were determined after automatic solid phase extraction with ASPEC (Gilson, Middleton, WI, USA) with HPLC-MS/MS Agilent 1200 HPLC-System (Agilent Technologies) and API 5500 MS/MS-System (AB Sciex) following TZW lab method. All substances had a Limit of Detection (LoD) of 0.01 µg/L except oxipurinol with a LoD of 0.025 µg/L. For analytic results below the LoD contaminant reduction rates and E_{EO} values were calculated by using the respective LoD.

The dissolved organic carbon (DOC) was analyzed with a TOC-V-CPH (Shimadzu, Kyoto, Japan) with integrated auto-sampler ASI-V (Shimadzu, Japan) following method DIN EN 1484:2019-04. The electrolytically produced Free Available Chlorine (FAC) and total chlorine was determined with AL410 photometer (Aqualytic, Dortmund, Germany) using Aqualytic DPD1 and DPD3 reagents with a measurement range between 0.01 and 6.0 mg/L. Electric conductivity (CDC401, Hach Düsseldorf, Germany) and pH (CDC401, Hach) were determined using a Hach HQ40d multimeter. THMs were analyzed following DIN EN ISO 10,301 using a 7890A GC/MS by Agilent Technologies (Santa Clara, CA, USA) with a detection limit of 0.1 µg/L.

3. Results and Discussion

3.1. Lab Test

Figure 4 shows the linear relation between current or current density applied at the MOX electrode and the formation of FAC at the given input chloride concentrations considering chloride concentration of 25, 50 and 100 mg/L in the synthetic water.

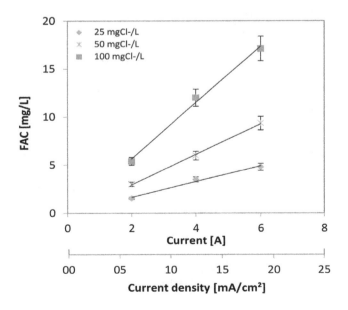

Figure 4. Relation between current and FAC concentration ($n = 9$) (Q = 75 L/h, EC = 400 µS/cm, T = 19 ± 1 °C).

Figure 5 shows the linear relation between the chloride concentration and the FAC as well as the FAC production rate achieved at the applied currents of 2, 4 and 6 A.

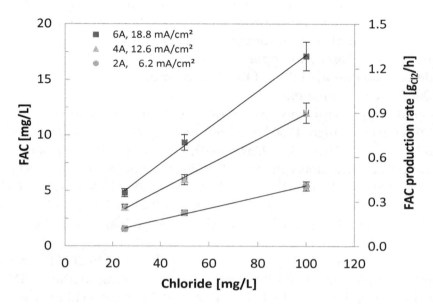

Figure 5. Relation between chloride concentration and FAC concentrations ($n = 9$) (Q = 75 L/h, EC = 400 µS/cm, T = 19 ± 1 °C).

Such relations were also derived in former studies as shown in [44]. The linearity largely simplifies the control of FAC generation, from an operational perspective. The currents at the MOX electrode can be easily adjusted in a treatment system within the here given current densities.

Figure 6 shows that charge specific chlorine production rates and current efficiencies are similar for currents of 2, 4, and 6 A with the applied cell. At higher currents, here shown with 8 A, the efficiency drops especially at higher chloride concentrations. This was related to the formation of gas bubbles at the anode surface.

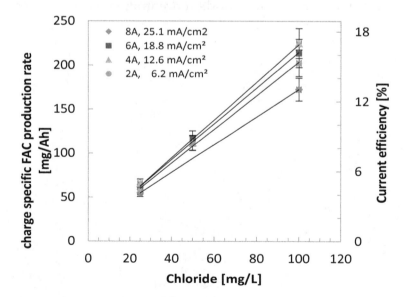

Figure 6. Relation of chloride concentration (with charge specific FAC production rate and current efficiency ($n = 9$) (Q = 75 L/h, EC = 400 µS/cm, T = 19 ± 1 °C).

Figure 7 shows the linear relation between the electrolytically produced FAC concentration and the resulting reduction in FAC when irradiated with 1 and 2 UV reactors.

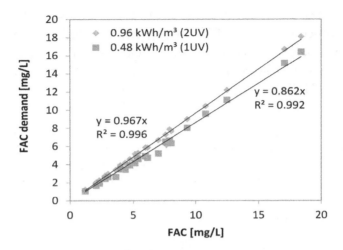

Figure 7. Chlorine consumption by UV treatment in dependence of FAC (Q = 75 L/h, EC = 400 µS/cm, T = 19 ± 1 °C).

At the given FAC concentrations any additional FAC is readily photolyzed independent of its concentration. Already with the energy input of one UV lamp (0.48 kWh/m³) 84% of the FAC is consumed. On average the chlorine concentration was reduced to 1.0 ± 0.5 mg/L. The trials with two UV lamps achieved a FAC reduction of 96% and the chlorine concentration was further reduced on average to 0.2 ± 0.1 mg/L, which would make FAC quenching as suggested by [24] dispensable. Whether the second UV lamp and with that the extra energy demanded is required, depends on the site-specific treatment targets, with regards to TOrC reduction, DBP formation potential, residual chlorine concentration and the design of the used UV-reactor.

Figure 8 shows the relation between FAC degradation and the radical formation by means of chlorine photolysis at the given pH values.

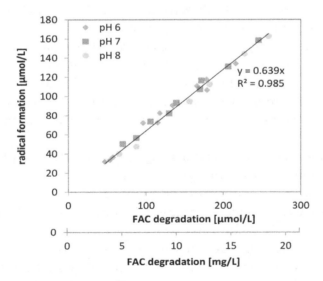

Figure 8. Radical formation in dependence of FAC demand at different pH (Q = 75 L/h, EC = 400 µS/L, T = 19 ± 1 °C) (*n* = 26).

The pH has no observable effect on the formation of radicals from photolyzed FAC. This relation has already been described by [28]. The average radical yield factor was constant over the here given FAC degradation range and with 64% (*n* = 26) significantly higher than the reported values of [45].

From an operational perspective the here achieved constant radical yield factor in the examined FAC concentration range offers the possibility to evaluate the degradation of TOrC by monitoring the chlorine degradation.

However, with rising pH the concentration of OCl⁻ is increasing (Equation (6)). As OCl⁻ has a slightly higher adsorption coefficient at 254 nm [28] slightly higher radical concentrations can be expected at higher pH values. On the other hand radical scavenger effects of HOCl and ClO⁻ may result in additional chlorine degradation and the formation of oxy-chlorine radicals. Such scavenger effects are stronger at basic pH [29].

3.2. Field Tests

The total chlorine and FAC concentration measured directly after the ECl₂ or the respective number of UV lamps (SP 3) during the treatment of Elbe river water is shown in Figure 9.

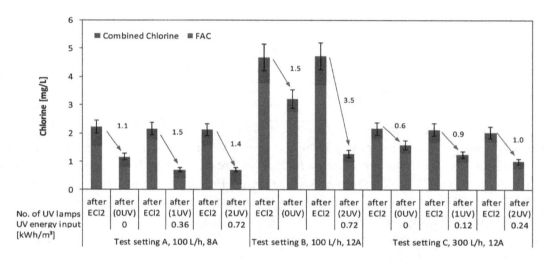

Figure 9. FAC and total chlorine concentrations measured in three different settings during field tests with Elbe river water.

The average concentrations of FAC directly after the ECl₂ reached 1.6 ($n = 3$), 3.8 ($n = 2$), and 1.7 mg/L ($n = 3$) for test settings A, B and C, respectively. Average total chlorine concentrations were 2.2 ($n = 3$), 4.7 ($n = 2$) and 2.1 mg/L ($n = 3$). The chlorine degradation was dependent on the initial chlorine concentration and the UV energy input into the water as already shown in Figure 7.

The reduction in chlorine concentration of the tested water during the trials without UV radiation indicated with 1.1, 1.4 and 0.6 mg/L a high DBP formation potential, typical for surface waters. The lower chlorine demand values during test setting C were related to the shorter HRT in the final storage tank.

The degradation of the tested TOrCs after ECl₂ only and after the photolysis of chlorine with the respective UV energy input is shown in Figure 10a–d.

The input concentrations for gabapentin and oxipurinol were 0.25 and 0.93 µg/L respectively. These substances were nearly completely degraded at all given test settings alone by the produced chlorine. After passing the first UV lamp in test setting A, eventually still available quantities of gabapentin and oxipurinol were degraded below the LoD. In contrast to that benzotriazole and 5-methylbenzotriazole could be degraded by only by 5% and 4% with ECl₂ alone during test setting A. Iomeprol and 4-methylbenzotriazole were degraded by 8% and 11%. Higher chlorine concentrations produced during test setting B did not substantially improve the degradation. However, in combination with UV radiation the degradation could be significantly increased. In test setting A the degradation of benzotriazole was increased to 49% and 73% with an energy input of 0.36 (1 UV) and 0.72 kWh/m³ (2 UV), respectively. In combination with UV radiation the additional chlorine made available during test setting B further increased the degradation of benzotriazole to 89%. Iomeprol degradation could be increased to 83% and 82% during test settings A and B, respectively, with the energy of two UV lamps (0.72 kWh/m³). The behavior of 4-methylbenzotriazole and 5-methylbenzotriazole degradation was similar.

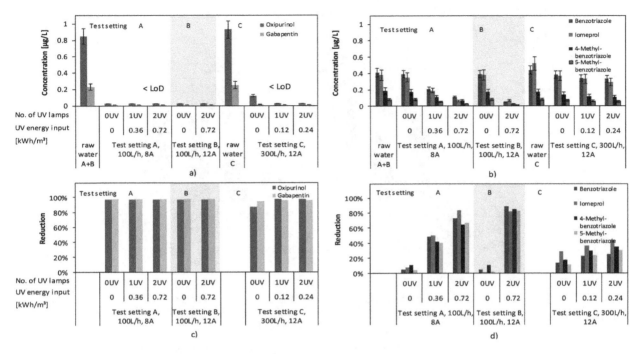

Figure 10. Concentrations of tested TOrCs (**a,b**) and degradation percentages (**c,d**) concentrations measured in three different settings during field tests with Elbe river water.

Even though nearly equal amounts of chlorine were available, the degradation of all substances susceptible for the UV/ECl$_2$ AOP was largely reduced during test setting C. As also smaller amounts of chlorine were degraded, it can be concluded that at a flow rate of 300 L/h the HRT and by that the UV energy input was not sufficient to produce equal amounts of radicals as in test setting A.

3.3. Behavior of DOC and Formation of Disinfection-by-Products and Metabolites (Toxicity) during Field Tests

The mean DOC of Elbe river water in Dresden, Germany, is about 5.2 mg/L (4.6–6.0 mg/L, $n = 325$) [46]. In test settings A and B the river water showed a slightly elevated DOC of 6.7 mg/L, and 5.6 mg/L in test setting C after ultrafiltration. The comparatively high DOC in test settings A and B was caused by runoff after heavy rainfall during the period of this test, whereas test setting C was performed a rainless period. DOC was reduced by 28% and 23% in test setting A and B and by only 3% in test setting C through ECl$_2$ alone as shown in Figure 11. The degradation of DOC without UV is explained by a direct oxidation of organics during the passage through the electrolytic cell as e.g., described in [47,48]. During test settings A and B the degradation of DOC was substantially higher compared to test setting C. This is related to the higher HRT and the nature of the DOC which is more easily degradable when containing runoff.

Figure 12 shows that DBPs measured as THMs have been formed after ECl$_2$ and for THMs the strict German threshold of 50 µg/L could not be adhered to for the majority of the tested settings. However, it was always possible to adhere to the EU guideline values of < 100 µg/L. The EU guideline values could always be maintained (Table 3). During the trials with UV applied the formation of THM was substantially lower compared to trials without UV radiation. Less THMs were formed because smaller amounts of chlorine were available after photolysis. During test setting A with 2 UV lamps the THM formation made up only 47% compared to the trial without UV radiation, by which the German guideline limits could be maintained. DBP formation is therefore related to residual chlorine levels rather than the AOP itself. This is confirmed by other studies where no significant increase of organic DBP formation during the application of UV/Cl$_2$ was found and most organic DBPs formed were related to the application of chlorine itself [49,50]. Also [24] found no significant quantities of THMs in UV/Cl$_2$ when adding 6 mg/L of chlorine prior to the UV lamp, using simulated wastewater with a DOC of 46 mg/L (COD ~120 mg/L). In this study, residual chlorine was completely quenched after the UV lamp.

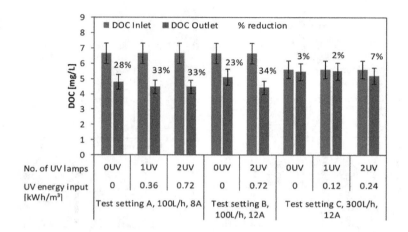

Figure 11. DOC concentrations measured in three different settings during field tests with Elbe river water.

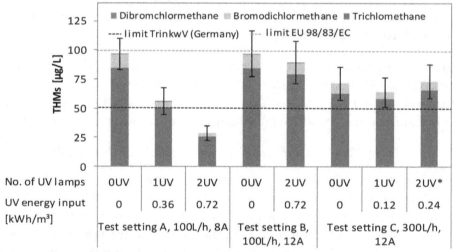

* insufficient quenching of sample caused further generation of THMs and AOX

Figure 12. THM concentrations measured in three different settings during field tests with Elbe river water.

Table 3. Selected guideline values concerning the formation of THMs during drinking water disinfection.

Parameter	Germany [51]	EU [52]	India [53]	WHO [54]
THM (µg/L)	10 [a]/50 [b]	100	-	
Bromoform (µg/L)			100	100
Dibromochloromethane (µg/L)			100	100
Bromodichloromethane (µg/L)			60	60
Chloroform (µg/L)			200	300

[a] At the end of treatment, [b] Point of use.

To prevent excessive THM formation it is therefore suggested to closely monitor the residual chlorine concentration directly after the UV lamps and reduce the concentration whenever required by increasing the UV radiation or the flow rate. The field test data in Figure 9 indicate that residual chlorine concentrations around 0.2 mg/L, as required by many drinking water guidelines, seems to be promising to meet strict German guidelines for THMs.

From an operational perspective such chlorine levels must be monitored by integrating online probes into the algorithm controlling the chlorine production.

3.4. Energy Efficiency

Figure 13a,b and Table A1 (in Appendix A) show the E_{EO} values achieved for the degradation of the analyzed substances under the given conditions differentiating between the energy input into the chlorine production by means of ECl_2 and photolysis of chlorine through UV radiation.

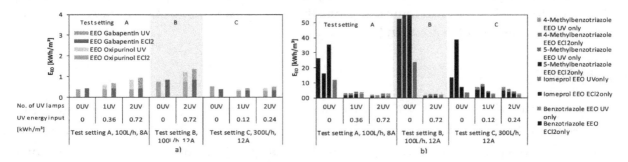

Figure 13. $E_{EO}s$ for gabapentin and oxipurinol (**a**) and 4-methylbenzotriazole, 5-methylbenzotriazole iomeprol and benzotriazole (**b**) calculated from three different settings during field tests with Elbe river water.

For test setting A the E_{EO} for oxipurinol and gabapentin increased from 0.38 to 0.84 and from 0.42 to 0.95 kWh m^{-3} order^{-1} by increasing the energy input for chlorine photolysis from 0 to 0.72 kWh/m^3. This increase is based on the fact that the additional energy input of the UV lamps does not substantially contribute to the degradation, as those substances are readily degraded by ECl_2 alone as shown in Figure 10a. In contrast to that, the chlorine photolysis substantially reduced the E_{EO} for benzotriazole, iomeprol, 5-methylbenzotriazole, and 4-methylbenzotriazole during all test settings. In test setting A the E_{EO} for e.g., 5-methylbenzotriazole was reduced from 35.6 to 2.7 kWh m^{-3} order^{-1} with an energy input for chlorine photolysis caused by the UV lamps of 0.72 kWh/m^3 (2 UV). For benzotriazole the E_{EO} was reduced from 26.5 to 2.3 kWh m^{-3} order^{-1} with the same energy input. During test setting B the UV lamps reduced the E_{EO} for benzotriazole from a rather theoretical value of 52.5 without UV lamps to 1.9 kWh m^{-3} order^{-1} with an UV energy input of 0.72 kWh/m^3 (2 UV). This shows the positive effect of adding a minor quantity of additional energy in the form of UV light into the setting in order to substantially reducing the E_{EO}.

Despite the substantial improvement of TOrC degradation through the photolysis of chlorine the achieved E_{EO} values are higher compared to the literature reported E_{EO} values for e.g., UV/H$_2$O$_2$ AOP. [24] reported an E_{EO} for benzotriazole for H$_2$O$_2$/UV of 0.52 kWh m^{-3} order^{-1}, neglecting hereby the energy required to produce and supply H$_2$O$_2$. However, even after removing the energy required for ECl_2 from the UV/ECl_2 AOP the resulting E_{EO} for e.g., benzotriazole was still higher with 1.0 and 1.2 kWh m^{-3} order^{-1} for test settings A and B. Whether this was related to the different water qualities tested cannot be evaluated here. Due to the limited data available the E_{EO} values calculated and the energy consumed can only be taken as an indication and further studies are required. A reduction of the energy demand by 30–70% by using UV/Cl$_2$ instead of UV/H$_2$O$_2$ as estimated by [24] seems not likely with the settings examined in this work.

3.5. Operational Experience and Optimization

The long term UV/ECl_2 trial was conducted over a period of 10 months during which 1023 m^3 (3.4 m^3/d) of water were treated. The system operated without any interruption caused by technical reasons. The only maintenance required was the manual cleaning of the quartz glass sleeves, from which deposits needed to be removed in a one-month interval, as those have reduced the UV dose. The components, and especially the electrolytic cells, did not show any sign of wear by the time the trial was finished. The concentration of total hardness found in Elbe river water (Table 2) have proven to be harmless with regards the formation of calcareous deposits in the cell surface with the selected polarity inversion intervals of three hours. In future tests these intervals should be extended. No chemicals for the operation and maintenance of the system were required.

Considering the high degradation rates of TOrCs, the comparably low formation of organic DBPs and the low E_{EO} for the setting with a flow rate of 100 L/h, a current at the MOX of 8 A and the UV radiation of 2×36 W lamps (0.72 kWh/m^3) was most promising and should constitute the base for further optimization. At this setting a total energy demand for the AOP process including pumping would sum up to 1.35 kWh/m^3. In remote regions this energy can be supplied easily with renewable PV Energy systems as already shown in [55].

4. Conclusions

The photolysis of chlorine (UV/Cl$_2$) was identified as promising AOP for the degradation of TOrCs present in drinking water sources. In the here presented work the chlorine required to run an UV/Cl$_2$ AOP was produced by means of inline electrolysis from the natural chloride content of the water and by that substituted the external supply of chlorine, allowing the operation of an AOP without any additional need of chemicals.

For FAC concentrations between 1 and 18 mg/L an UV energy input of 0.48 kWh/m^3 photolyzed 84% of the chlorine to form OH and Cl radicals. The degradation could be increased to 97% by doubling the UV energy input for the photolysis. The tests have further shown that the molar radical yield factor is about 64% and also very constant over the tested FAC range of 1–18 mg/L. In principle this allows to control the radical formation by adjusting the chlorine concentration simply through variation of the electric current at the MOX electrode.

By combining ECl$_2$ with UV the degradation of benzotriazole and iomeprol can be increased from 5% to 89% and from 8% to 84%, respectively, compared to the application of ECl$_2$ alone.

In most of the test settings the formation of organic DBPs measured as THMs have reached concentrations above the strict German guideline values for drinking water of 50 µg/L. EU guideline values could be adhered to with all of the tested settings. The formation of such DBPs was related to elevated residual chlorine concentrations after the photolysis of chlorine and can be controlled by reducing the residual chlorine concentrations to levels that are adequate for drinking water disinfection. This requires site specific adaption of the treatment process.

The calculated E_{EO} values have found to be higher than literature values of alternative AOPs. The identified optimization potential should be considered to reduce the overall energy requirement of this technology. The main advantage of the here presented approach is the application of an effective AOP to degrade TOrC independent of any external chemical input. By that the UV/ECl$_2$ AOP constitutes an alternative treatment approach especially for decentralized applications.

Further data and research is required to confirm the here presented results and the effect of the AOP on other TOrCs. Optimization potential is given by increasing the HRT in the UV reactors by adapting the reactor design.

Interesting for future application may also be further treatment of treated wastewater prior to its discharge into the environment.

Author Contributions: Conceptualization, methodology, investigation, P.O., F.-M.K., K.M., A.G., F.B., P.M. and T.G. (Tobias Gerhardt); writing—original draft preparation, P.O. and R.W.; writing—review and editing, P.O., T.G. (Thomas Grischek), F.-M.K., K.M. and T.G. (Tobias Gerhardt); supervision, T.G. (Thomas Grischek); funding acquisition, project administration, P.O. and F.-M.K. All authors have read and agreed to the published version of the manuscript.

Appendix A

Table A1. E_{EO} values achieved during field tests with Elbe river water considering ECl_2 and UV/ECl_2.

	$E_{EO\text{-}(ECl_2)}$ [kWh/(order·m³)]		Elimination [%]		Field test $E_{EO\text{-}(ECl_2 + UV)}$ [kWh/(Order·m³)]		Elimination [%]		Literature values $E_{EO}\text{-}UV/Cl_2$ [kWh/(Order·m³)] [24]
	min	max	min	max	min	max	min	max	
Benzotriazole	13.6	52.5	5%	14%	1.9	7.3	25%	89%	0.5
4-Methyl-benzotriazole	1.3	23.6	11%	18%	0.6	1.5	29%	85%	n.a.
5-Methyl-benzotriazole	7.2	213.9	1%	11%	0.9	1.6	24%	83%	n.a.
Iomeprol	16.1	996.2	0%	29%	1.7	9.3	37%	83%	n.a.
Gabapentin	0.4	≤0.8	94%	≥98%	≤0.4	1.4	≥96%	≥98%	n.a.
Oxipurinol	≤0.4	≤0.7	87%	≥97%	≤0.3	≤0.7	≥97%	≥97%	n.a.

References

1.   Kümmerer, K. *Pharmaceuticals in the Environment*; Springer Berlin Heidelberg: Berlin/Heidelberg, Germany, 2008; ISBN 978-3-540-74663-8.

2.   Ternes, T.; Joss, A. (Eds.) *Human Pharmaceuticals, Hormones and Fragrances. The Challenge of Micropollutants in Urban Water Management*; IWA Publ: London, UK, 2008; ISBN 9781843390930.

3.   Houtman, C.J. Emerging contaminants in surface waters and their relevance for the production of drinking water in Europe. *J. Integr. Environ. Sci.* **2010**, *7*, 271–295. [CrossRef]

4.   Montes-Grajales, D.; Fennix-Agudelo, M.; Miranda-Castro, W. Occurrence of personal care products as emerging chemicals of concern in water resources: A review. *Sci. Total Environ.* **2017**, *595*, 601–614. [CrossRef] [PubMed]

5.   Mohapatra, S.; Huang, C.-H.; Mukherji, S.; Padhye, L.P. Occurrence and fate of pharmaceuticals in WWTPs in India and comparison with a similar study in the United States. *Chemosphere* **2016**, *159*, 526–535. [CrossRef] [PubMed]

6.   Tran, N.H.; Reinhard, M.; Gin, K.Y.-H. Occurrence and fate of emerging contaminants in municipal wastewater treatment plants from different geographical regions-a review. *Water Res.* **2018**, *133*, 182–207. [CrossRef]

7.   Kibuye, F.A.; Gall, H.E.; Elkin, K.R.; Ayers, B.; Veith, T.L.; Miller, M.; Jacob, S.; Hayden, K.R.; Watson, J.E.; Elliott, H.A. Fate of pharmaceuticals in a spray-irrigation system: From wastewater to groundwater. *Sci. Total Environ.* **2019**, *654*, 197–208. [CrossRef]

8.   Balakrishna, K.; Rath, A.; Praveenkumarreddy, Y.; Guruge, K.S.; Subedi, B. A review of the occurrence of pharmaceuticals and personal care products in Indian water bodies. *Ecotoxicol. Environ. Saf.* **2017**, *137*, 113–120. [CrossRef]

9.   Hellauer, K.; Mergel, D.; Ruhl, A.; Filter, J.; Hübner, U.; Jekel, M.; Drewes, J. Advancing Sequential Managed Aquifer Recharge Technology (SMART) using different intermediate oxidation processes. *Water* **2017**, *9*, 221. [CrossRef]

10. Hirsch, R.; Ternes, T.; Haberer, K.; Kratz, K.-L. Occurrence of antibiotics in the aquatic environment. *Sci. Total Environ.* **1999**, *225*, 109–118. [CrossRef]

11. Kolpin, D.W.; Furlong, E.T.; Meyer, M.T.; Thurman, E.M.; Zaugg, S.D.; Barber, L.B.; Buxton, H.T. Pharmaceuticals, hormones, and other organic wastewater contaminants in U.S. streams, 1999-2000: A national reconnaissance. *Environ. Sci. Technol.* **2002**, *36*, 1202–1211. [CrossRef]

12. Schwab, B.W.; Hayes, E.P.; Fiori, J.M.; Mastrocco, F.J.; Roden, N.M.; Cragin, D.; Meyerhoff, R.D.; D'Aco, V.J.; Anderson, P.D. Human pharmaceuticals in US surface waters: A human health risk assessment. *Regul. Toxicol. Pharmacol.* **2005**, *42*, 296–312. [CrossRef]

13. Avisar, D.; Levin, G.; Gozlan, I. The processes affecting oxytetracycline contamination of groundwater in a phreatic aquifer underlying industrial fish ponds in Israel. *Environ. Earth Sci.* **2009**, *59*, 939–945. [CrossRef]

14. Ebele, A.J.; Abou-Elwafa Abdallah, M.; Harrad, S. Pharmaceuticals and personal care products (PPCPs) in the freshwater aquatic environment. *Emerg. Contam.* **2017**, *3*, 1–16. [CrossRef]

15. Bolong, N.; Ismail, A.F.; Salim, M.R.; Matsuura, T. A review of the effects of emerging contaminants in wastewater and options for their removal. *Desalination* **2009**, *239*, 229–246. [CrossRef]

16. Dolar, D.; Gros, M.; Rodriguez-Mozaz, S.; Moreno, J.; Comas, J.; Rodriguez-Roda, I.; Barceló, D. Removal of emerging contaminants from municipal wastewater with an integrated membrane system, MBR-RO. *J. Hazard. Mater.* **2012**, *239–240*, 64–69. [CrossRef] [PubMed]

17. Snyder, S.A.; Adham, S.; Redding, A.M.; Cannon, F.S.; DeCarolis, J.; Oppenheimer, J.; Wert, E.C.; Yoon, Y. Role of membranes and activated carbon in the removal of endocrine disruptors and pharmaceuticals. *Desalination* **2007**, *202*, 156–181. [CrossRef]

18. Yang, Y.; Pignatello, J.J.; Ma, J.; Mitch, W.A. Comparison of halide impacts on the efficiency of contaminant degradation by sulfate and hydroxyl radical-based advanced oxidation processes (AOPs). *Environ. Sci. Technol.* **2014**, *48*, 2344–2351. [CrossRef]

19. Rosenfeldt, E.J.; Linden, K.G.; Canonica, S.; Gunten, U. von. Comparison of the efficiency of *OH radical formation during ozonation and the advanced oxidation processes O_3/H_2O_2 and UV/H_2O_2. *Water Res.* **2006**, *40*, 3695–3704. [CrossRef]

20. Fang, J.; Fu, Y.; Shang, C. The roles of reactive species in micropollutant degradation in the UV/free chlorine system. *Environ. Sci. Technol.* **2014**, *48*, 1859–1868. [CrossRef]

21. Wang, W.-L.; Wu, Q.-Y.; Huang, N.; Wang, T.; Hu, H.-Y. Synergistic effect between UV and chlorine (UV/chlorine) on the degradation of carbamazepine: Influence factors and radical species. *Water Res.* **2016**, *98*, 190–198. [CrossRef]

22. Xiang, Y.; Fang, J.; Shang, C. Kinetics and pathways of ibuprofen degradation by the UV/chlorine advanced oxidation process. *Water Res.* **2016**, *90*, 301–308. [CrossRef]

23. Rott, E.; Kuch, B.; Lange, C.; Richter, P.; Kugele, A.; Minke, R. Removal of emerging contaminants and estrogenic activity from wastewater treatment plant effluent with UV/chlorine and UV/H_2O_2 advanced oxidation treatment at pilot scale. *Int. J. Environ. Res. Public Health* **2018**, *15*, 935. [CrossRef] [PubMed]

24. Sichel, C.; Garcia, C.; Andre, K. Feasibility studies: UV/chlorine advanced oxidation treatment for the removal of emerging contaminants. *Water Res.* **2011**, *45*, 6371–6380. [CrossRef] [PubMed]

25. Latimer, W.M. *The Oxidation States of the Elements and Their Potentials in Aqueous Solutions*, 2nd ed.; Prentice-Hall: New York, NY, USA, 1952.

26. Watts, M.J.; Linden, K.G. Chlorine photolysis and subsequent OH radical production during UV treatment of chlorinated water. *Water Res.* **2007**, *41*, 2871–2878. [CrossRef] [PubMed]

27. Baxendale, J.H.; Wilson, J.A. The photolysis of hydrogen peroxide at high light intensities. *Trans. Faraday Soc.* **1957**, *53*, 344. [CrossRef]

28. Feng, Y.; Smith, D.W.; Bolton, J.R. Photolysis of aqueous free chlorine species (HOCl and OCl) with 254 nm ultraviolet light. *J. Environ. Eng. Sci.* **2007**, *6*, 277–284. [CrossRef]

29. Kishimoto, N. State of the art of UV/chlorine Advanced Oxidation Processes: Their mechanism, byproducts formation, process variation, and applications. *J. Wat. Environ. Tech.* **2019**, *17*, 302–335. [CrossRef]

30. Grebel, J.E.; Pignatello, J.J.; Mitch, W.A. Effect of halide ions and carbonates on organic contaminant degradation by hydroxyl radical-based advanced oxidation processes in saline waters. *Environ. Sci. Technol.* **2010**, *44*, 6822–6828. [CrossRef]

31. Haaken, D.; Dittmar, T.; Schmalz, V.; Worch, E. Influence of operating conditions and wastewater-specific parameters on the electrochemical bulk disinfection of biologically treated sewage at boron-doped diamond (BDD) electrodes. *Desalin. Water Treat.* **2012**, *46*, 160–167. [CrossRef]

32. Haaken, D.; Dittmar, T.; Schmalz, V.; Worch, E. Disinfection of biologically treated wastewater and prevention of biofouling by UV/electrolysis hybrid technology: Influence factors and limits for domestic wastewater reuse. *Water Res.* **2014**, *52*, 20–28. [CrossRef]

33. Kraft, A. Electrochemical water disinfection: A short review. *Platinum Metals Review* **2008**, *52*, 177–185. [CrossRef]

34. Kraft, A.; Stadelmann, M.; Blaschke, M.; Kreysig, D.; Sandt, B.; Schröder, F.; Rennau, J. Electrochemical water disinfection: Part I: Hypochlorite production from very dilute chloride solutions. *J. Appl. Electrochem.* **1999**, *29*, 859–866. [CrossRef]

35. Buxton, G.V.; Subhani, M.S. Radiation chemistry and photochemistry of oxychlorine ions. Part 2. Photodecomposition of aqueous solutions of hypochlorite ions. *J. Chem. Soc. Faraday Trans.* **1972**, *68*, 958. [CrossRef]

36. Asmus, K.D.; Moeckel, H.; Henglein, A. Pulse radiolytic study of the site of hydroxyl radical attack on aliphatic alcohols in aqueous solution. *J. Phys. Chem.* **1973**, *77*, 1218–1221. [CrossRef]

37. Abdelraheem, W.H.; Nadagouda, M.N.; Dionysiou, D.D. Solar light-assisted remediation of domestic wastewater by NB-TiO_2 nanoparticles for potable reuse. *Applied Catalysis B Environ.* **2020**, *269*, 118807. [CrossRef]

38. LfULG, online data base iDA, Saxon State Authority for Environment, Agriculture and Geology. 2020. Available online: https://www.umwelt.sachsen.de/umwelt/infosysteme/ida/ (accessed on 5 August 2020).

39. Bolton, J.R.; Bircher, K.G.; Tumas, W.; Tolman, C.A. Figures-of-merit for the technical development and application of advanced oxidation technologies for both electric- and solar-driven systems. *Pure Appl. Chem.* **2001**, *2001*, 627–637. [CrossRef]

40. Bolton, J.R.; Stefan, M.I. Fundamental photochemical approach to the concepts of fluence (UV dose) and electrical energy efficiency in photochemical degradation reactions. *Res. Chem. Intermed.* **2002**, *28*, 857–870. [CrossRef]

41. Monod, A.; Chebbi, A.; Durand-Jolibois, R.; Carlier, P. Oxidation of methanol by hydroxyl radicals in aqueous solution under simulated cloud droplet conditions. *Atmos. Environ.* **2000**, *34*, 5283–5294. [CrossRef]

42. Nash, T. Colorimetric determination of formaldehyde under mild conditions. *Nature* **1952**, *170*, 976. [CrossRef]

43.   Payne, W.A.; Brunning, J.; Mitchell, M.B.; Stief, L.J. Kinetics of the reactions of atomic chlorine with methanol and the hydroxymethyl radical with molecular oxygen at 298 K. *Int. J. Chem. Kinet.* **1988**, *20*, 63–74. [CrossRef]
44.   Kraft, A.; Blaschke, M.; Kreysig, D.; Sandt, B.; Schröder, F.; Rennau, J. Electrochemical water disinfection. Part II: Hypochlorite production from potable water, chlorine consumption and the problem of calcareous deposits. *J. Appl. Electrochem.* **1999**, *29*, 895–902. [CrossRef]
45.   Jin, J.; El-Din, M.G.; Bolton, J.R. Assessment of the UV/chlorine process as an advanced oxidation process. *Water Res.* **2011**, *45*, 1890–1896. [CrossRef] [PubMed]
46.   Paufler, S.; Grischek, T.; Benso, M.; Seidel, N.; Fischer, T. The impact of river discharge and water temperature on manganese release from the riverbed during riverbank filtration: A case study from Dresden, Germany. *Water* **2018**, *10*, 1476. [CrossRef]
47.   Houk, L.L.; Johnson, S.K.; Feng, J.; Houk, R.S.; Johnson, D.C. Electrochemical incineration of benzoquinone in aqueous media using a quaternary metal oxide electrode in the absence of a soluble supporting electrolyte. *J. Appl. Electrochem.* **1998**, *28*, 1167–1177. [CrossRef]
48.   Johnson, S.K.; Houk, L.L.; Feng, J.; Houk, R.S.; Johnson, D.C. Electrochemical incineration of 4-chlorophenol and the identification of products and intermediates by mass spectrometry. *Environ. Sci. Technol.* **1999**, *33*, 2638–2644. [CrossRef]
49.   Wang, D.; Bolton, J.R.; Andrews, S.A.; Hofmann, R. Formation of disinfection by-products in the ultraviolet/chlorine advanced oxidation process. *Sci. Total Environ.* **2015**, *518–519*, 49–57. [CrossRef]
50.   Yang, X.; Sun, J.; Fu, W.; Shang, C.; Li, Y.; Chen, Y.; Gan, W.; Fang, J. PPCP degradation by UV/chlorine treatment and its impact on DBP formation potential in real waters. *Water Res.* **2016**, *98*, 309–318. [CrossRef]
51.   Bundesministerium für Justiz und Verbraucherschutz. Verordnung über die Qualität von Wasser für den menschlichen Gebrauch (Trinkwasserverordnung-TrinkwV). 2001. Gesetze im Internet. Available online: https://www.gesetze-im-internet.de/trinkwv_2001/BJNR095910001.html (accessed on 18 November 2020).
52.   European Commission. COUNCIL DIRECTIVE 98/83/EC of 3 November 1998 on the quality of water intended for human consumption. 1998. EUR-lex. Available online: https://eur-lex.europa.eu/legal-content/en/TXT/?uri=CELEX:52017PC0753 (accessed on 18 November 2020).
53.   IS 10500. *Drinking Water—Specification*; Indian Standard; Bureau of Indian Standards: New Delhi, India, (Second Revision); 2012; Law-resource; Available online: https://law.resource.org/pub/in/bis/S06/is.10500.2012.pdf (accessed on 18 November 2020)Law-resource.
54.   *Guidelines for Drinking-Water Quality*, 4th ed; World Health Organization: Geneva, Switzerland, Incorporating 1st Addendum; 2017; Available online: https://apps.who.int/iris/bitstream/handle/10665/44584/9789241548151_eng.pdf;jsessionid=592F2A53BA42E1A3A26CABA04706353E?sequence=1 ISBN 978-9241549950.
55.   Otter, P.; Malakar, P.; Jana, B.; Grischek, T.; Benz, F.; Goldmaier, A.; Feistel, U.; Jana, J.; Lahiri, S.; Alvarez, J. Arsenic removal from groundwater by solar driven inline-electrolytic induced co-precipitation and filtration—A long term field test conducted in West Bengal. *IJERPH* **2017**, *14*, 1167. [CrossRef]

3

A MATLAB-Based Application for Modeling and Simulation of Solar Slurry Photocatalytic Reactors for Environmental Applications

Raúl Acosta-Herazo [1],*, Briyith Cañaveral-Velásquez [2], Katrin Pérez-Giraldo [2], Miguel A. Mueses [2], María H. Pinzón-Cárdenas [1] and Fiderman Machuca-Martínez [1]

[1]   School of Chemical Engineering, Universidad del Valle, Cali A.A. 25360, Colombia;
maria.pinzon@correounivalle.edu.co (M.H.P.-C.); fiderman.machuca@correounivalle.edu.co (F.M.-M.)
[2]   Modeling and Applications of Advanced Oxidation Technologies Research Group, Photocatalysis and Solar
Photoreactors Engineering, Department of Chemical Engineering, Universidad de Cartagena,
Cartagena A.A. 1382-195, Colombia; briyithcarolina@hotmail.com (B.C.-V.); klpg_15@hotmail.com (K.P.-G.);
mmueses@unicartagena.edu.co (M.A.M.)
*   Correspondence: raul.acosta@correounivalle.edu.co

Abstract: Because of the complexity caused by photochemical reactions and radiation transport, accomplishing photoreactor modeling usually poses a barrier for young researchers or research works that focus on experimental developments, although it may be a crucial tool for reducing experimental efforts and carrying out a more comprehensive analysis of the results. This work presents PHOTOREAC, an open-access application developed in the graphical user interface of Matlab, which allows a user-friendly evaluation of the solar photoreactors operation. The app includes several solar photoreactor configurations and kinetics models as well as two variants of a radiation absorption-scattering model. Moreover, PHOTOREAC incorporates a database of 26 of experimental solar photodegradation datasets with a variety of operational conditions (model pollutants, photocatalyst concentrations, initial pollutant concentrations); additionally, users can introduce their new experimental data. The implementation of PHOTOREAC is presented using three example cases of solar photoreactor operation in which the impact of the operational parameters is explored, kinetic constants are estimated according to experimental data, and comparisons are made between the available models. Finally, the impact of the application on young researchers' projects in photocatalysis at the University of Cartagena was investigated. PHOTOREAC is available upon request from Professor Miguel Mueses.

Keywords: computer-based learning; solar photocatalysis; water contaminants; kinetic modeling; photoreactor design

1. Introduction

Heterogeneous photocatalysis is an example of an emerging environmental technology with a variety of promising applications, such as air and water disinfection and decontamination, clean fuel production and green product manufacturing [1–3].

Modeling and computer simulation of photoreactors are crucial for their design, scale-up and technology transfer; since they allow engineers and researchers to understand the role of the design parameters and operational conditions without performing an excessive number of experiments. However, modeling a solar photoreactor is a very complicated task, because it requires a combination of knowledge in applied solar energy, geometric optics, radiative transfer, materials science and photochemical reaction engineering.

The implementation of commercial packages for photoreactor simulations is limited. Simulation packages for chemical plants, such as Aspen HYSYS® (Aspen Technology, Inc., Bedford, MA, USA) or Aspen plus® (Aspen Technology, Inc., Bedford, MA, USA), do not incorporate photocatalytic reactors. On the other hand, modeling and simulation of photoreactors can be carried out in Computational Fluid Dynamics (CFD) packages, such as COMSOL Multiphysics® (COMSOL, Inc., Burlington, MA, USA) and ANSYS® Fluent (ANSYS, Inc., Southpointe, PA, USA). However, they do not have modules dedicated to photoreactor engineering. Therefore, the simulations are performed by adapting the existing simulation modules for the simulation of photocatalytic reactors. This configuration of the CFD modules must be carried out manually by the user, which may result in an approach not intuitive enough for non-experts in photoreactor engineering. Another alternative is to perform the direct coding of the photoreactor model in a programming language. Still, this may result in a challenge for researchers that have not taken advanced courses in programming and numerical methods.

For the above reasons, the direct coding or the use of CFD simulators to implement a photoreactor model could be found inconvenient by non-expert researchers in photoreactors engineering, such as young researchers or those focused on experimental developments. However, implementing a photoreactor model may be a crucial tool for reducing the experimental efforts and carrying out a more comprehensive analysis of the results.

In this work, we present PHOTOREAC, an open-access computational application developed in the graphical user interface of Matlab wholly dedicated to the modeling and simulation of large-scale slurry solar photocatalytic reactors for environmental applications. It is based on the experience gathered by our research groups at Cartagena University (Cartagena de Indias, Colombia) and the Universidad del Valle (Cali, Colombia) during the last twenty years of research in heterogeneous solar photocatalysis, and also on extensive literature research in photoreactor engineering.

The application aims to provide non-expert researchers in photoreactors engineering a user-friendly, dedicated and efficient tool for the modeling and simulation of solar photoreactors, providing them with valuable information without implementing very sophisticated methods.

By employing PHOTOREAC, the users will be able to explore the role of critical parameters of the system on the radiation absorption performance of the photoreactor and the overall kinetic behavior of the photocatalytic process; parameters include the photoreactor geometry, the photoreactor dimensions, the model pollutant, the kinetic expression, the photocatalyst concentration, the photocatalysts optical properties, the initial pollutant concentration, the volume of treated water and the incident radiation. Additionally, PHOTOREAC incorporates a database of experimental information collected in our laboratory regarding the solar photodegradation of a variety of model pollutants under different operational conditions. Therefore, users will have empirical data available to carry out analyses and comparisons with their data.

2. Solar Photoreactors Modeling by PHOTOREAC

PHOTOREAC performs the modeling and simulation of the photoreactors following the general algorithm described in Figure 1. The algorithm considers mathematical simplifications to maintain the approach as rigorously and computationally efficient as possible, and thus it provides the users with valuable information without implementing sophisticated numerical methods that, although they can improve the quantitative results, may not affect the qualitative analysis. These assumptions and simplifications will be described and discussed in the upcoming sections.

The basis of the PHOTOREAC approach is that the radiation field modeling can be carried out independently of the photocatalytic kinetics modeling since the radiation balance in the photoreactor is not a function of the concentration of the chemical species. Therefore, the radiation balance is decoupled from the mass and momentum balances of the system. Besides, the radiation field described by the local volumetric rate of photon absorption (LVRPA) profile inside the photoreactor is considered to be in a steady-state, i.e., it does not vary along the reaction rime its reaction time does not change [4,5]. On the other hand, to carry out a kinetic analysis independent of the radiation absorption effects, i.e., the optimized kinetic parameters are not a function of the irradiation conditions, it is mandatory to know the radiation field in the photoreactor beforehand [6,7].

Thus, PHOTOREAC considers two modules: (i) the photon absorption-scattering module, in which the user will be able to determine the radiation field of the available photoreactor configurations by following the procedure described by the red box in Figure 1; and (ii) the kinetic modeling module, in which the user will be able to estimate the radiation-independent kinetic parameters for the four available kinetics expressions following the procedure described in the blue box in Figure 1.

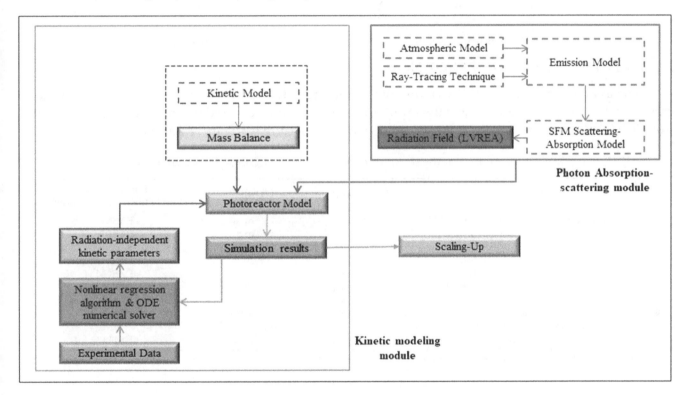

Figure 1. General algorithm for the modeling and simulation of solar slurry photoreactors in PHOTOREAC. ODE: Ordinary Differential Equation; SFM: Six Flux Model.

2.1. The Photoreactors Set-Up in PHOTOREAC

PHOTOREAC includes three configurations of pilot-scale solar photoreactors: a flat plate photoreactor (FPP), a compound parabolic collector photoreactor (CPCP) and a tubular-type photoreactor (TTP). These are the most common configurations for solar-pilot applications of heterogeneous photocatalysis; a detailed description of them can be found in the literature [3,8,9]. For the TTP, a novel prototype is also included, the offset multi-tubular photoreactor (OMTP) [10]. All of the photoreactors operate in recirculation, a flow-through mode with the water passing through an external tank, as shown in Figure 2. The photoreactor is exposed to the sunlight, facing the sun, while the reservoir tank is in the dark. The flow consists of an aqueous suspension of photocatalyst powder and the dissolved contaminant. The Evonik TiO_2 P25 was selected as the model photocatalyst in PHOTOREAC because it is considered the most promising alternative for commercial applications due to its low cost, photochemical stability, and high oxidation power [3]. Therefore, it is widely studied, and its physicochemical and optical properties are well known in the literature [11].

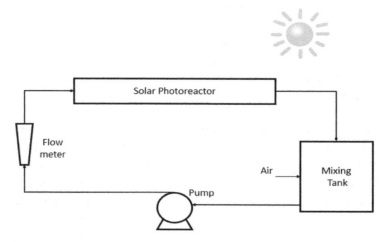

Figure 2. Scheme of a solar pilot photoreactor set-up.

2.2. The Input Data for the Use of PHOTOREAC

The availability and reliability of the input data provided to PHOTOREAC are crucial for good results. Table 1 shows a summary of the input information that is required to simulate the photoreactors in PHOTOREAC. Additionally, it is indicated in which module is the information used.

Table 1. Summary of the input information for the PHOTOREAC modules.

Parameter	Symbol	Units	Belonging to PHOTOREAC Module [a]
Photoreactor radius (CPCP and OMTP)	R	m	PASM
Water film thickness (FFP)	δ	m	PASM
Photoreactor length	L	m	PASM
Solar incident radiation	I_0	W/m^2	PASM
Reaction volume	V_R	L	PASM/KMM
Photocatalyst concentration	C_{cat}	g/L	PASM/KMM
Total volume	V_T	L	KMM
Number of experimental photodegradation data	N	Dimensionless	KMM
Concentration vs. accumulated energy data [b]	C_i vs. ξ_{AE}	ppm vs. J/m^2	KMM
Concentration vs. standard time data [b]	C_i vs. t	ppm vs. min	KMM

[a] PASM: photon absorption-scattering module; KMM: kinetic modeling module. [b] If it is a multicomponent mixture, C_i is replaced by *TOC*.

At the same time, the experimental photodegradation data for kinetic analysis in PHOTOREAC deserves special attention. The effects of the adsorption must be carefully considered in the solar photocatalytic experimental test. The photodegradation data used to feed PHOTOREAC must be reported at the zero-point of photodegradation, where adsorption has already been allowed to homogenize, which is usually achieved by allowing the system to recirculate under darkness for 30 min to establish adsorption–desorption equilibrium conditions before being exposed to solar light. Thus, although the kinetic models in PHOTOREAC do not contemplate the competitive effects of molecular adsorption, the data used will already be corrected with that effect. Therefore, there is no problem with the application of the models [10,12].

During the exposure time to sunlight, the data should be reported as the pollutant concentration C_i against the accumulated ultraviolet (UV) energy ξ_{AE}. The experiments finish when the desired accumulated UV total energy in J/m^2 is reached. Additionally, it is required to record the corresponding standard time for each sample.

2.3. The PHOTOREAC Photon Absorption-Scattering Module

The PHOTOREAC photon absorption-scattering module performs the radiation field modeling of the three available configurations of solar photoreactors: FPP, CPCP and OMTP. It provides the

LVRPA spatial distribution inside the photoreactor and the overall volumetric rate of photon absorption (OVRPA), which corresponds to the LVRPA averaged over the entire volume of the reactor. The latter is a critical magnitude for the kinetic assessment [12].

The PHOTOREAC modeling approach is focused on the six-flux absorption-scattering model (SFM). SFM is an analytical equation in which the leading hypothesis is that scattering only occurs in the six Cartesian directions [5]. Despite being a simplified model, it retains the key aspects of the radiation field modeling in photoreactors and has been implemented successfully at the solar pilot scale [13,14]. Other modeling approaches for solar photoreactors, such as the discrete ordinate method (DOM) or the Monte Carlo simulation, offer a more accurate description of the radiation transport phenomena. However, they are more time-consuming in the computations and their mathematical formulation is of high complexity. The SFM short computation times are ideal for exploring the impact of operational parameters, including the photocatalyst concentration, photoreactor dimensions and incident radiation, in particular for users that are dabbling in photoreactor engineering, to which PHOTOREAC is oriented. Independently of the photoreactor configuration, the central equation of SFM is given by [14]:

$$LVRPA = \frac{I_0}{\lambda_{\omega corr} \omega_{corr}(1-\gamma)}\left[\left(\omega_{corr} - 1 + \sqrt{1-\omega_{corr}^2}\right)e^{-\frac{r_p}{\lambda_{\omega corr}}} + \gamma\left(\omega_{corr} - 1 - \sqrt{1-\omega_{corr}^2}\right)e^{\frac{r_p}{\lambda_{\omega corr}}}\right] \quad (1)$$

where I_0 is the incident solar radiation in W/m^2 and r_p is a spatial coordinate in the reactor domain whose definition depends on the reactor geometry. Finally, the corrected photon path length $\lambda_{\omega corr}$ in m, the dimensionless corrected scattering albedo ω_{corr} and the dimensionless parameter γ are all parameters derived from the SFM formulation. PHOTOREAC also includes a more recent variant of the SFM, the Six Flux Model coupled to the Henyey-Greenstein scattering phase function (SFM-HG). In it, the Henyey–Greenstein (HG) scattering phase function is used to describe the optical properties of the TiO$_2$ P25 photocatalyst. By contrast, the SFM describes TiO$_2$ based on a diffuse reflectance scattering phase function [15]. By incorporating both variants of SFM, the users will be able to observe the role of the scattering phase function. The parameters and implementation of Equation (1) are detailed in the literature, and the modeling details for the FFP are given in previous work [16].

On the other hand, for the CPCP and the OMTP, a ray-tracing technique together with Equation (1) must be implemented, since, besides the incident radiation, the direction with which solar rays impact the photoreactor is crucial. A complete description of the SFM implementation for CPCP and OMTP is reported elsewhere [10,13,14].

2.4. The PHOTOREAC Kinetic Modeling Module

The PHOTOREAC kinetic modeling module estimates the kinetic parameters from the photodegradation experimental data provided. Table 2 shows the photocatalytic kinetic models in PHOTOREAC. These models explicitly consider the effect of the radiation absorption on the average reaction rate in $\langle -r_i \rangle_{V_R}$ by including the E_g, and the overall rate of photon absorption (OVRPA) in W/m^3, which corresponds to the LVRPA averaged over the entire volume of the reactor. Additionally, C_i is the concentration of the water contaminant in mol/m^3, $\kappa_P = 2/S_g\,C_{cat}$ is the particle constant in m^3/m^2, S_g is the catalyst specific surface area m^2/kg, C_{Cat} is the photocatalyst concentration kg/m^3, C_{O2} is the oxygen concentration in mol/m^3 and ϕ_g^{eff} is the effective quantum yield in mol/(s watts). Finally, k^{L-H}, K_{kin}, α_1 and α_2 are the kinetic constants of the models, which are independent of the irradiation conditions.

Table 2. Photocatalytic kinetic models in PHOTOREAC.

Kinetic Model	Mathematical Expression	Fitting Parameters	Refs.
Langmuir–Hinshelwood	$\langle -r_i \rangle_{V_R} = -\dfrac{K_{Kin} k^{L-H} C_i}{1 + k^{L-H} C_i} \left(E_g \right)^{0.5}$	k^{L-H} (L/mol), K_{kin} (mol L^{-1} s^{-1} $W^{-0.5}$)	[13]
Zalazar et al.	$\langle -r_i \rangle_{V_R} = -\dfrac{\phi_g^{eff} E_g}{\dfrac{1}{2} + \left[\dfrac{1}{4} + K_{kin} \dfrac{\phi_g^{eff} E_g}{2 C_{cat}^2 C_i C_{O_2}} \right]^{0.5}}$	ϕ_g^{eff} (mol s^{-1} watts^{-1}), K_{kin} (mole s kg^2 m^{-9})	[17]
Ballari et al.	$\langle -r_i \rangle_{V_R} = -2 \dfrac{\alpha_1}{\kappa_P} \left[-1 + \sqrt{1 + \kappa_P \dfrac{\alpha_2 E_g}{C_i}} \right] C_i$	α_1 (cm s^{-1}), α_2 (mol watts^{-1} cm^{-1})	[18]
Mueses et al.	$\langle -r_i \rangle_{V_R} = -2 \dfrac{\alpha_1}{\kappa_P} \left[-1 + \sqrt{1 + \dfrac{\kappa_P}{\alpha_1} \phi_g^{eff} E_g} \right] \dfrac{k^{L-H} C_i}{1 + k^{L-H} C_i}$	α_1 (mol m^{-2} s^{-1}), ϕ_g^{eff} (mol s^{-1} watts^{-1}), k^{L-H}(m^3 mol^{-1})	[12]

Each of these previous expressions has its features and limitations, from either a phenomenological or a numerical point of view. For instance, the Langmuir–Hinshelwood expression is a semi-empirical model. By contrast, the other models were deduced from a detailed reaction mechanism. Zalazar et al. and Mueses et al.'s kinetic expressions consider the effect of the effective quantum yield ϕ_g^{eff} explicitly, a critical parameter to evaluate photocatalytic reactions. However, the expression proposed by Mueses et al. is the only one with three fitting parameters, unless the effective quantum yield of the system is previously known [12].

To determine the kinetics parameters, it is necessary to follow a rigorous approach to account for the effects of the diffusion and convection in the material balance of the photoreactor. Although the inclusion of these effects will provide more accurate results for the kinetic parameters (such parameters will be independent of the diffusion and convection), it also implies the implementation of more advanced numerical techniques, e.g., finite differences and orthogonal collocation [7,19]. PHOTOREAC considers the photoreactor-tank system as a batch mode reactor; therefore, the effects of the diffusion and convection are lumped in the kinetic parameters, which simplifies the numerical approach.

The following assumptions are established for the mass balance of the system (represented by Figure 2): (i) the system is perfectly mixed; (ii) there are no mass transport limitations; (iii) the conversion per pass in the reactor is differential; and (iv) parallel dark reactions can be neglected. The mass balance in the reservoir tank can then be expressed as follows [7,18]:

$$\frac{dC_i}{dt} = \frac{V_R}{V_T} \langle -r_i \rangle_{V_R} \tag{2}$$

where C_i is the concentration of the water contaminant in mol/m^3 at time t, t is time in s, $\langle -r_i \rangle_{V_R}$ is the average reaction rate in (mol m^3 s^{-1}), and V_R and V_T are the volumes of the photoreactor and the total reaction volume in m^3, respectively. However, for solar photoreactors, the standard time may not be the more appropriate magnitude for following the concentration of the water pollutant due to the fluctuation of the incident solar irradiance because of the atmospheric phenomena and the time of day. Therefore, a change of variable is proposed as follows [10]:

$$\frac{dC_i}{dt} = \left(\frac{dC_i}{d\xi_{AE}} \right) \left(\frac{d\xi_{AE}}{dt} \right) \tag{3}$$

$$\frac{dC_i}{d\xi_{AE}} = \frac{\beta}{\xi_t} \langle -r_i \rangle_{V_R} \tag{4}$$

With the initial condition, C_i ($\xi_{AE} = 0$) $= C_{i,0}$, where C_i is the water contaminant concentration for a given ξ_{AE} is the accumulated energy in J/m^2, $\xi_t = \left(\frac{d\xi_{AE}}{dt} \right)$ in J/m^{2}s is the slope of the straight line resulting from the experimental data relationship of the accumulative incident solar radiation vs. time for each experimental test, and the dimensionless factor $\beta = V_R/V_T$.

The search for the best values for the kinetic parameters of the model is carried out using a non-linear regression procedure, as is shown in Figure 1. It starts with an initial guess and follows an optimization criterion until the required convergence is reached. The error function is given by the sum of the squared errors of the experimental water contaminant concentration $C_{i,\exp}$ and the value determined from the numerical solution of Equation (4) $C_{i,calc}$:

$$F_{obj} = \sum_{i=1}^{N} \left(C_{i,\exp} - C_{i,calc} \right)^2 \qquad (5)$$

where N is the number of experimental data. The Matlab function fminsearch, which uses the Nelder–Mead algorithm, is implemented as the optimization solver together with the Matlab function ode 45 for solving the ordinary differential equation (ODE) given by Equation (4).

For the photodegradation of multicomponent mixtures, the concentration C_i may be replaced by a global concentration parameter such as total organic carbon (TOC) [12]. Therefore, Equation (4) is written as:

$$\frac{dTOC}{d\xi_{AE}} = \frac{\beta}{\xi_t} \langle -r_{TOC} \rangle_{V_R} \qquad (6)$$

with the initial condition $TOC (\xi_{AE} = 0) = TOC_0$, where TOC is the total organic carbon of the mixture mol/m^3 for a given ξ_{AE}, ξ_{AE} is the accumulated energy in J/m^2, TOC_0 is TOC of the mixture measured at the starting point of the experiment, and V_R and V_T are the volumes of the photoreactor and the total reaction volume in m^3, respectively. $\langle -r_{TOC} \rangle_{V_R}$ is the average reaction rate of the TOC of the mixture in (mol m^3 s^{-1}). The mathematical expressions for $\langle -r_{TOC} \rangle_{V_R}$ are the same given in Table 2, replacing $\langle -r_i \rangle_{V_R}$ by $\langle -r_{TOC} \rangle_{V_R}$ and C_i by TOC.

Similarly, Equation (5) is rewritten as:

$$F_{obj} = \sum_{i=1}^{N} \left(TOC_{i,\exp} - TOC_{i,calc} \right)^2 \qquad (7)$$

Then, for multicomponent mixtures, the TOC of the mixture must be provided to PHOTOREAC as a function of the accumulated energy instead of the concentration of a pure component water contaminant. This approach is particularly useful in real environmental applications because in such cases the most usual situation is that the content of the wastewater is unknown, and it would be tough and resource-consuming to determine it. Therefore, it is easier to establish a global parameter such as the TOC, which shows the mineralization of both intermediates and the precursor compounds in the wastewater. By contrast, the monitoring of each initial pure component in the mixture does not consider the formation of intermediates.

The Kinetic Modeling Module Database

In the kinetic modeling module, PHOTOREAC incorporates a database that consists of 26 datasets of the solar photocatalytic degradation of water contaminants using TiO$_2$ P25 Evonik as a photocatalyst. The information was collected by the Modeling and Applications of Advanced Oxidation Technologies Research Group at Cartagena University (Cartagena de Indias, Colombia) and the Research Group on Advanced Processes for Biological and Chemical Treatments (GAOX) at the Universidad del Valle (Cali, Colombia). Table 3 details the information available in the database: two solar photoreactor configurations (CPCP and OMTP) and five model pollutants at different initial concentrations and photocatalyst concentrations. By selecting the dataset to perform the kinetic analysis, PHOTOREAC loads the information about the experimental test: the pollutant concentration vs. accumulated energy data, the OVRPA and the $\beta = V_R/V_T$ factor.

Table 3. PHOTOREAC database of the solar photodegradation of water contaminants.

Water Contaminant	Photoreactor Configuration	Initial Concentration of the Contaminant, ppm	Photocatalyst Concentration, g/L
Dichloroacetic acid (DCA)	CPCP	30 60 120	0.1, 0.5 0.1, 0.35 0.1, 0.35, 0.5
	OMTP	60 120	0.35 0.35
Phenol (PH)	CPCP	60 120	0.1 0.1
	OMTP	60 120	0.1 0.1
4-chlorophenol (4-CP)	CPCP	60 120	0.5 0.5
	OMTP	60 120	0.5 0.5
Methylene Blue (MB)	CPCP	10	0.25
	OMTP	10	0.2, 0.25, 0.3, 0.35
Amoxicillin (AMX)	CPCP	20	0.3, 0.6, 0.9, 1.0

3. Implementation of PHOTOREAC in Solar Photoreactors

In this section, three example cases to demonstrate the use of the PHOTOREAC application are presented. All of the cases are based on an experimental test already performed in the solar photoreactor platforms of our research groups in Cartagena, Colombia (10°25′25″ N, 75°31′31″ W) and Cali, Colombia (3°27′00″ N, 76°32′00″ W). Further information about the set-up and operation of the experimental solar tests can be found in previous works [10,12].

3.1. Example Case I: Solar Photodegradation of Dichloroacetic Acid (DCA) in a CPCP

This example shows the implementation of PHOTOREAC for an analysis of the solar photocatalytic degradation of DCA in a CPCP. The photoreactor consists of ten borosilicate tubes with radius $R = 0.016$ m and length $L = 1.2$ m providing a reaction volume of $V_R = 9.7$ L. The DCA initial concentration was $C_i = 30$ ppm using a TiO_2 P25 Evonik concentration of $C_{cat} = 0.5$ g/L. The main objective of the example case was to determine the radiation-independent kinetic parameters of the system from the experimental data provided to the application using the SFM as the radiative model.

First, the radiation field is determined by the photon absorption-scattering module. Figure 3 shows the main screen of the PHOTOREAC GUI: (1) the photoreactor panel, where the photoreactor configuration was selected; (2) the system properties panel, where the input data were introduced for the simulation; (3) the SFM model panel, where the SFM variant for the simulation is selected; (4) the SFM scattering phase function probabilities are displayed according to the SFM variant that was selected, in this case, the SFM; (5) the resulting LVRPA spatial distribution in the cross-section of the CPCP tube is plotted; (6) the resulting OVRPA of the system is displayed; (7) the options menu. Together with the main screen shown in Figure 3, PHOTOREAC generates a secondary screen with the results of the ray-tracing simulation (Figure 4).

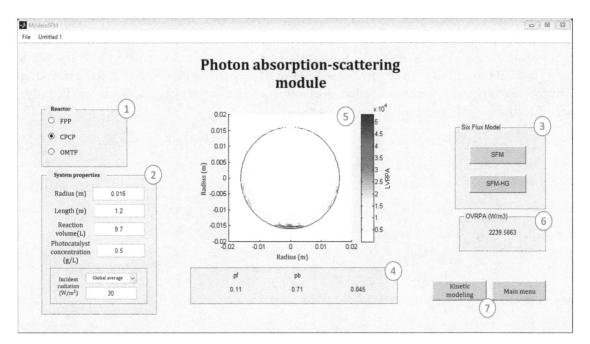

Figure 3. Radiation field simulation for a CPCP in the photon absorption-scattering module. (1) photoreactor panel; (2) system properties panel; (3) the SFM panel; (4) display the corresponding SFM scattering probabilities used in the simulation; (5) LVRPA spatial distribution plot; (6) OVRPA; (7) options menu.

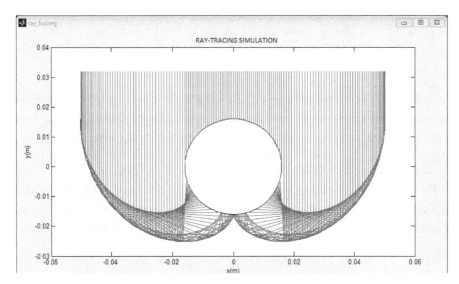

Figure 4. Ray-tracing simulation for the CPCP.

From the results presented by the photon absorption-scattering module, the user will be able to extract essential findings regarding the impact of variables on the photocatalyst concentration. For instance, for this example case, in the LVRPA distribution plot shown in Figure 3, it is observed that the highest values of the LVRPA are around $y = -0.015$ m and $y = -0.005$ m. This result is due to the fact that at these coordinates there is a high concentration of rays that come from the CPCP reflectors, as can be observed in Figure 4. Additionally, it is observed that the LVRPA is concentrated near to the CPCP wall, and the center of the tube shows very low LVRPA values, as a result of the relatively high photocatalyst concentration used in the simulation ($C_{cat} = 0.5$ g/L). This behavior is well-known in the literature: at high concentrations of the photocatalyst, the photons cannot penetrate deeply into the tube and the absorbed energy is concentrated around the boundary wall [14].

Once the radiation field for the CPCP is determined, the application proceeds to the kinetic modeling module. Figure 5 shows the input panel displayed by PHOTOREAC. The application loads the system parameters determined previously, such as the TiO_2 concentration and the OVRPA. The remaining system parameters must be provided manually by the user. Similarly, the photodegradation vs. accumulated energy data should be introduced in the experimental data panel. Finally, the user may proceed to the kinetic modeling module's main screen.

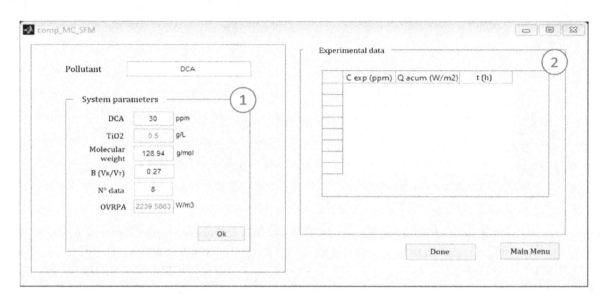

Figure 5. Input panel for the kinetic modeling module. (1) System parameters panel; (2) experimental data panel.

Figure 6 shows the main screen of the PHOTOREAC GUI at the kinetic modeling module: (1) kinetic models panel, where the user can choose the kinetic models to be fitted; (2) experimental data and models simulations plot, where the experimental data and the fitting curves of the models that were previously selected are displayed; (3) fitted kinetic parameters panel, where the values of the fitting parameters of each model chosen are displayed; (4) the x-axis magnitude panel, where the user can determine if the displayed data are presented in accumulated energy or standard time as the x-axis magnitude; (5) correlation coefficient panel, which displays the higher R^2 among the selected kinetic models; (6) correlation coefficient panel, which shows the kinetic model with the highest R^2 value among the chosen ones; (7) export data button, which exports the results of the fitting curve to a Microsoft Excel file; (8) options menu panel.

From the PHOTOREAC kinetic modeling module screen in Figure 6, it is observed that the best fitting is achieved for the Ballari et al. model with $R^2 = 0.97392$. The other models reported $R^2 = 0.97365$ for Mueses et al., $R^2 = 0.63843$ for Langmuir–Hinshelwood and $R^2 = 0.$ 63585 for Zalasar et al. Due to PHOTOREAC only displaying the model with the higher value for the correlation coefficient R^2, it selected the Ballari et al. model. However, Mueses et al.'s expression showed an almost identical R^2, and it should not be discarded without further analysis. From Table 2, it is observed that the mathematical structure of the Ballari et al. and Mueses et al. expressions are very similar; indeed, the Ballari et al. expression is considered a particular case of the Mueses et al. model for systems with high molecular adsorption [12]. Therefore, it is expected that both models performed similarly, as is the case for the DCA photodegradation. The Langmuir–Hinshelwood and Zalasar et al. expressions may not lead to successful results due to the fact that they do not describe the effects of the absorbed radiation (OVRPA) accurately. On the other hand, Ballari et al. and Mueses et al. may perform better since they include an OVRPA squared root correction factor. the same can be said for Ballari et al. and Mueses et al. regarding the OVRPA squared root correction factor.

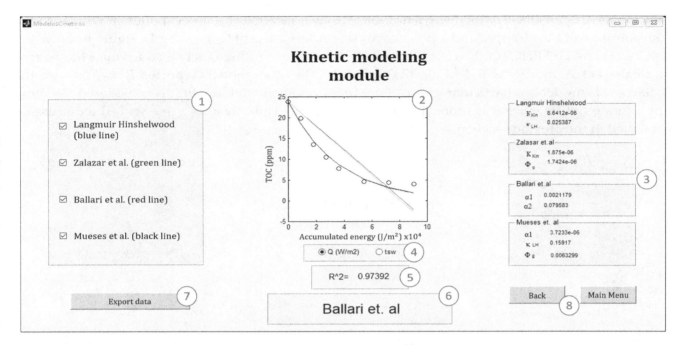

Figure 6. DCA fitting for the available kinetics models in the kinetics modeling module. (1) Kinetic models panel; (2) experimental data and models simulations plot; (3) fitted kinetic parameters panel; (4) x-axis magnitude panel; (5) correlation coefficient panel; (6) display of the kinetic model with the highest R^2 value among the ones selected; (7) export data button; (8) options menu panel.

3.2. Example Case II: Solar Photodegradation of Methylene Blue in an OMTP

In the previous example case, the user must provide all the required information to perform the computations. In this example, the use of the database incorporated in the PHOTOREAC kinetic modeling module is shown. Figure 7 shows the PHOTOREAC screen of the kinetic modeling module: (1) the photoreactor configuration panel, for selecting the photoreactor to be studied; (2) the model pollutant panel, for choosing the water contaminants from the five available options in the database; (3) the photocatalyst-pollutant panel, for choosing the photocatalyst concentration-initial pollutant concentration combination from the available options in the database; (4) the experimental data panel, for loading the pollutant concentration vs. accumulated energy (or standard time); (5) the system parameters panel, which displays the OVRPA and the $\beta = V_R/V_T$ factor charged.

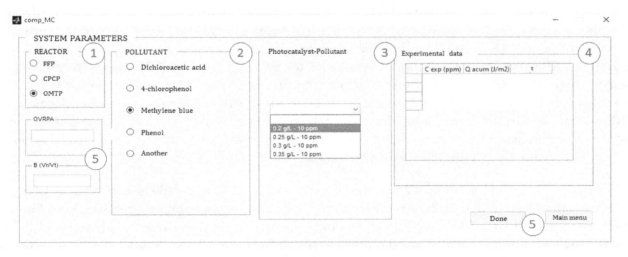

Figure 7. Database in the kinetic modeling module. (1) photoreactor configuration panel; (2) model pollutants panel; (3) photocatalyst-pollutant panel; (4) experimental data panel; (5) system parameters panel.

In this case, an OMTP with methylene blue (MB) was selected as a model pollutant with an initial concentration of $C_i = 10$ ppm and a photocatalyst concentration of $C_{cat} = 0.2$ g/L. Figure 8 shows the results obtained by PHOTOREAC. It is observed that the best fitting is achieved for the Mueses et al. and Ballari et al. models with $R^2 = 0.99737$ for both. The other models reported $R^2 = 0.001$. As the Ballari et al. model is a particular case of the Mueses et al. model, the first is considered the more appropriate option since it is more specific for this case. These results agree with the discussion presented in the previous section.

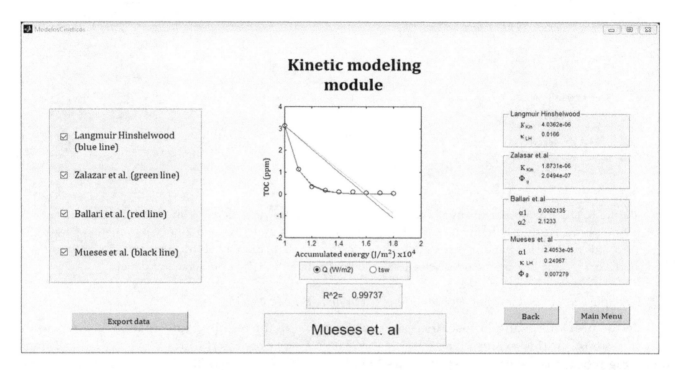

Figure 8. Methylene blue solar photodegradation fitting in an offset multi-tubular photoreactor (OMTP) for the four available kinetics models in the kinetic modeling module.

3.3. Example Case III: Radiation Field Modeling in a Flat Plate Photoreactor (FPP)

In this case, the objective was to compare the radiation field simulation for an FPP using SFM and SFM-HG. The photoreactor consists of a titled squared flat plate of length $L = 1$ m, which is placed facing the sun and uniformly irradiated. A water film of 1 cm thickness flows over its surface, providing a reaction volume of $V_R = 10$ L. The TiO$_2$ P25 Evonik concentration is $C_{cat} = 0.2$ g/L. Figures 9 and S1 show the LVRPA profile in the FPP calculated with the SFM-HG and the SFM, respectively. In both cases, the highest LVRPA values are found near to the surface of the water film (thickness = 0–0.2 cm) because this is the boundary that the solar light irradiates. After 0.2 cm, exponential decay in the LVRPA occurs as a result of the absorption and scattering of photons by the suspended photocatalyst. Due to the photoreactor being considered as uniformly irradiated, the changes in the LVRPA profile are only significant along with the water film thickness.

In Figure S1, which uses the SFM, a shaper exponential LVRPA profile is observed, with higher values near the irradiated boundary (at thickness = 0–0.2 cm) when comparing to values in Figure 9, which uses the SFM-HG. These results are due to the difference in the scattering phase function; the SFM-HG uses a predominantly forward scattering phase function, which causes photons to penetrate deeper into the water film. By contrast, the SFM uses a predominantly backward phase function, which causes that photons to be redirected toward the irradiated boundary and be mostly absorbed in the beginning of the film or escape from the system [15].

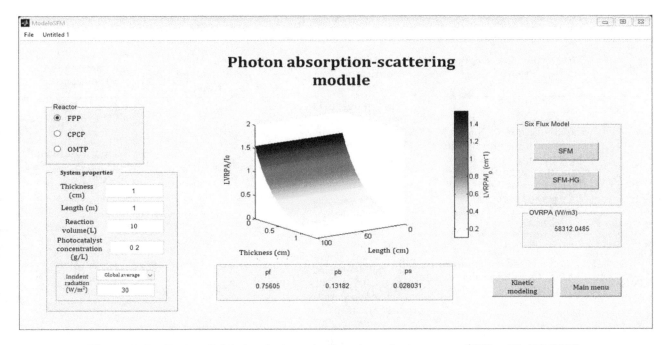

Figure 9. Radiation field simulation of a flat plate photoreactor (FPP) with SFM-HG.

4. PHOTOREAC Implementation in Research Projects in Heterogeneous Photocatalysis and Photoreactor Engineering by Chemical Engineering Undergraduates

Since the year 2015, different versions of PHOTOREAC have supported the final degree projects of chemical engineering students belonging to the Modeling and Applications of Advanced Oxidation Technologies Research Group at Cartagena University. The students developed research on heterogeneous photocatalysis and photoreactor engineering. A survey was done amongst them to determine the perceived impact of PHOTOREAC on their final degree projects. Table 4 shows the results of the survey. Between 2015–2018, ten final degree projects were developed in the research group, with an average impact of 37.5% perceived by the students. The use of PHOTOREAC can be summarized as follows: in four of the degree projects, both modules of PHOTOREAC were implemented since they performed the radiation field simulation and kinetic modeling of model pollutants; in two other projects, the photon-scattering module was used to determine the radiation field in photoreactors; finally, in four projects, the application was used in the learning process for modeling solar photoreactors. As a relevant outcome, two of the degree projects supported publications in high-impact journals. In all of the projects, the authors highlighted the use of PHOTOREAC as a user-friendly tool that allows them to reach the main objective of the projects or to achieve a fast advance in the learning curve, therefore allowing them to focus on more complex research.

Table 4. PHOTOREAC impact on final degree projects in chemical engineering.

Year	Title of the Final Degree Project	Related Publication/Ref.	PHOTOREAC Impact on the Project Perceived by the Students	PHOTOREAC Implementation in the Project
2015	Design and evaluation of a modified compound parabolic collector solar reactor	A Novel Prototype Offset Multi Tubular Photoreactor (OMTP) for solar photocatalytic degradation of water contaminants/ref. [10]	30%	Modeling the radiation field and kinetics of methylene blue for both the CPCP and OMTP
2015	Effect of oxygen transfer from the air on the photocatalytic degradation of dichloroacetic acid using a flat plate reactor	–	50%	Modeling the radiation field and kinetics of dichloroacetic acid in an FFP

Table 4. *Cont.*

Year	Title of the Final Degree Project	Related Publication/Ref.	PHOTOREAC Impact on the Project Perceived by the Students	PHOTOREAC Implementation in the Project
2016	Radiant field modeling in heterogeneous photoreactors implementing Monte Carlo simulation: Modification of the Six Flux Model to new phase functions	Coupling the Six Flux Absorption-Scattering Model to the Henyey–Greenstein scattering phase function: Evaluation and optimization of radiation absorption in solar heterogeneous photoreactors/ref. [15]	60%	Modeling the radiation field for the FFP and a CPCP
2016	Evaluation of the temperature effect on the heterogeneous photocatalytic degradation kinetics	Modeling and experimental evaluation of a non-isothermal photocatalytic solar reactor: temperature effect on the reaction rate kinetics/ref. [20]	20%	The learning process for modeling CPCP
2016	Solar heterogeneous photocatalytic degradation of organic pollutants in a pilot-scale modified tubular collector	A Novel Prototype Offset Multi Tubular Photoreactor (OMTP) for solar photocatalytic degradation of water contaminants/ref. [10]	50%	Modeling the radiation field and kinetics of DCA, PH and 4-CP for both the CPCP and OMTP
2016	Simulation of in series and in parallel arrangements of solar reactors (CPCP) for wastewater treatment	–	30%	The learning process for modeling CPCP
2016	Experimental evaluation and mathematical modeling of the performance of TiO_2-P25 reuse in heterogeneous solar photocatalytic degradation of acetaminophen	–	30%	The learning process for modeling CPCP
2017	Solar photocatalytic ozonation applied to amoxicillin degradation in wastewater at pilot-plant scale	–	30%	The learning process for modeling CPCP
2018	Mathematical modeling and simulation of photocatalytic hydrogen production	–	35%	Radiation field modeling of an FFP
2018	Experimental evaluation and mathematical modeling of the regeneration of commercial TiO_2 by the photocatalytic degradation of glyphosate	–	40%	Modeling the radiation field and the kinetics glyphosate in a CPCP

5. Analysis of the Overall Performance of PHOTOREAC

PHOTOREAC was shown to be a useful tool for modeling and simulation of solar photoreactors, and in particular for a non-expert public. Its user-friendly interface developed in the graphical user interface of Matlab proved to be intuitive enough to be used successfully by chemical engineering undergraduates, which develop research in heterogeneous photocatalysis.

In Section 3, the application was evaluated for disparate operational conditions, showing that it can fit and simulate the photodegradation experimental data provided for the two cases evaluated: CPCP-DCA (example case I) and OMTP-MB (example case II). These example cases were very different from each other, mainly because of the different photoreactor geometries: CPCP can capture more solar radiation per length than the OMTP as a result of being equipped with reflectors. However, OMTP has more volume than the CPCP [10]. Additionally, the employed model pollutant, its initial concentration, and the photocatalyst concentration were different. In both examples, PHOTOREAC performed successfully, allowing the user to evaluate different kinetic expressions and extract relevant findings from it. Finally, example case III focused on the photon absorption-scattering module, in which the impact of the radiation model was evaluated and discussed for an FFP. In this case, PHOTOREAC shows its versatility for researchers with an interest in studying the energy absorption behaviors of photoreactors.

The dedicated interface of PHOTOREAC for photoreaction engineering, together with its numerical algorithm, allowed the evaluation of the performance of large scale solar photoreactors without time-consuming computations and a complex mathematical formulation. The time invested in preparing and launching a simulation in PHOTOREAC is between 5–10 min, and the calculation time does not exceed 45 s. In contrast to commercial CFD simulators in which preparing and starting a first-time simulation may take a couple of hours needed for generating the photoreactor geometry in the system (or it importing it from CAD software), preparing the simulation modules and their models and selecting the proper meshing and numerical algorithms; besides, the computational time for each simulation is, generally, measured in hours [21–23]. Nevertheless, the results obtained by CFD simulators are much more complete and accurate than the results that PHOTOREAC may offer; for instance, CFD simulators provide detailed flow patterns for studying the hydrodynamics in the photoreactor. However, its high computational time may result in a barrier when exploring the impact of numerous parameters on a wide range of values.

Moreover, the most common CFD commercial simulators used in modeling photoreactors are very expensive licensed software. At the same time, PHOTOREAC is an open-access application that is available on-demand, by email to one of the authors of the paper, professor Miguel A. Mueses (mmueses@unicartagena.edu.co).

In conclusion, PHOTOREAC is recommended for the following cases: (i) for an introduction to photoreactor engineering; (ii) when a quantitative margin of error is still acceptable in the calculations; (iii) when qualitative results are the main objective of the work; and (iv) when the parametric space in the study is extensive, i.e., it is required to study the impact of numerous variables in broad ranges. In this case, PHOTOREAC may be employed to reduce the parametric space and then to implement a CFD simulator.

6. Limitations and Future Work

As with every modeling software, PHOTOREAC is limited by the availability and reliability of the input data provided by the users. Additionally, the computational application is limited to Titanium Dioxide P25 Evonik as photocatalyst. Although TiO_2 P25 is the most common photocatalyst, the capability of performing simulations for any photocatalyst will be crucial for the software, since an area of intensive research in heterogeneous photocatalysis is the development and testing of new photoactive materials. On the other hand, expanding the available kinetic models would also be a considerable improvement, because it will allow users to make a more comprehensive analysis by comparing the results of the kinetic models' fitting. Moreover, it is necessary to implement the option that users introduces their own kinetic expression, since some pollutants will require concrete mathematical expression because their kinetic mechanism may not follow the most common postulates. These drawbacks are expected to be overcome in the upcoming version of PHOTOREAC.

In the authors' opinion, some important challenges for PHOTOREAC and, in general, for photoreaction engineering at the pilot-solar scale are that the models account for the variability of the incident radiation on the solar photoreactor caused by fluctuations in atmospheric conditions. This improvement will allow more accurate quantification of the energy absorbed by the suspended photocatalyst, and therefore better quantification of the chemical species produced by the photoactivation of the photocatalyst.

Author Contributions: Conceptualization, B.C.-V., K.P.-G. and M.A.M.; data curation, B.C.-V., and K.P.-G.; formal analysis, R.A.-H.; funding acquisition, M.H.P.-C. and F.M.-M.; investigation, B.C.-V. and K.P.-G.; methodology, B.C.-V. and K.P.-G.; project administration, M.H.P.-C. and F.M.-M.; supervision, R.A.-H. and M.A.M.; validation, B.C.-V. and K.P.-G.; writing—original draft, R.A.-H.; writing—review and editing, R.A.-H., M.H.P.-C., M.A.M. and F.M.-M. All authors have read and agreed to the published version of the manuscript.

Acknowledgments: The authors gratefully thank Universidad de Cartagena (Cartagena, Colombia) with the project 017-2018: "Plan de Fortalecimiento del Grupo de Investigación Modelado y Aplicación de Procesos Avanzados de Oxidación" and Universidad del Valle (Cali, Colombia) with the Project CI. 21022 "Estudio del efecto hidrodinámico y de transporte de energía radiante en el diseño y optimización de reactores fotocatalíticos heterogéneos solares. CI 21022." for financial support. Acosta-Herazo thanks the CEIBA foundation with the program "Bolivar Gana con Ciencia" for financing his doctoral studies.

References

1.   Byrne, C.; Subramanian, G.; Pillai, S.C. Recent advances in photocatalysis for environmental applications. *J. Environ. Chem. Eng.* **2018**, *6*, 3531–3555. [CrossRef]

2.   Kumaravel, V.; Mathew, S.; Bartlett, J.; Pillai, S.C. Photocatalytic hydrogen production using metal doped TiO2: A review of recent advances. *Appl. Catal. B Environ.* **2019**, *244*, 1021–1064. [CrossRef]

3.   Spasiano, D.; Marotta, R.; Malato, S.; Fernandez-Ibañez, P.; Di Somma, I. Solar photocatalysis: Materials, reactors, some commercial, and pre-industrialized applications. A comprehensive approach. *Appl. Catal. B Environ.* **2015**, *170–171*, 90–123. [CrossRef]

4.   Cassano, A.E.; Alfano, O.M. Reaction engineering of suspended solid heterogeneous photocatalytic reactors. *Catal. Today* **2000**, *58*, 167–197. [CrossRef]

5.   Li Puma, G.; Brucato, A. Dimensionless analysis of slurry photocatalytic reactors using two-flux and six-flux radiation absorption-scattering models. *Catal. Today* **2007**, *122*, 78–90. [CrossRef]

6.   Grčić, I.; Li Puma, G. Photocatalytic degradation of water contaminants in multiple photoreactors and evaluation of reaction kinetic constants independent of photon absorption, irradiance, reactor geometry, and hydrodynamics. *Environ. Sci. Technol.* **2013**, *47*, 13702–13711. [CrossRef]

7.   Marugán, J.; van Grieken, R.; Pablos, C.; Satuf, M.L.; Cassano, A.E.; Alfano, O.M. Rigorous kinetic modelling with explicit radiation absorption effects of the photocatalytic inactivation of bacteria in water using suspended titanium dioxide. *Appl. Catal. B Environ.* **2011**, *102*, 404–416. [CrossRef]

8.   Colina-Marquez, J.; Castilla-Caballero, D.; Machuca-Martinez, F. Modeling of a falling-film photocatalytic reactor: Fluid dynamics for turbulent regime. *Appl. Math. Model.* **2016**, *40*, 4812–4821. [CrossRef]

9.   Malato, S.; Maldonado, M.I.; Fernández-Ibáñez, P.; Oller, I.; Polo, I.; Sánchez-Moreno, R. Decontamination and disinfection of water by solar photocatalysis: The pilot plants of the Plataforma Solar de Almeria. *Mater. Sci. Semicond. Process.* **2016**, *42*, 15–23. [CrossRef]

10.   Ochoa-Gutiérrez, K.S.; Tabares-Aguilar, E.; Mueses, M.Á.; Machuca-Martínez, F.; Li Puma, G. A Novel Prototype Offset Multi Tubular Photoreactor (OMTP) for solar photocatalytic degradation of water contaminants. *Chem. Eng. J.* **2018**, *341*, 628–638. [CrossRef]

11.   Manassero, A.; Satuf, M.L.; Alfano, O.M. Evaluation of UV and visible light activity of TiO2 catalysts for water remediation. *Chem. Eng. J.* **2013**, *225*, 378–386. [CrossRef]

12.   Mueses, M.A.; Machuca-Martinez, F.; Li Puma, G. Effective quantum yield and reaction rate model for evaluation of photocatalytic degradation of water contaminants in heterogeneous pilot-scale solar photoreactors. *Chem. Eng. J.* **2013**, *215–216*, 937–947. [CrossRef]

13.   Colina-Márquez, J.; Machuca-Martínez, F.; Li Puma, G. Photocatalytic mineralization of commercial herbicides in a pilot-scale solar CPC reactor: Photoreactor modeling and reaction kinetics constants independent of radiation field. *Environ. Sci. Technol.* **2009**, *43*, 8953–8960. [CrossRef] [PubMed]

14.   Colina-Marquez, J.; Machuca-Martínez, F.; Puma, G.L. Radiation absorption and optimization of solar photocatalytic reactors for environmental applications. *Environ. Sci. Technol.* **2010**, *44*, 5112–5120. [CrossRef] [PubMed]

15.   Acosta-Herazo, R.; Monterroza-Romero, J.; Mueses, M.A.; Machuca-Martínez, F.; Li Puma, G. Coupling the Six Flux Absorption-Scattering Model to the Henyey-Greenstein scattering phase function: Evaluation and optimization of radiation absorption in solar heterogeneous photoreactors. *Chem. Eng. J.* **2016**, *302*, 86–96. [CrossRef]

16.   Otálvaro-Marín, H.L.; Mueses, M.A.; Machuca-Martínez, F. Boundary layer of photon absorption applied to heterogeneous photocatalytic solar flat plate reactor design. *Int. J. Photoenergy* **2014**, *2014*. [CrossRef]

17. Zalazar, C.S.; Romero, R.L.; Martín, C.A.; Cassano, A.E. Photocatalytic intrinsic reaction kinetics I: Mineralization of dichloroacetic acid. *Chem. Eng. Sci.* **2005**, *60*, 5240–5254. [CrossRef]

18. De Los Ballari, M.M.; Alfano, O.O.; Cassano, A.E. Photocatalytic degradation of dichloroacetic acid. A kinetic study with a mechanistically based reaction model. *Ind. Eng. Chem. Res.* **2009**, *48*, 1847–1858. [CrossRef]

19. Otálvaro-Marín, H.L.; González-Caicedo, F.; Arce-Sarria, A.; Mueses, M.A.; Crittenden, J.C.; Machuca-Martinez, F. Scaling-up a heterogeneous H2O2/TiO2/solar-radiation system using the DamkÖhler number. *Chem. Eng. J.* **2019**, *364*, 244–256. [CrossRef]

20. Molano, M.; Mueses, M.A.; Fiderman, M.M. Modelado y evaluación experimental de un reactor solar fotocatalítico no isotérmico: Efecto de la temperatura sobre la cinética de la velocidad de reacción. *Ing. Compet.* **2017**, *19*, 143–154. [CrossRef]

21. Casado, C.; García-Gil, Á.; van Grieken, R.; Marugán, J. Critical role of the light spectrum on the simulation of solar photocatalytic reactors. *Appl. Catal. B Environ.* **2019**, *252*, 1–9. [CrossRef]

22. Casado, C.; Marugán, J.; Timmers, R.; Muñoz, M.; van Grieken, R. Comprehensive multiphysics modeling of photocatalytic processes by computational fluid dynamics based on intrinsic kinetic parameters determined in a differential photoreactor. *Chem. Eng. J.* **2017**, *310*, 368–380. [CrossRef]

23. Boyjoo, Y.; Ang, M.; Pareek, V. Some aspects of photocatalytic reactor modeling using computational fluid dynamics. *Chem. Eng. Sci.* **2013**, *101*, 764–784. [CrossRef]

4

Recent Trends in Removal Pharmaceuticals and Personal Care Products by Electrochemical Oxidation and Combined Systems

Khanh Chau Dao [1,2], **Chih-Chi Yang** [2]🆔, **Ku-Fan Chen** [2]🆔 and **Yung-Pin Tsai** [2,*]🆔

[1] Department of Health and Applied Sciences, Dong Nai Technology University, Bien Hoa,
    Dong Nai 810000, Vietnam; daokhanhchau07@gmail.com
[2] Department of Civil Engineering, National Chi Nan University, Nantou Hsien 54561, Taiwan;
    chi813@gmail.com (C.-C.Y.); kfchen@ncnu.edu.tw (K.-F.C.)
*   Correspondence: yptsai@ncnu.edu.tw

Abstract: Due to various potential toxicological threats to living organisms even at low concentrations, pharmaceuticals and personal care products in natural water are seen as an emerging environmental issue. The low efficiency of removal of pharmaceuticals and personal care products by conventional wastewater treatment plants calls for more efficient technology. Research on advanced oxidation processes has recently become a hot topic as it has been shown that these technologies can effectively oxidize most organic contaminants to inorganic carbon through mineralization. Among the advanced oxidation processes, the electrochemical advanced oxidation processes and, in general, electrochemical oxidation or anodic oxidation have shown good prospects at the lab-scale for the elimination of contamination caused by the presence of residual pharmaceuticals and personal care products in aqueous systems. This paper reviewed the effectiveness of electrochemical oxidation in removing pharmaceuticals and personal care products from liquid solutions, alone or in combination with other treatment processes, in the last 10 years. Reactor designs and configurations, electrode materials, operational factors (initial concentration, supporting electrolytes, current density, temperature, pH, stirring rate, electrode spacing, and fluid velocity) were also investigated.

Keywords: advanced oxidation processes; electrochemical advanced oxidation processes; pharmaceuticals and personal care products; electrochemical oxidation; anodic oxidation

---

1. Introduction

The concern for pharmaceuticals and personal care products (PPCPs) as toxic substances in the environment and the essential to assess their environmental risks have significantly increased recently. PPCPs are defined as a group of compounds that is including pharmaceutical drugs, cosmetic ingredients, food supplements, and ingredients in other consumer products (e.g., shampoos, lotions) [1]. Pharmaceuticals are used to prevent or treat diseases on humans and animals, whereas personal care products (PCPs) are used mostly to improve the quality of daily life [2]. They are considered as emerging pollutants (new products or chemicals without regulatory status) and whose effects on the environment and human health are unidentified [3]. Due to the widespread occurrence in water bodies, regardless of the low concentrations (normally ranging from ng/L to µg/L), residues of PPCP can harm human and animal health when it enters and accumulates in the food chain, causing unknown long-term effects [2,4].

During wastewater treatment (WWT) processes, many PPCPs experience microbial mediated reactions [5] in the environment. Thus, transformation products are formed. The transformation of PPCPs can occur during WWT, depending on the compound's physicochemical properties and conditions, where PPCPs can be destroyed or partially transformed or remained unchanged [6]. In this review, it can be seen that the effect of PPCPs in the environment does not only depend on concentration but also persistence, bioaccumulation, biotransformation, and elimination. Some PPCPs produce metabolites or by-products more harmful than the parent compounds. Toxicity evaluation is an important environmental pollution control factor since the degradation by-products from the initial structure can be more toxic.

Biodegradation, photodegradation, and other processes of abiotic transformation, such as hydrolysis [7], can reduce environmental concentrations of PPCPs and result in partial loss and mineralization of these compounds. Chiron et al. [8] revealed that acridine is a photodegradation product of carbamazepine under artificial estuarine water conditions, whereas tetracycline could not be photodegraded due to its sediment adsorption [9].

The electrochemical oxidation process (EOP) can be described as an electrochemical technology capable of achieving oxidation of contaminants from water or wastewater, either by direct or mediated oxidation processes originating on the anode surface of the electrochemical cell. This means that these oxidative processes should not actually be carried out on the anode, but only on its surface. As a consequence, this technique incorporates two main types of processes [10]: heterogeneous and homogeneous oxidation. Direct anodic oxidation or electrolysis occurs directly on the anode (M) with direct charge transfer reactions between the surface of the anode and the organic contaminants involved. The mechanism requires only the mediation of electrons that are capable of oxidizing such organic compounds at defined potentials more negative the oxygen evolution potential [11]. The indirect electrochemical oxidation by reactive oxygen species is based on the electro-generation of adsorbed *OH ($E° = 2.8$ V/SHE) onto the anode surface as an intermediate of the OEP [10,12].

This paper intends to be a powerful tool for researchers in the pursuit of comprehensive information on the removal of PPCPs from liquid solutions by EOP, alone or in combination with other treatment processes. The remediation of aqueous or real wastewater was assessed, regarding many features like the configuration of the electrochemical reactor, anode and cathode characteristics, and operational parameters such as initial PPCPs concentration, supporting electrolytes, current density (j), temperature, pH, temperature, stirring rate, electrode spacing, and fluid velocity.

2. Origins and Classification of PPCPs

Direct and indirect pathways can introduce PPCPs into the environment. PPCPs may enter surface water by direct discharge into surface water from factories, hospitals, households, and WWTPs, as well as through land runoff in the case of biosolids distributed over agricultural land that may touch groundwater by leaching or bank filtration. Sediment can adsorb PPCPs within the surface water compartment because of various binding sites [13]. Soil may also be one of the PPCPs sinks. PPCPs can pass through irrigation into the soil with PPCPs containing treated and untreated wastewater. These can also be moved to the soil through an atmospheric wet deposition for some PPCPs [14].

Wastewater, including domestic, municipal, and hospital wastewater, are the primary sources that bring pharmaceuticals into the environment (both point- and nonpoint-sources) from various activities such as wastes (human and animal), landfill leachate, biosolid, and direct disposal of pharmaceuticals. Such pharmaceuticals then can not be biodegradable ultimately in WWTPs and enter the receiving waters [15–17]. In WWTPs, activated sludge is the main process for secondary treatment which can remove various kinds of PPCPs from wastewater. However, the removal rate depends greatly on physiochemical characteristics, reactors applied, and operational conditions (hydraulic retention time, sludge retention time, and pH) as well [18]. Table 1 summarizes the target PPCPs selected for this study and their structures, Table 2 updates the removal efficiency of PPCPs by combining biological treatment with other processes.

Table 1. Structures, chemical abstracts service registry number (CAS), and classification for the target pharmaceuticals and personal care products (PPCPs) selected for this study.

Compounds (CAS) Classification	Structure	Compounds (CAS) Classification	Structure
Aspirin (50-78-2) Nonsteroidal anti-inflammatory drugs (NSAIDs)		Lamivudine (134678-17-4) Antivirals	
Atenolol (29122-68-7) Beta-blockers		Levodopa (59-92-7) Antiparkinson Agents	
Berberine (2086-83-1) Antibiotics		Methotrexate (59-05-2) Antineoplastics	
Caffeine (58-08-2) Stimulant		Metronidazole (443-48-1) Antibiotics	
Carbamazepine (298-46-4) Anticonvulsants		Musk ketone (81-14-1) Fragrances	

Table 1. *Cont.*

Compounds (CAS) Classification	Structure	Compounds (CAS) Classification	Structure
Carboplatin (41575-94-4) Antineoplastics		Naproxen (22204-53-1) NSAIDs	
Ceftazidime (78439-06-2) Antibiotics		N,N-diethyl-m Toluamide (134-62-3) Insect repellents	
Ceftriaxone sodium (104376-79-6) Antibiotics		Norfloxacin (70458-96-7) Antibiotics	
Cephalexin (15686-71-2) Antibiotics		Ofloxacin (82419-36-1) Antibiotics	
Chloramphenicol (56-75-7) Antibiotics		Omeprazole (73590-58-6) Antibiotics	

Table 1. *Cont.*

Compounds (CAS) Classification	Structure	Compounds (CAS) Classification	Structure
Ciprofloxacin (85721-33-1) Antibiotics		Methyl Paraben (99-76-3) Preservatives	
Clofibric acid (882-09-7) Blood lipid regulators		Paracetamol (103-90-2) NSAIDs	
Diclofenac (15307-86-5) NSAIDs		Rifampicin (13292-46-1) Antibiotics	
Enrofloxacin (93106-60-6) Antibiotics		Salicylic acid (69-72-7) NSAIDs	
Estrone (53-16-7) Hormones		Sulfamethoxazole (723-46-6) Antibiotics	

Table 1. *Cont.*

Compounds (CAS) Classification	Structure	Compounds (CAS) Classification	Structure
Ibuprofen (15687-27-1) NSAIDs		Sulfachloropyrida-zine (80-32-0) Antibiotics	
Iohexol (66108-95-0) Radiological Non-Ionic Contrast Media		Sulfadiazine (68-35-9) Antibiotics	
2-methyl-4-isothiazolin-3-one (2682-20-4) Preservatives		Tetracycline (60-54-8) Antibiotics	
Ketoprofen (22071-15-4) NSAIDs			

Table 2. The removal efficiency of PPCPs by combining biological treatment with other processes.

Compounds	Initial Concentration	Treatment Processes	Removal Efficiency (%)	Ref.
Aspirin	930 ng/L	Modified Bardenpho process	92	[19]
Atenolol	255 ng/L	Grit tanks|primary sedimentation|bioreactor|clarifiers	47.1	[19]
Atenolol	1197 ng/L	Pretreatment|primary (settling)|secondary activated sludge (AS)	14.4	[20]
Berberine	2.3 ± 2.0	Grit removal|primary clarifier|denitrification|nitrification|second clarifier	84	[21]
Berberine	75.0–375.0 mg/L	Upflow anaerobic sludge blanket (UASB)—membrane bioreactor (MBR)	99	[22]
Caffeine	82 ± 36 µg/L	Grit removal|primary clarifier|denitrification|nitrification|second clarifier	99.7	[21]
Caffeine	22,849 ng/L	Anaerobic/Anoxic/Oxic (A2O)	94.9	[23]
Caffeine	208–416 ng/L	A series of different waste stabilization ponds	73	[24]
Carbamazepine	129 ng/L	Pretreatment|primary (settling)|secondary AS	9.5	[20]
Carbamazepine	2.0 ± 1.3 µg/L	Grit removal|primary clarifier|denitrification|nitrification|second clarifier	0	[21]
Carboplatin	4.7 to 145 µg/L	Adsorption to AS	70%	[25]
Ceftazidime	40 mg/L	Coupling ultraviolet (UV)|algae-algae treatment	97.26	[26]
Ceftriaxone	14 µg/L	AS process	<1	[27]
Cephalexin	4.6 mg/L	Grit channels|primary clarifies|conventional AS|Final settling	87	[28]
Chloramphenicol	206 ± 56 ng/L	Preliminary screening|primary sedimentation|conventional AS treatment	>70	[29]
Chloramphenicol	31 ± 16 ng/L	Screen|primary clarifier|AS system for denitrification and nitrification	50	[30]
Ciprofloxacin	2200 ng/L	Grit channels|primary clarifies|conventional AS	−88.6	[31]
Ciprofloxacin	5524 ng/L	Pretreatment|primary (settling)|secondary AS	57	[20]
Ciprofloxacin	2 mg/L	Aerobic sequencing batch reactors (SBRs) with mixed microbial cultures	51	[32]
Clofibric acid	0.25 ± 0.09 µg/L	Grit removal|primary clarifier|denitrification|nitrification|second clarifier	52	[21]
Clofibric acid	26 ng/L	Pretreatment|primary (settling)|secondary AS	54.2	[20]
Clofibric acid	20–70 mg/L	Primary treatment|Orbal oxidation ditch|UV disinfection	10–60	[33]
Diclofenac	2.0 ± 1.5 µg/L	Grit removal|primary clarifier|denitrification|nitrification|second clarifier	96	[21]
Diclofenac	232 ng/L	Pretreatment|primary (settling)|secondary AS	5	[20]
Diclofenac	9–170 ng/L	Conventional AS|UV disinfection	65	[34]
Estrone	57 ng/L	Grit channels|primary clarifies|conventional AS	93.7	[31]
Estrone	4500 ng/L	Grit channels|primary clarifies|conventional AS	99.7	[31]
Ibuprofen	3.4 ± 1.7 µg/L	Grit removal|primary clarifier|denitrification|nitrification|second clarifier	96	[21]
Ibuprofen	2687 ng/L	Pretreatment|primary (settling)|secondary AS	95	[20]
Iohexol	9.0 ± 2.0 µg/L	Grit removal|primary clarifier|denitrification|nitrification|second clarifier	89	[21]
2-methyl-4-isothiazolin-3-one	1–3 mg/L	Aerobic process	80–100	[35]
Ketoprofen	441 ng/L	Anaerobic/Anoxic/Oxic (A2O)	11.2	[23]
Lamivudine	210 ± 13 ng/L	Screen|aerated grit-removal|primary clarifier|nitrification/denitrification	>76	[36]

Table 2. *Cont.*

Compounds	Initial Concentration	Treatment Processes	Removal Efficiency (%)	Ref.
Methotrexate	7.30–55.8 ng/L	Pretreatment|primary (settling)|secondary AS	100	[20]
Metronidazole	90 ng/L	Anaerobic/Anoxic/Oxic (A2O)	38.7	[23]
Musk ketone	0.640 ± 0.395 µg/L	Primary gravitational settling|AS	91.0 ± 5.2	[37]
Naproxen	3000 ng/L	Grit channels|primary clarifies|conventional AS	96.2	[31]
	2363 ng/L	Pretreatment|primary (settling)|secondary AS	60.9	[20]
DEET	503 ng/L	Primary|secondary treatment with AS	19.2–46.2	[38]
Norfloxacin	229 ± 42 ng/L	Screen|primary clarifier|AS system for denitrification and nitrification	66	[30]
	2100 ng/L	Grit channels|primary clarifies|conventional AS	124.2	[31]
Ofloxacin	2275 ng/L	Pretreatment|primary (settling)|secondary AS	64.1	[20]
Omeprazole	365 ng/L	Pretreatment|primary (settling)|secondary AS	8.5	[20]
Methyl Paraben	801 ng/L	Conventional biological treatment with P and N removal	100	[39]
	218,000 ng/L	Modified Bardenpho process	99	[19]
Paracetamol	23,202 ng/L	Pretreatment|primary (settling)|secondary AS	100	[20]
Rifampicin	0–31 ng/L	Secondary treatment process: AS, biological filtration oxygenated reactor, anoxic/oxic (A/O), cyclic AS technology (CAST), and A2O	0–100	[40]
Salicylic acid	5.866 µg/L	Primary|secondary treatment: trickling filter beds|final clarification.	>98	[41]
	7400 ng/L	Grit channels|primary clarifies|conventional AS	−35.8	[31]
Sulfamethoxazole	0.82 ± 0.23 µg/L	Grit removal|primary clarifier|denitrification|nitrification|second clarifier	24	[21]
	524 ng/L	Pretreatment|primary (settling)|secondary AS	31.2	[20]
	118 ± 17 ng/L	Screen|primary clarifier|AS system for denitrification and nitrification	64	[30]
Sulfachloropyridazine	0.19 µg/L	Conventional AS	62	[42]
Sulfadiazine	72 ± 22 ng/L	Screen|primary clarifier|AS system for denitrification and nitrification	50	[30]
Tetracycline	257 ± 176 ng/L	Preliminary screening|primary sedimentation|conventional AS treatment	69	[29]

3. Analytical Methods of PPCPs

Figure 1 shows the analytical method that is essential to investigate the occurrence of PPCPs in the environment, whichs consists of several main steps. This includes selecting appropriate analytical instruments (Table 3), which depend on the characteristics of PPCPs; extracting and purifying the samples by using techniques such as solid-phase extraction (SPE), liquid-liquid extraction (LLE), liquid-liquid micro-extraction (LLME), and solid-phase micro-extraction (SPME) that was introduced in various studies [43,44]; and optimizing of measurement parameters.

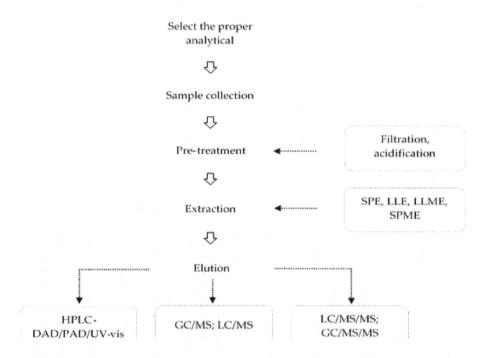

Figure 1. PPCPs analytical method procedure. Solid-phase extraction: SPE; liquid-liquid extraction: LLE; liquid-liquid micro-extraction: LLME; solid-phase micro-extraction: SPME; HPLC: High-performance liquid chromatography; DAD: Diode array detector; PAD: photodiode detector; UV-vis: ultraviolet-visible detector; GC/MS: Gas chromatography–mass spectrometry; LC/MS: Liquid chromatography–mass spectrometry.

Table 3. The analytical methods of PPCPs in the literature.

Analytical Methods	PPCPs
GC-MS	Ciprofloxacin, Chloramphenicol, Methyl paraben
HPLC	Lamivudine, Ceftazidime, Carboplatin, Aspirin, Cephalexin, Musk ketone, Norfloxacin, Ceftriaxone sodium, Levodopa, N,N-diethyl-m-Toluamide (DEET)
HPLC-DAD	Acetaminophen, Diclofenac, Sulfamethoxazole, Chloramphenicol, Ofloxacin, Berberine, Tetracycline
HPLC-UV/HPLC-UV vis/UV-vis	Ciprofloxacin, Rifampicin, Carbamazepine, Caffeine, Enrofloxacin, Sulfamethoxazole, Diclofenac, Isothdiazolin-3-ones, Metronidazole, Estrone, Paracetamol, Diclofenac, Methyl paraben, Clofibric acid, Sulfonamides
HPLC-HR-MS/HPLC-MS/HPLC-MS-MS	Carbamazepine, Iohexol, Ceftazidime, Methotrexate, Ibuprofen, Clofibric acid
HPLC-PDA	Atenolol, Paracetamol, Salicylic acid, Parabens, Sulfachloropyridazine, Omeprazole, Ibuprofen, Naproxen, Carbamazepine

4. Removal of PPCPs from Liquid Solutions by EOP

4.1. Electrochemical Reactor Designs and Configurations

There are two types of electrodes: two-dimensional and three-dimensional. Compared to two-dimensional, three-dimensional electrodes ensures a high electrode surface-to-cell volume ratio value. Due to the ease of scale up to a larger electrode size, more electrode pairs, or an increased number of cell stacks, cell designs using the parallel plate geometry in a filter press arrangement are widely used [45].

In the configuration of the reactor, the cell arrangement (divided and undivided cells) must be considered. The anolyte and catholyte are separated into divided cells by a porous diaphragm or ion-conducting membrane. Choosing the separating diaphragm or membrane is as critical for divided cells as choosing the correct electrode materials for proper electrolyte system functioning. Generally, the use of divided cells should be avoided wherever possible regarding the cost of separators, the complexity of reducing the electrode gap and the problems of the mechanic, and corrosion [46]. Undivided cells working in batch mode are often under magnetic stirring for mixing at a thermostatically controlled temperature (Figure 2). The number of electrodes can increase the active area per volume unit.

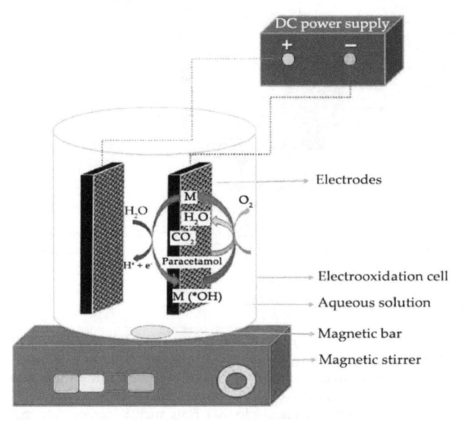

Figure 2. Diagram of the electrochemical reactor, using a glass beaker. The solution was stirred continuously throughout the process with a magnetic bar on a magnetic stirrer. The graphite anode was used as a working anode and a distance of 2 mm.

Most of the studies were conducted in undivided electrochemical reactors, usually using solution volumes ranging from 100 to 500 mL, although 1 L or larger volumes were sometimes used [48–50]. Divided cells use a separator between anolyte and catholyte, which makes the treatment process more costly and challenging due to the penalty overvoltage of the separator. The investigation of norfloxacin degradation in an electrochemical reactor with the presence and absence of an ion-exchange membrane proved the use of the membrane is highly advantageous as it enhances the anodic reaction kinetics and improves the current efficiency. This leads to an improvement in the degradation of norfloxacin, mineralization, and the consequent mineralization current efficiency [51]. Moreover, Chen et al. [52] used successfully divided and thermostated cells and a Nafion 212 ion-exchange membrane separator to perform electrodegradation of DEET with total removal.

Since the metal deposition occurs on the surface of the cathode to boost the space-time yield, it is required to increase the surface area. Therefore, the fluidized bed electrode was developed, with granular graphite and glass beads for filling the gap between the main electrodes and used as the third electrode [32].

Filter-press cells have been used by coupling to a pump and a reservoir (Figure 3). One module including an anode, a cathode, and a membrane (if necessary) makes it relatively easy to operate and maintain the reactor.

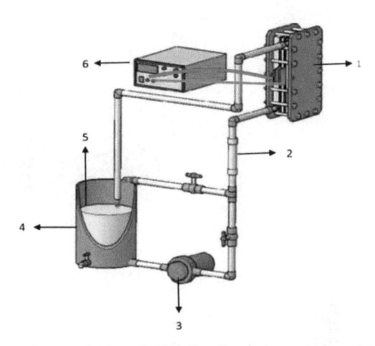

Figure 3. Experimental setup of 4 L undivided filter flow press reactor used for the treatment of paracetamol and diclofenac. 1. flow electrolytic cell, 2. flow meter, 3. peristaltic pump, 4. reservoir, 5. sampling, and 6. power supply.

4.2. Electrode Materials

It has also been shown that the anodes with high over-potential O_2 yield better electrochemical oxidation results [53–56]. Consequently, the electrode material (M) has a significant impact on the performance of PPCPs in oxidative degradation. Accordingly, an interesting issue is a systematic research on the comparative performance of electrode materials.

Sopaj et al. [57] tested on different electrode materials such as carbon felt, carbon fiber, carbon graphite, Platinum (Pt), lead dioxide, dimensionally stable anode (DSA) [58], (Ti/RuO$_2$–IrO$_2$), and boron-doped diamond (BDD) for removing of amoxicillin in aqueous media. BDD anode was more effective in oxidizing and mineralizing amoxicillin in water than the DSA. Moreover, it can be obtained very high electrolysis efficiency for the BDD electrode during the initial stage, even for high current densities.

Barışçı et al. [59] showed the performance of electrodes was significantly different for the anti-cancer drug carboplatin degradation with various mixed metal oxide (MMO) electrodes and BDD electrode (Figure 4). CV voltammograms unveiled that BDD, Ti/IrO$_2$-RuO$_2$, Ti/RuO$_2$, and Ti/IrO$_2$-Ta$_2$O$_5$ anodes had the highest levels of oxygen evolution and the poorest anodes were SnO$_2$/Pt, Ti/Pt and Ti/Ta$_2$O$_5$-SnO$_2$-IrO$_2$. Besides, higher oxygen evolution overpotential explained the formation of OH* on the surface of anode instead of molecular oxygen, which improved the efficiency.

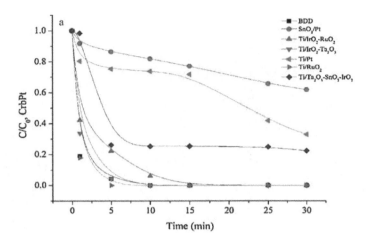

Figure 4. The effect of electrode material on anti-cancer drug carboplatin degradation under conditions: supporting electrolyte, 200 mg/L Na_2SO_4; pH 7; current density, 30 mA cm^{-2}.

4.2.1. Lead and Lead Dioxide

Because of the stability, low cost, and high oxygen evolution potential, lead and lead dioxide have been used as anode materials [60] (Table 4). Recent studies have paid considerable efforts to improve the performance, including the addition of a new intermediate layer between the substrate and the oxidation layer, doping metal, or non-metallic ions and the adoption of new preparation methods [61,62].

Dai et al. [55] found the catalytic effect of La–Gd–PbO_2 showed the highest performance followed by that of La–PbO_2, Gd–PbO_2, PbO_2, respectively, in levodopa degradation. Moreover, compared to the pure PbO_2 electrode, the PbO_2 electrode with 1% Mo had a higher oxygen evolution potential and higher current of reduction and oxidation peaks, which led to increasing in electrochemical activity and decreasing of energy consumption [63].

Porous Ti plays an essential role in improving lead dioxide electrode performance compared to the traditional planar Ti substrate. Zhao et al. [64] found that compared to the traditional PbO_2 electrode, Ti/SnO_2-Sb_2O_3/PbO_2 had higher stability, safety, and removal performance of musk ketone. Xie et al. [65] developed a TiO_2-based SnO_2-Sb/polytetrafluoroethylene resin-PbO_2 electrode based on TiO_2 nanotubes and demonstrated the growing of TiO_2 nanotubes on Ti material led to an increase in current efficiency. Before electrons flow, the electrode needs a large overpotential that minimizes the oxygen evolution, decreases the production of hydrogen peroxide and ozone, and favors the creation of *OH, with the electron efficiency of 88.45%. The degradation of ibuprofen demonstrated the degradation rate constant over Ti/SnO_2-Sb/Ce-PbO_2 was two times of the value over Ti/Ce-PbO_2 [66].

4.2.2. DSA

In recent decades, MMO electrodes, known as DSA, have been made commercially available (Table 5). These consist of the corrosion-resistant base material, such as titanium or tantalum, coated with a metal oxide layer. DSA is catalytic oxide electrodes that, due to their low Cl_2 overpotential, can effectively produce active chlorine species [67].

Studies verify the performance of three-dimensional (3D) was much better, more cost-effective, and saved more energy than traditional two-dimensional (2D). The highest efficiency was recorded in the 3D process for removing carbamazepine compared to a 2D electrochemical process [68]. Furthermore, using a 3D electrode reactor to treat estriol, in batch mode, exhibited reaction rate per unit area was significantly higher and lower energy consumption than conventional 2D electrode reactor with indirect oxidation as the main contributor to the degradation in the batch 3D electrode reactor at all electrode distances [69]. Over 80% of the removal efficiency was attributed to indirect oxidation at an electrode distance of 2 cm (Figure 5).

Table 4. Selected results reported for PPCPs removal by electrochemical oxidation process (EOP) with lead and lead dioxide anodes.

PPCPs	Initial C	Electrolyte	j/mA cm^{-2}	Reactors/Operational Parameters	Electrodes		pH	Reaction Time (min)	Removal (%)	Ref.
					Anode	Cathode				
Lamivudine	5 mg/L	20 mM Na$_2$SO$_4$	$\geq$10	Undivided cell, V 450 mL, current density (j) (6–14 mA cm^{-2})	Ti/SnO$_2$-Sb/Ce-PbO$_2$; 7 cm × 10 cm × 1 mm	Stainless steel (SS); 7 cm × 10 cm × 1 mm, gap 2 cm	3–11	240	70 (TOC)	[70]
Ciprofloxacin	50 mg/L	0.1 mol/L Na$_2$SO$_4$	30	Filter-press flow reactor; pH (3, 7, and 10), flow rate (qV = 2.5, 4.5, and 6.5 L min^{-1}), j (6.6, 20, and 30 mA cm^{-2}), and T = 10, 25, and 40 °C	Ti-Pt/β- PbO$_2$; 3.1 cm × 2.0 cm, 3.1 cm × 2.7 cm	AISI 304 SS plate	10	120	100	[71]
Ofloxacin	20 mg/L	Na$_2$SO$_4$	30	Differential column batch reactor, fluid velocity: 0.003 and 0.048 m/s, detention time: 10.3–0.54 min.	TiO$_2$-based SnO$_2$-Sb/FR-PbO$_2$; 2 cm × 5 cm	SS foil; Same shape and size, gap 0.5 and 3 cm	6.25	90	99.00	[65]
Enrofloxacin	10 mg/L	20 mM Na$_2$SO$_4$	8	Undivided electrolytic cell, V 30 mL, j (2–10 mA/cm^2), pH (~3–11)	Ti/SnO$_2$-Sb-La-PbO$_2$; 25 cm^2	Ti; Same area; gap 5 mm	3–11	30	95.1 (TOC)	[72]
Musk ketone	50 mg/L	0.06 mol/L Na$_2$SO$_4$	40	Cylindrical single compartment cell, V 100 mL, stirring rate 800 rmin^{-1}, j (10–50 mA cm^{-2}), pH (3–11)	Ti/SnO$_2$-Sb$_2$O$_3$/PbO$_2$; 1 cm × 1 cm	Stainless copper foil; (2 cm × 2 cm), gap 1.5 cm	7	120	99.93	[64]
Levodopa	100 mg/L	0.1 mol/L Na$_2$SO$_4$	50	Electrochemical system, V 250 mL, j (15–70 mA cm^{-2})	La-Gd- PbO$_2$; 12 cm × 2 cm, thickness: 1 mm, 14 cm^2	Ti; The same area; gap 4 cm	5.9	120	100.00	[55]

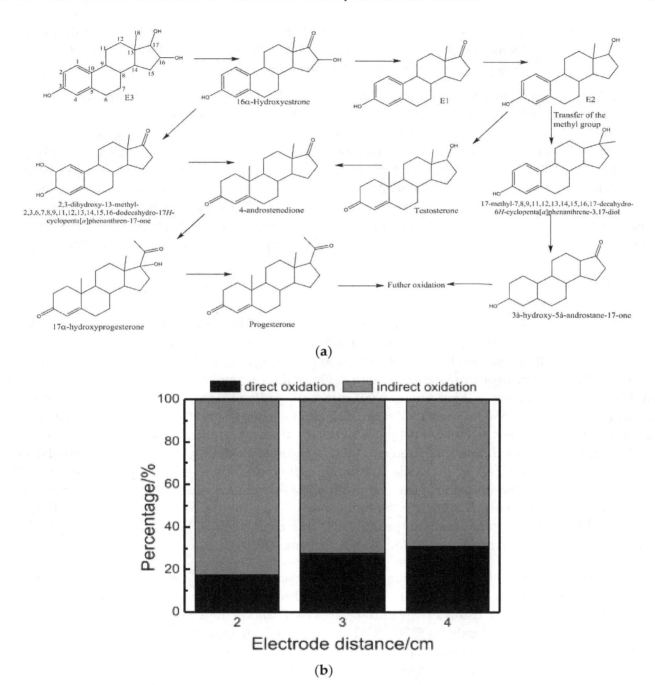

Figure 5. Proposed simplified pathways for estriol (E3) degradation in batch 3D electrolysis (**a**) and (**b**) the contribution of direct and indirect oxidation at various distances under operating conditions: C_0 = approximately 1000 μg/L, j = 2.2 mAcm^{-2}, constant current, Q = 150 mL/min.

By adding powder activated carbon (PAC) or metal particles, the conductivity, mass transfer, or adsorption may also be increased in the 3D process [73]. The possibility of catalytic reaction and more reactive sites for adsorption are advantages of the 3D process that lead to better removal performance [74].

The SnO$_2$ electrode has been widely used in wastewater treatment because of its high oxidation activity, it lower toxicity than PbO$_2$, and it being more cost-effective than BDD. Sadly, it also contains limitations due to high energy consumption and instability. Adding TiO$_2$ could reduce the electrode's internal passivation and charge transfer resistance, improving its stability and efficiency in oxidation when Cu limits the growth of crack morphology and offer more effective active sites. They accelerated electronic transfer and decreased SnO_2 surface potential, improved the OEP, and increased the response current peak, which increased the electrode's oxidative degradation capacity [75]. Ti/SnO_2-Cu showed better stability and higher corrosion resistance than the conventional Ti/SnO_2-Sb electrode [76].

RuO_2/IrO_2-coated titanium anode with improved electrocatalytic behavior and stability are readily available in practical mesh geometries and have extended the lifetime and lower costs compared to BDD electrodes. Various DSA such as Ti/RuO_2, Ti/Pt, Ti/IrO_2-RuO_2, Ti/IrO_2-Ta_2O_5, Ti/Ta_2O_5-SnO_2-IrO_2, and Pt/SnO_2 used for removing X-ray contrast iohexol demonstrated that Ti/RuO_2 provided the highest degradation efficiency [77]. Barışçı et al. [59] found that Ti/RuO_2 could reach complete degradation of carboplatin anti-cancer drug in just 5 min and obtained zero toxicity at the end of the process. However, the use of IrO_2, RuO_2 on large scale is restricted by low abundance, high cost, and difficulty in their separation. Ir/IrO_2 nanoparticles could be immobilized on Fe_3O_4 core/ SiO_2 shell via surface-modified NH_2 functional groups resulted in high catalytic activity, high stability, and efficient recyclability.

4.2.3. Boron-Doped Diamond

The BDD anode showed high performance on various kinds of PPCPs, as seen in Table 6. The low-pressure conversion of carbon to diamond crystals has allowed a thin layer of diamond film to develop on suitable substrates like silicon, niobium, tungsten, molybdenum, and titanium [78]. He et al. [79] examined aspirin degradation with PbO_2, BDD, and porous Ti/BDD as the anode. On BDD electrodes, the electrochemical process involves direct and indirect electrochemical oxidation, whereas, on the PbO_2 electrode, only indirect oxidation. The kinetic results can be explained by the mechanism of aspirin degradation, which may take place in two distinct forms: direct oxidation at the electrode surface and indirect oxidation mediated by *OH. In indirect oxidation, the initial step involves the formation of *OH from water molecule discharge. The oxidation is indirectly mediated by *OH contributing to the mineralization of organic pollutants. Aspirin mineralization is mainly performed by reaction with *OH. Porous Ti/BDD is the highest excellent potential for aspirin relative to flat BDD and PbO_2 electrode when niobium-supported BDD thin film (Nb/BDD) anode could be applied in a wide range of pH, reducing chemicals for pH adjustment [48].

In various systems, BDD allowed for higher removal rates of PPCPs than other anodes as higher quantities of *OH produced. Sirés et al. [80] indicated that the performance was demonstrated to be much more productive using a large surface area BDD anode than a Pt one, explained by a large number of active hydroxyl radicals BDD (*OH) and minimizing their parasitic reactions. Compared to the Pt and glassy carbon anodes, the BDD anode showed better efficiency for isothiazolin-3-one degradation [81]. BDD physisorbed *OH was observed to cause the combustion of ketoprofen into CO_2 and H_2O. The poor mineralization was attributed to the formation of chlorinated organic compounds that are refractory at both BDD and Pt anodes [82]. Omeprazole was primarily oxidized by *OH formed from water oxidation at the surface of the Pt or BDD [54]. It also can be seen that the BDD anode was superior to the Pt and PbO_2 electrodes for DEET abatement. At the same j value and temperature, the DEET abatement degradation in the order BDD, PbO_2, and Pt [52]. It also can be seen the higher oxidation power of BDD became evident in removing estrone than β-PbO_2 anode [83].

BDD electrode in a single compartment filter-press flow cell represented the conversion of cephalexin and its hydroxylated intermediates to CO_2 depended solely on their diffusion to the BDD surface. Due to the different types and quantities of electrogenerated oxidants, the oxidation rate of cephalexin using distinct salts as supporting electrolytes showed distinct rates; however, none of them were able to mineralize cephalexin and its intermediates, which only occurred through a diffusion mechanism on the surface of the BDD [84]. Due to the high concentration of *OH generated on the BDD surface, with the release of NH_4^+ and NO_3^- ions, nearly 50% of mineralization of paracetamol and diclofenac is always achieved [50].

Table 5. Selected results reported for PPCPs removal by EOP with dimensionally stable anode (DSA) anodes.

PPCPs	Initial C	Electrolyte	j/mA cm^{-2}	Reactors/Operational Parameters	Electrodes		pH	Reaction Time (min)	Removal (%)	Ref.
					Anode	Cathode				
Ceftazidime	5 mg/L	1 g/L Na$_2$SO$_4$	1.25	V reactor and electrolytic wastewater was 150 mL and 120 mL, respectively	Ti/TiO$_2$/SnO$_2$-Sb-Cu; (50 mm × 30 mm × 2 mm)	Pt wire; gap 4 cm	6	-	97.65	[75]
Iohexol	0.525 mg/L	0.1 M Na$_2$SO$_4$	38.1–45	Batch experiments, V 350 mL, pH 7.2, iohexol concentration 0.525 mg/L; j = 15, 30, and 45 mA/cm^2; pH (4.0, 7.0 ± 02, and 9.0)	Ti/RuO$_2$; 25 cm^2	SS; 0.5-mm gap	7.1	19.8–30	>90	[77]
Carboplatin	0.5 mg/L	0.1 M Na$_2$SO$_4$	30	One-compartment cell 350 mL; pH range 4–9; j = 15, 30 and 45 mA/cm^2;	Ti/RuO$_2$; 25 cm^2	SS plate; 25 cm^2 gap 0.5 cm	7	5	100.00	[59]
Methotrexate	0.5 mg/L	200 mg/L Na$_2$SO$_4$	30	One-compartment cell, V 350 mL, Na$_2$SO$_4$ (100, 200, 300 mg/L), pH range of 4–9; j = 15, 30 and 45 mA cm^{-2}	Ti/IrO$_2$-RuO$_2$; 25 cm^2	SS plate; 0.5 cm gap	7	5	95.00	[85]
Estriol	1000 µg/L	0.1M Na$_2$SO$_4$	20	Batch 3D electrolysis, an undivided rectangular reactor, V 300 mL, filled with approximately 50 g granular graphite particles and 70 g glass beads	Ti/IrO$_2$-RuO$_2$; 5 × 10 cm	Ti; 5 × 10 cm; gap could be adjusted	3–7	50	80.00	[69]
Sulfamethoxazole	200 mg/L	0.1 mol/L NaCl	≥20	Single compartment filter press-type flow cell reactor, flow rate: 425 mL/min	Ti/Ru$_{0.3}$Ti$_{0.7}$O; 14 cm^2	Ti plate; The same geometric area	3	30	>98	[86]
Ceftriaxone sodium	10 mg/L	0.1 mol/L Na$_2$SO$_4$	(The external potential of +2.0 V)	A cylindrical glass reactor made, fused and sealed at one end	TiO$_2$(40)/Nano-G	Titanium mesh; gap 2 cm	-	120	97.70	[87]
Clofibric acid	50 mg/L	50mM Na$_2$SO$_4$	33.6	250 mL undivided glass beaker containing 200 mL solution, T constant at 20 °C, constant current	Plate mixed metal oxide (DSA, Ti/RuO$_2$-IrO$_2$); 5.0 cm × 11.9 cm	SS; Same dimension; gap 4.0 cm	4	180	64.70 (TOC)	[88]

4.2.4. Other Electrodes

The Pt electrode showed better performance in sulfamethoxazole and diclofenac degradation wit electrolyte supports under the same conditions as the carbon electrode [89]. Compared to RuO_2/Ti, IrO_2/Ti, and $RuIrO_2/Ti$ electrodes, Pt/Ti demonstrated that the removal efficiency of berberine was considerably higher [90].

Carbon nanotubes are recognized in wastewater as an advanced anode material for recalcitrant antibiotics for electrocatalysis oxidation. Cyclic voltammetry analysis of La_2O_3-CuO_2/carbon nanotube (CNT) showed a stronger catalytic activity of the modified electrode and stable working life with an efficiency of 90% to 1 mg/L ceftazidime within 30 min, which is much higher than that of pristine CNTs and DSA [91]. The addition of TiO_2 could promote the electron transfer and reusability of the CeO_2-ZrO_2/TiO_2/CNT electrode [92]. In three electrodes promoted by multiwall carbon nanotubes (MWCNTs) (MWCNT, MWCNT-COOH, and MWCNT-NH_2), concerning the electrode surface chemistry, MWCNT-NH_2, with the highest isoelectric point (4.70), is the most promising material due to improved reactant interactions [93].

4.3. Influence of Operational Parameters

4.3.1. Initial PPCPs Concentration

The initial drug concentration significantly influenced the rate of electrochemical decomposition and the process efficiency for both drugs, ifosfamide, and cyclophosphamide [94]. The higher degradation rate of ibuprofen achieved at relative lower initial concentrations at the initial ibuprofen concentrations ranges from 1.0 to 20.0 mg/L [66]. The concentration of parabens in the aqueous matrix was the element that, regardless of the aqueous matrix under investigation, exerts a more extraordinary effect on the target variable. An increase in the initial parabens concentration resulted in a decrease in the efficiency of removal [95] and the mineralization rate decreases when salicylic acid concentration rose from 200 mg/L. During bulk electrolyzes at a low j value and high salicylic acid concentration, salicylic acid was oxidized to aromatic compounds due to a low local concentration on the anode surface of electrogenerated *OH relative to salicylic acid. As bulk electrolyzes at a high j value and low salicylic acid concentration, the product was directly combusted to CO_2 due to a high local concentration on the anode surface of electrogenerated *OH relative to salicylic acid [96].

Interestingly, it could be seen that the efficiency improved with the increased concentration of paracetamol and diclofenac due to the gradual increase in the concentration of *OH to oxidize contaminants before participating in non-oxidizing reactions [50]. The removal of caffeine had two stages, depending on its concentration. At low concentrations, the efficiency significantly increased with j value, suggesting a crucial role of mediated oxidation processes [97].

4.3.2. Supporting Electrolytes

In the presence of NaCl as the supporting electrolyte, the degradation rate of PPCPs was favored. Experiments on RuO_2/Ti, IrO_2/Ti, $RuIrO_2/Ti$, and Pt Ti electrodes showed a constant reaction rate in NaCl solution three to five times higher than in Na_2SO_4 and the oxidation rate of berberines increased due to active chlorine formation [90]. Ambuludi et al. [98] indicated that the pseudo-first-order rate constant increased when NaCl replaced Na_2SO_4 as the electrolyte support and it was almost unaffected by the concentration of ibuprofen. Otherwise, the poor mineralization of ketoprofen was due to the formation of chlorinated organic compounds, which are refractory, at both BDD and Pt anodes in the presence of NaCl as supporting electrolyte while total mineralization using Na_2SO_4 as an electrolyte was achieved [82]. Indermuhle et al. [97] found using NaCl, compared to Na_2SO_4, caffeine could reach a faster degradation but more reaction intermediates are formed and the mechanism is consistent with other proposed (Figure 6).

Table 6. Selected results reported for PPCPs removal by EOP with BDD anodes.

PPCPs	Initial C	Electrolyte	j/mA cm^{-2}	Reactors/Operational Parameters	Anode	Cathode	pH	Reaction Time (min)	Removal (%)	Ref.
Atenolol	0.19 mmol/L	14 mmol/L Na$_2$SO$_4$	30	Double-jacket glass, one-compartment flow filter-press reactor, V 0.002 m^3, pH: 3 and 10, flow rate 3.33×10^{-5} m^3 s^{-1}, j (5, 10, 20 and 30 mAcm^{-2}), T = 25 °C	Nb/BDD500; 0.01 m^2	AISI 304L; Gap 0.02 m	10	120	100.00	[48]
Rifampicin	200 mg/L	0.5 mol/L Na$_2$SO$_4$	90	250-mL undivided open cell, equipped with magnetic stirring at 30 °C	BDD; 3.0 × 2.5 cm	Ti/Ru$_{0.3}$Ti$_{0.7}$O$_2$; 4.0 × 4.0 cm	3	180	95.00	[99]
Norfloxacin	100 mg/L	0.1 mol/L Na$_2$SO$_4$	10	One-compartment filter-press flow reactor, pH (3, 7, 10, and without specific control) j (10, 20, and 30 mA cm^{-2}), T (10, 25, and 40 °C)	BDD; Thickness of 2.9 μm	SS; area of 3.54 cm × 6.71 cm	not pH dependent	300	100.00	[100]
Estrone	500 μg/L	0.1 mol/L Na$_2$SO$_4$	10	A filter-press electrochemical reactor, 0.5L solution, flow rate (2.0, 3.0, 4.0, 5.0, 6.0, and 7.0 L/min), j (5, 10, and 25 mA cm^{-2}), pH (3.0, 7.0, and 10.0)	BDD; each face was 2.5 cm × 3.0 cm 15 cm^2	SS; (3.0 cm × 4.0 cm)	<=7	30	98.00	[83]
Paracetamol Diclofenac	50 mg/L, 100 mg/L	0.05 M Na$_2$SO$_4$	1.56–6.25	4L undivided filter flow press reactor, j (1.56 to 6.25 mAcm^{-2}), flow rate kept constant at 2 L/min	BDD; 64 cm^2	SS; Gap 2 cm	3	60	50.00 (TOC)	[50]
Methyl paraben	100 mg/L	0.05 mol/L K$_2$SO$_4$	10.8	One-compartment pyrex cell (400 mL) operated at 25 ± 1 °C in batch mode, j (1.35 to 21.6 mA cm^{-2})	BDD; 9.68 cm^{-2}	Titanium foil; The same area	5.7	300	100.00	[101]
Sulfonamides	50 mg/L	6.1 g/L Na$_2$SO$_4$	15	Undivided electrolytic cell, V 100 mL, pH (from 2.0 to 7.4), T (from 25 to 60 °C), and j (from 0.05 to 15 mA cm^{-2})	Si/BDD; 10 cm^2	SS; Gap 1cm	6.4	180	92.00	[102]

Table 6. *Cont.*

PPCPs	Initial C	Electrolyte	j/mA cm^{-2}	Reactors/Operational Parameters	Electrodes		pH	Reaction Time (min)	Removal (%)	Ref.
					Anode	Cathode				
Tetracycline	100 mg/L	5 g/L Na$_2$SO$_4$ or NaCl	25 to 300 A m^{-2}	Up-flow electrochemical cell, 20 cm^3, batch mode with recirculation; pH (2 to 12), j (25 to 300 A m^{-2})	BDD; 20 cm^2	SS; Gap 1 cm	5.6	30 min	100.00	[103]
Sulfachloropy-ridazine	0.2 mM	0.05 M Na$_2$SO$_4$	350 mA	An open, cylindrical and undivided glass cell 250 mL with magnetic stirring	BDD; 25 cm^2	Carbon-felt; 77 cm^2 (14.0 cm × 5.5 cm)	4.5	8h	95.00	[104]
Omeprazole	169 mg/L	0.05 Na$_2$SO$_4$	100	Undivided and cylindrical glass cell of 150 mL, with a double jacket; j = 33.3–150 mA cm^{-2}, T = 35 °C, stirred with 800 rpm	BDD; 3 cm^2	Carbon-PTFE air-diffusion; Gap 1 cm	7	360	78.00 (TOC)	[54]
Ibuprofen	0.2 mM	0.05 M Na$_2$SO$_4$	50–500 mA	Cylindrical, open, one-compartment cell 200 mL, at T (20 ± 2 °C)	BDD; 25 cm^2	Carbon-felt; 14 cm×5 cm each side, 0.5 cm width	3	480	>96 (TOC)	[98]

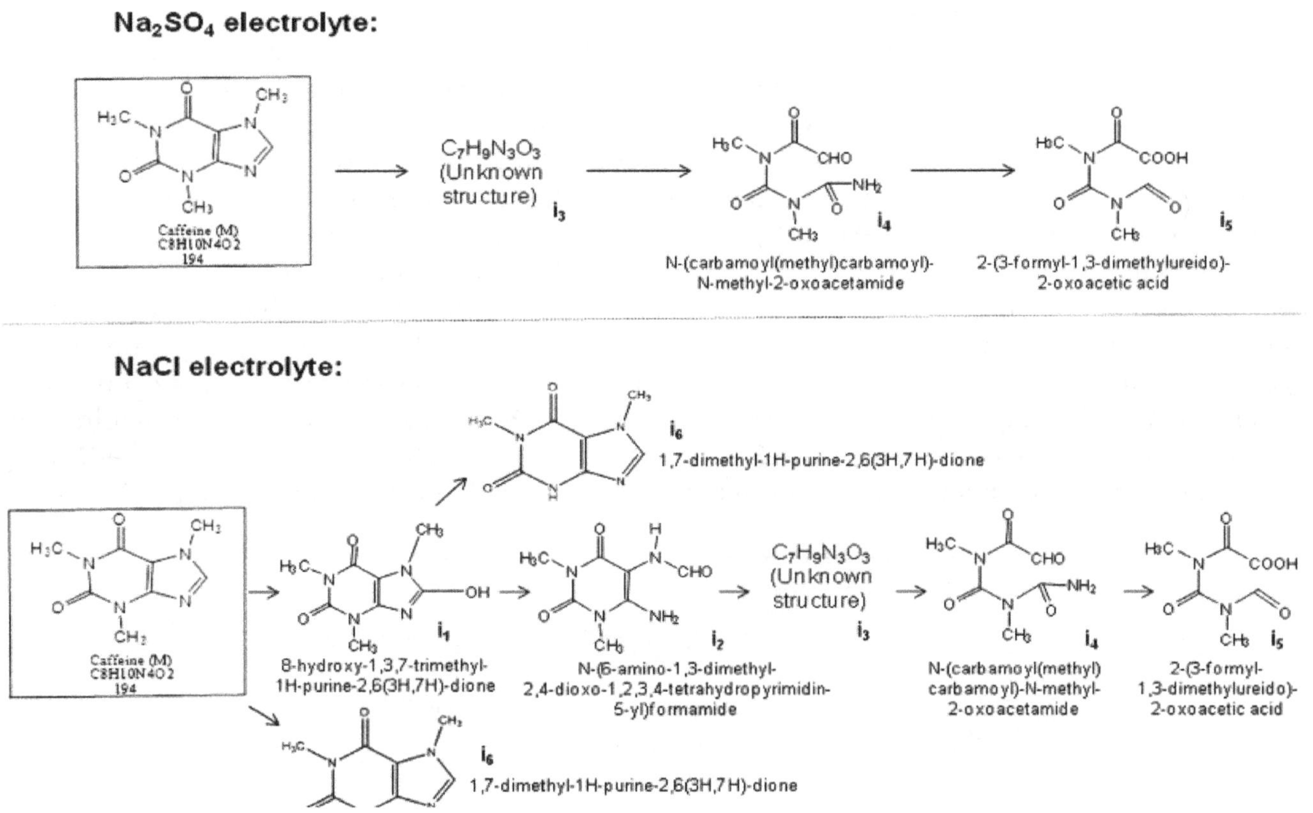

Na₂SO₄ electrolyte:

NaCl electrolyte:

Figure 6. The mechanism model proposed for caffeine degradation by electrochemical oxidation with conductive-diamond electrodes using Na_2SO_4 or $NaCl$ as the electrolyte.

In the presence of Na_2SO_4, the increasing concentration of Na_2SO_4 provided a higher rate of degradation of the anti-cancer drug carboplatin but further increased the concentration of Na_2SO_4, which did not offer a higher rate of degradation due to SO_4^{2-} excess [59]. Moreover, 0.1 M electrolyte-supporting Na_2SO_4 was found to be more active for sulfamethoxazole and diclofenac mineralization, with an efficiency of 15%–30% higher than 0.1 M electrolyte-supporting phosphate buffer on Pt and carbon electrodes [89].

Various inorganic ions have significant effects on removing certain PPCPs that were compared with a higher removal rate in the presence of chloride species than other ions. Acetaminophen, diclofenac, and sulfamethoxazole degradation showed high removal efficiencies, and faster reaction rates may correlate with the presence of chloride species, which may be due to the involvement of hypochlorite ions. Although all of the drugs were degraded by indirect electrochemical oxidation, cyclic voltammograms suggested that chloride species may have coexisted with *OH and have been converted into by-products of degradation [49], whereas ions Cl^- and PO_4^{3-} significantly increased the decomposition rate of ifosfamide [94].

4.3.3. Current Density, pH, Temperature, and Stirring Rate

Current density (j), pH, and temperature also among parameters that have been optimized and investigated in the EOP. Which factor most crucial for efficiency removal depends on the kinds of PPCPs, the material of electrodes and the nature of electrolytes applied. For naproxen removal, the current influence was the greatest among these variables, and the second was the salt concentration, the third flow rate and the fourth pH [105]. Domínguez et al. [106] also proved that the influence of the current was the greatest, then the concentration of salt and the flow rate, respectively, on carbamazepine degradation.

The j value shows a vital role in the removal efficiency with increasing removal efficiency when j increased in most cases of PPCPs [66,97,104] and other factors are dependent or not significant under certain operating conditions. Isothiazolin-3-ones degradation rate was faster as the j value applied increased but nearly independent of electrolyte pH [81]. Moreover, the complete removal of norfloxacin is dependent on pH. However, the removal increased with the temperature at 10 °C, 25 °C, and 40 °C may result from a gradual increase in the diffusion coefficient and the oxidation of byproducts under temperature conditions [100]. Interestingly, DEET degradation increased with increasing current density but was moderately affected by temperature (25–75 °C) [52]. Similarly, the salicylic acid mineralization rate increased at 25 °C with an increase of applied current, the pH impact was not significant [96]. This also can be seen in the case of ketoprofen [82], ifosfamide, and cyclophosphamide [94]. Interestingly, the carboplatin degradation rates increased significantly in the initial phases of electrolysis as j value increased on the Ti/RuO$_2$ electrode. However, a further increase in j did not affect the rate of degradation [59]. Sun et al. [107] found that pH decreased, the efficiency of chloramphenicol degradation increased, and maximum degradation was achieved at pH 2, Figure 7.

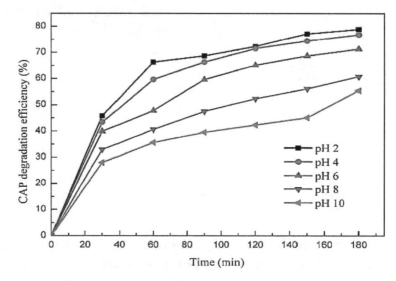

Figure 7. Effect of initial pH of wastewater on the chloramphenicol degradation efficiency of particle electrodes.

Stirring increased the rate of mass transfer and PPCPs formed a contentious relationship on the electrode surface to increase the efficiency of removal. When the stirring speed was too slow, the mass transfer resistance would be the limitation. With the free radical produced from the electrode surface, PPCPs were unable to react quickly. It was also not possible to transfer the hydroxyls produced to the solution in time. On the other hand, the high stirring speed turned leads to short time for PPCPs touching the electrode surface, PPCPs could not be wholly oxidized and soon left the electrode surface. O$_2$ and H$_2$ bubbles produced from H$_2$O electrolysis would be more competitive to access molecule surface with extreme disturbance, resulting in reducing removal efficiency. The kinetic study of naproxen degradation at fix potential indicates that the rate of degradation increases with the stirring speed at 250 and 500 rpm [93]. For diffusion reactions, the stirring rate is an essential factor. The stirring rate showed a definite increase in the removal of ceftazidime and then decreased as the stirring speed between 150 and 200 rad min^{-1} [76].

4.3.4. Electrode Spacing and Fluid Velocity

The changes in the spacing of the electrodes would affect not only the mass transfer limitations but also the electron transport and electric resistance [108]. The effect of electrode spacing, however, depends on the direct or indirect oxidation. In the latter case, the electrode spacing should be matched with the diffusion length of *OH species. Duan, et al. [76] found that the oxidation of ceftazidime decreased as the spacing of the Ti/SnO$_2$-Cu electrode changed at 1, 2, and 3 cm under the current of 20 mA. As the spacing increased, the electrochemical resistance also increased while the charge in the electrolyte decreased. Xie et al. [65] (Figure 8) tested ofloxacin removal with the changes in electrode spacing. The reaction rate increased with the first-order pseudo constant changed as the distance decreased from 3 cm to 0.5 cm and the mass transfer coefficient increased.

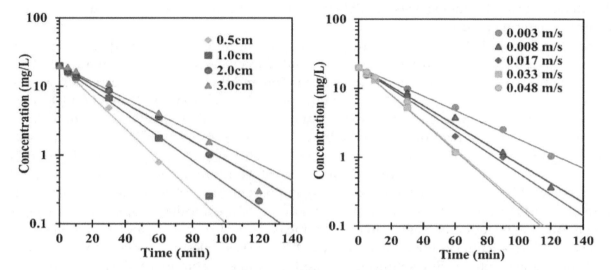

Figure 8. Effect of electrode spacing and fluid velocity on ofloxacin degradation. Anode surface area 10 cm^2, electrolyte concentration = 0.05 M Na$_2$SO$_4$ solution, j = 30 mA cm^{-2}, initial ofloxacin concentration 20 mg/L, voltage 6.2–6.3 V, initial pH 6.25, temperature 25 °C.

It can be seen that the electrocatalytic oxidation process relied primarily on the high potential for direct oxidation on the electrode surface and the generation of free radicals for indirect oxidation of PPCPs. Consequently, the spacing increases, which leads to a loss in *OH production and oxidation power on anode surfaces. Diffusion efficiency also affects removal efficiency and so at a larger electrode spacing, the electrolysis process needs more time because of longer diffusion distance. Both electrode spacing and fluid velocity are critical since increasing velocities that lead to an increase in the rate of mass transfer while decreasing electrode spacing increases the surface area available for mass transfer [65].

4.4. Applications for Real Water and Wastewater Containing PPCPs

EOP is a promising technique with different degradation rates for the removal of PPCPs from water and WWTP effluents under optimal conditions concerning the ecological system [98,107,109,110].

Because of the presence of chloride ions in the effluent, oxidation in secondary treated wastewater was faster than in pure water [111]. Carbamazepine electrodegradation is feasible for WWT in several aqueous matrices [112], after 50 min of electrolysis time, caffeine was removed entirely in DIW and was almost removed in the wastewater sample may be related to the organic matter in wastewater. Having regard to these results, EOP is an effective method for further removal of caffeine from effluent from aerobic or anaerobic reactors that treat municipal wastewater, even though a high concentration of caffeine was used compared to low concentration in natural water. Compared to conventional methods for removing caffeine from urban wastewater, this approach appears to be more feasible for the following reasons: ease of operating, rapid removal of caffeine, and the effective efficiency of treatment [110]. The caffeine elimination obtained in real wastewater was found to be higher than in synthetic wastewater due to the contribution of electrogenerated oxidant species, such as hypochlorite [113], when sulfonamides and DEET removal were most efficient in the presence of municipal wastewater treatment plant (MWWTP) effluents [52,102].

4.5. Combined Systems

While EOP has been widely demonstrated for their ability to remove trace and persistent PPCPs in water and wastewater, complex water matrices could be found that inhibit their efficient operation. As a result, they may potentially reduce or fully retard the efficiency, requiring longer hydraulic retention time or higher volume capacity for compensation. System hybridization or combination of EOP with other water technologies is possible to overcome the operational problems associated with the complex water matrices.

Zaghdoudi et al. [114] investigated the possibility of coupling an electroreduction pretreatment before a biological process for dimetridazole removal. Direct electrolysis was initially conducted at the low potential to reduce amino derivatives formation and then azo dimer formation with a total degradation of dimetridazole achieved and the ratio of biochemical oxygen demand (BOD_5)/chemical oxygen demand (COD) increased. As mineralization yields of all electrolyzed solutions increased significantly, the enhancement of biodegradability was demonstrated during biological treatment. Nevertheless, the real mineralization yields should most likely be significantly higher if the contribution of titanocene, which is possibly biorecalcitrant, is not taken into account in the amount of TOC. Belkheiri et al. [115] examined the biodegradability improvement of tetracycline-containing solutions after an electrochemical pretreatment, as a large amount of the applied drugs are not metabolized and, therefore, can be found in wastewater. BOD_5 measurements verified biodegradability increased with the oxidation potential as the ratio of BOD_5/COD increased. Despite its chemical transformation, none of the reduced tetracycline solutions are biodegradable. Yahiaoui et al. [116] found after 5 h of electrochemical pre-treatment of tetracycline, the BOD_5/COD ratio increased considerably and confirmed during biological treatment, with 76% of dissolved organic carbon (DOC) removed.

Pharmaceutical degradation in conventional WWTPs is a problem because industrial sewage and hospital effluents contain low-concentration pharmaceuticals. Rodríguez-Nava et al. [117] found high efficiencies in removal without affecting activated sludge performance of integrating EOP with a biological system for simultaneous removal from wastewater of recalcitrant drugs (bezafibrate, gemfibrozil, indomethacin, and sulfamethoxazole). Drugs contained in wastewater without electrochemical pretreatment was persistent in the biological process and encouraged bulking formation. García-Gómez et al. [118] proved membrane bioreactor (MBR) high capacity to remove COD and low capacity for degradation (20%) of carbamazepine after 120 days, which presumably suggests that given the weak degradation and carbamazepine was not toxic to microorganisms. The EOP, on the other hand, was able to degrade carbamazepine completely.

In an exciting study for investigating pre- and post-treatment in one system to remove synthetic hospital wastewater fortified with four drug pollutants including carbamazepine, ibuprofen, estradiol, and venlafaxine by the combination of MBR and EOP, MBR alone treatment of wastewater showed a high percentage of ibuprofen and estradiol removal (about 90%), while carbamazepine and venlafaxine performed a low elimination (at around 10%). EOP as post-treatment, this allowed high removal (about 97%) of the four pharmaceutical pollutants and far more successful compared to EOP as pre-treatment [119]. The integration of electrochemical processes into MBR systems can utilize the mechanism of biodegradation, sorption, hydrolysis, and filtration on conventional MBR and electrocoagulation, electroosmosis, and electrophoresis on electrochemical processes that improve both the performance and the control of membrane fouling for eliminating recalcitrant micropollutants [120,121].

5. Conclusions

EOP is a promising technique with different degradation rates for the removal of PPCPs from water and wastewater, from synthetic or real, concerning the ecological system. There are numerous studies that have recently focused on the finding of electrode materials and optimal conditions, including initial PPCPs concentration, supporting electrolytes, j value, pH, temperature, stirring rate, and electrode spacing that are effective for removing a certain or groups of PPCPs with considering reduce operating cost. In terms of operational parameters, it was shown that the current influence was the greatest among these variables in some mentioned studies. Although the electrochemical process has recorded several influential factors, only some of them show a significant impact on real systems.

Studies showed that the EOP system depends heavily on the type of anode. BDD anode shows high performance on various kinds of PPCPs. The BDD anodes have been reported to produce higher organic oxidation rates and higher current efficiencies than other metal oxides commonly used. The development of BDD anodes and the enormous advantages of this electrode compared to others make this material was investigated on most of the works published in the literature. The performance of 3D electrolysis is much better, more cost-effective, and saves more energy consumption than traditional 2D electrolysis. The results validate 3D electrolysis in pretreatment or advanced treatment applications as a promising alternative method to remove PPCPs from secondary effluents.

Real field samples may contain other species of radical electrolytes that may participate in the electrochemical process and therefore act as interferences within the EOP system. It is therefore recommended that the electrochemical degradation process be the last step in the domestic water treatment since the technique also largely depends on the electrolytes in the water.

Toxicity evaluation is an essential environmental pollution control factor since the degradation by-products from the initial structure can be more toxic. It can be seen that in some kinds of PPCPs, intermediates are more toxic than the molecule of the parent, while others are less harmful. By evaluating toxicity, it helps significantly in optimizing treatment conditions to achieve the elimination of adverse effects of by-products.

EOP has widely demonstrated their ability to remove trace and persistent PPCPs in water and wastewater. Further, complex water matrices could be found that inhibit their efficient operation. System hybridization or combination of EOP with other water technologies is possible to overcome the operational problems associated with the complex water matrices.

Author Contributions: Conceptualization, K.C.D., Y.-P.T. and K.-F.C.; resources, Y.-P.T. and K.-F.C; writing—original draft preparation, K.C.D.; writing—review and editing, K.C.D., and C.-C.Y.; supervision, Y.-P.T. and K.-F.C.; project administration, C.-C.Y.; funding acquisition, Y.-P.T. All authors have read and agreed to the published version of the manuscript.

References

1. Shen, R.; Andrews, S.A. Demonstration of 20 pharmaceuticals and personal care products (PPCPs) as nitrosamine precursors during chloramine disinfection. *Water Res.* **2011**, *45*, 944–952. [CrossRef] [PubMed]
2. Boxall, A.B.; Rudd, M.A.; Brooks, B.W.; Caldwell, D.J.; Choi, K.; Hickmann, S.; Innes, E.; Ostapyk, K.; Staveley, J.P.; Verslycke, T. Pharmaceuticals and personal care products in the environment: What are the big questions? *Environ. Health Perspect.* **2012**, *120*, 1221–1229. [CrossRef]
3. Deblonde, T.; Cossu-Leguille, C.; Hartemann, P. Emerging pollutants in wastewater: A review of the literature. *Int. J. Hyg. Environ. Health* **2011**, *214*, 442–448. [CrossRef]
4. Rajapaksha, A.U.; Vithanage, M.; Lim, J.E.; Ahmed, M.B.M.; Zhang, M.; Lee, S.S.; Ok, Y.S. Invasive plant-derived biochar inhibits sulfamethazine uptake by lettuce in soil. *Chemosphere* **2014**, *111*, 500–504. [CrossRef] [PubMed]
5. Helbling, D.E.; Hollender, J.; Kohler, H.-P.E.; Singer, H.; Fenner, K. High-throughput identification of microbial transformation products of organic micropollutants. *Environ. Sci. Technol. Water Treat.* **2010**, *44*, 6621–6627. [CrossRef] [PubMed]
6. Xia, K.; Bhandari, A.; Das, K.; Pillar, G. Occurrence and fate of pharmaceuticals and personal care products (PPCPs) in biosolids. *J. Environ. Qual.* **2005**, *34*, 91–104. [CrossRef]
7. Blair, B.D.; Crago, J.P.; Hedman, C.J.; Klaper, R.D. Pharmaceuticals and personal care products found in the Great Lakes above concentrations of environmental concern. *Chemosphere* **2013**, *93*, 2116–2123. [CrossRef]
8. Chiron, S.; Minero, C.; Vione, D. Photodegradation processes of the antiepileptic drug carbamazepine, relevant to estuarine waters. *Environ. Sci.* **2006**, *40*, 5977–5983. [CrossRef]
9. Tolls, J. Sorption of veterinary pharmaceuticals in soils: A review. *Environ. Sci. Technol. Water Treat.* **2001**, *35*, 3397–3406. [CrossRef]

10. Panizza, M.; Cerisola, G. Direct and mediated anodic oxidation of organic pollutants. *Chem. Rev.* **2009**, *109*, 6541–6569. [CrossRef]

11. Hollender, J.; Zimmermann, S.G.; Koepke, S.; Krauss, M.; McArdell, C.S.; Ort, C.; Singer, H.; von Gunten, U.; Siegrist, H. Elimination of organic micropollutants in a municipal wastewater treatment plant upgraded with a full-scale post-ozonation followed by sand filtration. *Environ. Sci. Technol. Water Treat.* **2009**, *43*, 7862–7869. [CrossRef]

12. Martinez-Huitle, C.A.; Ferro, S. Electrochemical oxidation of organic pollutants for the wastewater treatment: Direct and indirect processes. *Chem. Soc. Rev.* **2006**, *35*, 1324–1340. [CrossRef]

13. Kaestner, M.; Nowak, K.M.; Miltner, A.; Trapp, S.; Schaeffer, A. Classification and modelling of nonextractable residue (NER) formation of xenobiotics in soil–a synthesis. *Crit. Rev. Environ. Sci. Technol. Water Treat.* **2014**, *44*, 2107–2171. [CrossRef]

14. Kallenborn, R. *Perfluorinated Alkylated Substances (PFAS) in the Nordic Environment*; Nordic Council of Ministers: Copenhagen, Denmark, 2004.

15. Daughton, C.G.; Ternes, T.A. Pharmaceuticals and personal care products in the environment: Agents of subtle change? *Environ. Health Perspect.* **1999**, *107*, 907–938. [CrossRef] [PubMed]

16. Golet, E.M.; Alder, A.C.; Hartmann, A.; Ternes, T.A.; Giger, W. Trace determination of fluoroquinolone antibacterial agents in urban wastewater by solid-phase extraction and liquid chromatography with fluorescence detection. *Anal. Chem.* **2001**, *73*, 3632–3638. [CrossRef]

17. Lishman, L.; Smyth, S.A.; Sarafin, K.; Kleywegt, S.; Toito, J.; Peart, T.; Lee, B.; Servos, M.; Beland, M.; Seto, P. Occurrence and reductions of pharmaceuticals and personal care products and estrogens by municipal wastewater treatment plants in Ontario, Canada. *Sci. Total Environ.* **2006**, *367*, 544–558. [CrossRef]

18. Roberts, J.; Kumar, A.; Du, J.; Hepplewhite, C.; Ellis, D.J.; Christy, A.G.; Beavis, S.G. Pharmaceuticals and personal care products (PPCPs) in Australia's largest inland sewage treatment plant, and its contribution to a major Australian river during high and low flow. *Sci. Total Environ.* **2016**, *541*, 1625–1637. [CrossRef] [PubMed]

19. Yu, Y.; Wu, L.; Chang, A.C. Seasonal variation of endocrine disrupting compounds, pharmaceuticals and personal care products in wastewater treatment plants. *Sci. Total Environ.* **2013**, *442*, 310–316. [CrossRef] [PubMed]

20. Martín, J.; Camacho-Muñoz, D.; Santos, J.L.; Aparicio, I.; Alonso, E. Occurrence and ecotoxicological risk assessment of 14 cytostatic drugs in wastewater. *Water Air Soil Pollut.* **2014**, *225*, 1896. [CrossRef]

21. Ternes, T.A.; Bonerz, M.; Herrmann, N.; Teiser, B.; Andersen, H.R. Irrigation of treated wastewater in Braunschweig, Germany: An option to remove pharmaceuticals and musk fragrances. *Chemosphere* **2007**, *66*, 894–904. [CrossRef]

22. Qiu, G.; Song, Y.-H.; Zeng, P.; Duan, L.; Xiao, S. Characterization of bacterial communities in hybrid upflow anaerobic sludge blanket (UASB)–membrane bioreactor (MBR) process for berberine antibiotic wastewater treatment. *Bioresour. Technol.* **2013**, *142*, 52–62. [CrossRef]

23. Rosal, R.; Rodríguez, A.; Perdigón-Melón, J.A.; Petre, A.; García-Calvo, E.; Gómez, M.J.; Agüera, A.; Fernández-Alba, A.R. Occurrence of emerging pollutants in urban wastewater and their removal through biological treatment followed by ozonation. *Water Res.* **2010**, *44*, 578–588. [CrossRef]

24. Leclercq, M.; Mathieu, O.; Gomez, E.; Casellas, C.; Fenet, H.; Hillaire-Buys, D. Presence and fate of carbamazepine, oxcarbazepine, and seven of their metabolites at wastewater treatment plants. *Arch. Environ. Contam. Toxicol.* **2009**, *56*, 408. [CrossRef] [PubMed]

25. Lenz, K.; Hann, S.; Koellensperger, G.; Stefanka, Z.; Stingeder, G.; Weissenbacher, N.; Mahnik, S.N.; Fuerhacker, M. Presence of cancerostatic platinum compounds in hospital wastewater and possible elimination by adsorption to activated sludge. *Sci. Total Environ.* **2005**, *345*, 141–152. [CrossRef]

26. Liu, Y.; Wang, Z.; Yan, K.; Wang, Z.; Torres, O.L.; Guo, R.; Chen, J. A new disposal method for systematically processing of ceftazidime: The intimate coupling UV/algae-algae treatment. *Chem. Eng. J.* **2017**, *314*, 152–159. [CrossRef]

27. Junker, T.; Alexy, R.; Knacker, T.; Kümmerer, K. Biodegradability of 14C-labeled antibiotics in a modified laboratory scale sewage treatment plant at environmentally relevant concentrations. *Environ. Sci. Technol.* **2006**, *40*, 318–324. [CrossRef] [PubMed]

28. Watkinson, A.J.; Murby, E.J.; Costanzo, S.D. Removal of antibiotics in conventional and advanced wastewater treatment: Implications for environmental discharge and wastewater recycling. *Water Res.* **2007**, *41*, 4164–4176. [CrossRef]

29. Leung, H.W.; Minh, T.B.; Murphy, M.B.; Lam, J.C.; So, M.K.; Martin, M.; Lam, P.K.; Richardson, B.J. Distribution, fate and risk assessment of antibiotics in sewage treatment plants in Hong Kong, South China. *Environ. Int.* **2012**, *42*, 1–9. [CrossRef]

30. Xu, W.; Zhang, G.; Li, X.; Zou, S.; Li, P.; Hu, Z.; Li, J. Occurrence and elimination of antibiotics at four sewage treatment plants in the Pearl River Delta (PRD), South China. *Water Res.* **2007**, *41*, 4526–4534. [CrossRef]

31. Blair, B.; Nikolaus, A.; Hedman, C.; Klaper, R.; Grundl, T. Evaluating the degradation, sorption, and negative mass balances of pharmaceuticals and personal care products during wastewater treatment. *Chemosphere* **2015**, *134*, 395–401. [CrossRef]

32. Salgado, R.; Oehmen, A.; Carvalho, G.; Noronha, J.P.; Reis, M.A. Biodegradation of clofibric acid and identification of its metabolites. *J. Hazard. Mater.* **2012**, *241*, 182–189. [CrossRef]

33. Sun, Q.; Lv, M.; Hu, A.; Yang, X.; Yu, C.P. Seasonal variation in the occurrence and removal of pharmaceuticals and personal care products in a wastewater treatment plant in Xiamen, China. *J. Hazard. Mater.* **2014**, *277*, 69–75. [CrossRef] [PubMed]

34. He, K.; Soares, A.D.; Adejumo, H.; McDiarmid, M.; Squibb, K.; Blaney, L. Detection of a wide variety of human and veterinary fluoroquinolone antibiotics in municipal wastewater and wastewater-impacted surface water. *J. Pharm. Biomed. Anal.* **2015**, *106*, 136–143. [CrossRef]

35. Voets, J.P.; Pipyn, P.; Van Lancker, P.; Verstraete, W. Degradation of microbicides under different environmental conditions. *J. Appl. Bacteriol.* **1976**, *40*, 67–72. [CrossRef]

36. Prasse, C.; Schlusener, M.P.; Schulz, R.; Ternes, T.A. Antiviral drugs in wastewater and surface waters: A new pharmaceutical class of environmental relevance? *Environ. Sci. Technol.* **2010**, *44*, 1728–1735. [CrossRef] [PubMed]

37. Simonich, S.L.; Federle, T.W.; Eckhoff, W.S.; Rottiers, A.; Webb, S.; Sabaliunas, D.; De Wolf, W. Removal of fragrance materials during US and European wastewater treatment. *Environ. Sci. Technol.* **2002**, *36*, 2839–2847. [CrossRef] [PubMed]

38. Nakada, N.; Tanishima, T.; Shinohara, H.; Kiri, K.; Takada, H. Pharmaceutical chemicals and endocrine disrupters in municipal wastewater in Tokyo and their removal during activated sludge treatment. *Water Res.* **2006**, *40*, 3297–3303. [CrossRef]

39. Molins-Delgado, D.; Díaz-Cruz, M.S.; Barceló, D. Ecological risk assessment associated to the removal of endocrine-disrupting parabens and benzophenone-4 in wastewater treatment. *J. Hazard. Mater.* **2016**, *310*, 143–151. [CrossRef]

40. Zhang, X.; Zhao, H.; Du, J.; Qu, Y.; Shen, C.; Tan, F.; Chen, J.; Quan, X. Occurrence, removal, and risk assessment of antibiotics in 12 wastewater treatment plants from Dalian, China. *Environ. Sci. Pollut. Res. Int.* **2017**, *24*, 16478–16487. [CrossRef]

41. Kosma, C.I.; Lambropoulou, D.A.; Albanis, T.A. Occurrence and removal of PPCPs in municipal and hospital wastewaters in Greece. *J. Hazard. Mater.* **2010**, *179*, 804–817. [CrossRef]

42. Verlicchi, P.; Al Aukidy, M.; Zambello, E. Occurrence of pharmaceutical compounds in urban wastewater: Removal, mass load and environmental risk after a secondary treatment—A review. *Sci. Total Environ.* **2012**, *429*, 123–155. [CrossRef]

43. Huddleston, J.G.; Willauer, H.D.; Swatloski, R.P.; Visser, A.E.; Rogers, R.D. Room temperature ionic liquids as novel media for 'clean' liquid–liquid extraction. *Chem. Commun.* **1998**, 1765–1766. [CrossRef]

44. Mohammadhosseini, M.; Tehrani, M.S.; Ganjali, M.R. Preconcentration, determination and speciation of chromium (III) using solid phase extraction and flame atomic absorption spectrometry. *J. Chin. Chem. Soc.* **2006**, *53*, 549–557. [CrossRef]

45. Rajeshwar, K.; Ibanez, J.G. *Environmental Electrochemistry: Fundamentals and Applications in Pollution Sensors and Abatement*; Elsevier: Amsterdam, The Netherlands, 1997.

46. Wendt, H.; Kreysa, G. *Electrochemical Engineering: Science and Technology in Chemical and Other Industries*; Springer: Berlin/Heidelberg, Germany, 1999.

47. Periyasamy, S.; Muthuchamy, M. Electrochemical oxidation of paracetamol in water by graphite anode: Effect of pH, electrolyte concentration and current density. *J. Environ. Chem. Eng.* **2018**, *6*, 7358–7367. [CrossRef]

48. Da Silva, S.W.; do Prado, J.M.; Heberle, A.N.A.; Schneider, D.E.; Rodrigues, M.A.S.; Bernardes, A.M. Electrochemical advanced oxidation of Atenolol at Nb/BDD thin film anode. *J. Electroanal. Chem.* **2019**, *844*, 27–33. [CrossRef]

49. Liu, Y.-J.; Hu, C.-Y.; Lo, S.-L. Direct and indirect electrochemical oxidation of amine-containing pharmaceuticals using graphite electrodes. *J. Hazard. Mater.* **2019**, *366*, 592–605. [CrossRef]

50. García-Montoya, M.F.; Gutiérrez-Granados, S.; Alatorre-Ordaz, A.; Galindo, R.; Ornelas, R.; Peralta-Hernandez, J.M.; Chemistry, E. Application of electrochemical/BDD process for the treatment wastewater effluents containing pharmaceutical compounds. *J. Ind. Eng. Chem.* **2015**, *31*, 238–243. [CrossRef]

51. Mora-Gomez, J.; Ortega, E.; Mestre, S.; Pérez-Herranz, V.; García-Gabaldón, M. Electrochemical degradation of norfloxacin using BDD and new Sb-doped SnO_2 ceramic anodes in an electrochemical reactor in the presence and absence of a cation-exchange membrane. *Sep. Purif. Technol.* **2019**, *208*, 68–75. [CrossRef]

52. Chen, T.-S.; Chen, P.-H.; Huang, K.-L. Electrochemical degradation of N, N-diethyl-m-toluamide on a boron-doped diamond electrode. *J. Taiwan Inst. Chem. Eng.* **2014**, *45*, 2615–2621. [CrossRef]

53. Brillas, E.; Sires, I.; Arias, C.; Cabot, P.L.; Centellas, F.; Rodriguez, R.M.; Garrido, J.A. Mineralization of paracetamol in aqueous medium by anodic oxidation with a boron-doped diamond electrode. *Chemosphere* **2005**, *58*, 399–406. [CrossRef]

54. Cavalcanti, E.B.; Garcia-Segura, S.; Centellas, F.; Brillas, E. Electrochemical incineration of omeprazole in neutral aqueous medium using a platinum or boron-doped diamond anode: Degradation kinetics and oxidation products. *Water Res.* **2013**, *47*, 1803–1815. [CrossRef]

55. Dai, Q.; Xia, Y.; Sun, C.; Weng, M.; Chen, J.; Wang, J.; Chen, J. Electrochemical degradation of levodopa with modified PbO_2 electrode: Parameter optimization and degradation mechanism. *Chem. Eng. J.* **2014**, *245*, 359–366. [CrossRef]

56. Radjenovic, J.; Bagastyo, A.; Rozendal, R.A.; Mu, Y.; Keller, J.; Rabaey, K. Electrochemical oxidation of trace organic contaminants in reverse osmosis concentrate using RuO_2/IrO_2-coated titanium anodes. *Water Res.* **2011**, *45*, 1579–1586. [CrossRef]

57. Sopaj, F.; Rodrigo, M.A.; Oturan, N.; Podvorica, F.I.; Pinson, J.; Oturan, M.A. Influence of the anode materials on the electrochemical oxidation efficiency. Application to oxidative degradation of the pharmaceutical amoxicillin. *Chem. Eng. J.* **2015**, *262*, 286–294. [CrossRef]

58. Oaks, J.L.; Gilbert, M.; Virani, M.Z.; Watson, R.T.; Meteyer, C.U.; Rideout, B.A.; Shivaprasad, H.; Ahmed, S.; Chaudhry, M.J.I.; Arshad, M. Diclofenac residues as the cause of vulture population decline in Pakistan. *Nature* **2004**, *427*, 630. [CrossRef]

59. Barışçı, S.; Turkay, O.; Ulusoy, E.; Soydemir, G.; Seker, M.G.; Dimoglo, A. Electrochemical treatment of anti-cancer drug carboplatin on mixed-metal oxides and boron doped diamond electrodes: Density functional theory modelling and toxicity evaluation. *J. Hazard. Mater.* **2018**, *344*, 316–321. [CrossRef]

60. El-Ashtoukhy, E.-S.; Amin, N.; Abdelwahab, O. Treatment of paper mill effluents in a batch-stirred electrochemical tank reactor. *Chem. Eng. J.* **2009**, *146*, 205–210. [CrossRef]

61. Wang, Q.; Jin, T.; Hu, Z.; Zhou, L.; Zhou, M. TiO_2-NTs/SnO_2-Sb anode for efficient electrocatalytic degradation of organic pollutants: Effect of TiO_2-NTs architecture. *Sep. Purif. Technol.* **2013**, *102*, 180–186. [CrossRef]

62. Wu, W.; Huang, Z.-H.; Lim, T.-T. Recent development of mixed metal oxide anodes for electrochemical oxidation of organic pollutants in water. *Appl. Catal. A Gen.* **2014**, *480*, 58–78. [CrossRef]

63. Dai, Q.; Zhou, J.; Meng, X.; Feng, D.; Wu, C.; Chen, J. Electrochemical oxidation of cinnamic acid with Mo modified PbO_2 electrode: Electrode characterization, kinetics and degradation pathway. *Chem. Eng. J.* **2016**, *289*, 239–246. [CrossRef]

64. Zhao, W.; Xing, J.; Chen, D.; Jin, D.; Shen, J. Electrochemical degradation of Musk ketone in aqueous solutions using a novel porous Ti/SnO_2-Sb_2O_3/PbO_2 electrodes. *J. Electroanal. Chem.* **2016**, *775*, 179–188. [CrossRef]

65. Xie, R.; Meng, X.; Sun, P.; Niu, J.; Jiang, W.; Bottomley, L.; Li, D.; Chen, Y.; Crittenden, J. Electrochemical oxidation of ofloxacin using a TiO_2-based SnO_2-Sb/polytetrafluoroethylene resin-PbO_2 electrode: Reaction kinetics and mass transfer impact. *Appl. Catal. B Environ.* **2017**, *203*, 515–525. [CrossRef]

66. Wang, C.; Yu, Y.; Yin, L.; Niu, J.; Hou, L.-A. Insights of ibuprofen electro-oxidation on metal-oxide-coated Ti anodes: Kinetics, energy consumption and reaction mechanisms. *Chemosphere* **2016**, *163*, 584–591. [CrossRef] [PubMed]

67. Brillas, E.; Martínez-Huitle, C.A. Decontamination of wastewaters containing synthetic organic dyes by electrochemical methods. An updated review. *Appl. Catal. B Environ.* **2015**, *166*, 603–643. [CrossRef]

68. Alighardashi, A.; Aghta, R.S.; Ebrahimzadeh, H. Improvement of Carbamazepine Degradation by a Three-Dimensional Electrochemical (3-EC) Process. *Int. J. Environ. Res. Public Health* **2018**, *12*, 451–458. [CrossRef]

69. Shen, B.; Wen, X.-H.; Huang, X. Enhanced removal performance of estriol by a three-dimensional electrode reactor. *Chem. Eng. J.* **2017**, *327*, 597–607. [CrossRef]

70. Wang, Y.; Zhou, C.; Chen, J.; Fu, Z.; Niu, J. Bicarbonate enhancing electrochemical degradation of antiviral drug lamivudine in aqueous solution. *J. Electroanal. Chem.* **2019**, *848*, 113314. [CrossRef]

71. Wachter, N.; Aquino, J.M.; Denadai, M.; Barreiro, J.C.; Silva, A.J.; Cass, Q.B.; Rocha-Filho, R.C.; Bocchi, N. Optimization of the electrochemical degradation process of the antibiotic ciprofloxacin using a double-sided β-PbO$_2$ anode in a flow reactor: Kinetics, identification of oxidation intermediates and toxicity evaluation. *Environ. Sci. Pollut. Res.* **2019**, *26*, 4438–4449. [CrossRef]

72. Wang, C.; Yin, L.; Xu, Z.; Niu, J.; Hou, L.-A. Electrochemical degradation of enrofloxacin by lead dioxide anode: Kinetics, mechanism and toxicity evaluation. *Chem. Eng. J.* **2017**, *326*, 911–920. [CrossRef]

73. Wei, L.; Guo, S.; Yan, G.; Chen, C.; Jiang, X. Electrochemical pretreatment of heavy oil refinery wastewater using a three-dimensional electrode reactor. *Electrochim. Acta* **2010**, *55*, 8615–8620. [CrossRef]

74. Fortuny, A.; Font, J.; Fabregat, A. Wet air oxidation of phenol using active carbon as catalyst. *Appl. Catal. B Environ.* **1998**, *19*, 165–173. [CrossRef]

75. Li, X.; Duan, P.; Lei, J.; Sun, Z.; Hu, X. Fabrication of Ti/TiO$_2$/SnO$_2$-Sb-Cu electrode for enhancing electrochemical degradation of ceftazidime in aqueous solution. *J. Electroanal. Chem.* **2019**, *847*, 113231. [CrossRef]

76. Duan, P.; Hu, X.; Ji, Z.; Yang, X.; Sun, Z. Enhanced oxidation potential of Ti/SnO$_2$-Cu electrode for electrochemical degradation of low-concentration ceftazidime in aqueous solution: Performance and degradation pathway. *Chemosphere* **2018**, *212*, 594–603. [CrossRef] [PubMed]

77. Turkay, O.; Barisci, S.; Ulusoy, E.; Dimoglo, A. Electrochemical Reduction of X-ray Contrast Iohexol at Mixed Metal Oxide Electrodes: Process Optimization and By-product Identification. *Water Air Soil Pollut.* **2018**, *229*, 170. [CrossRef]

78. Chen, X.; Chen, G. Fabrication and application of Ti/BDD for wastewater treatment. *Synth. Diam. Film. Prep. Electrochem. Charact. Appl.* **2011**, 353–371. [CrossRef]

79. He, Y.; Huang, W.; Chen, R.; Zhang, W.; Lin, H.; Li, H. Anodic oxidation of aspirin on PbO$_2$, BDD and porous Ti/BDD electrodes: Mechanism, kinetics and utilization rate. *Sep. Purif. Technol.* **2015**, *156*, 124–131. [CrossRef]

80. Sirés, I.; Oturan, N.; Oturan, M.A. Electrochemical degradation of β-blockers. Studies on single and multicomponent synthetic aqueous solutions. *Water Res.* **2010**, *44*, 3109–3120. [CrossRef] [PubMed]

81. Kandavelu, V.; Yoshihara, S.; Kumaravel, M.; Murugananthan, M. Anodic oxidation of isothiazolin-3-ones in aqueous medium by using boron-doped diamond electrode. *Diam. Relat. Mater.* **2016**, *69*, 152–159. [CrossRef]

82. Murugananthan, M.; Latha, S.; Raju, G.B.; Yoshihara, S. Anodic oxidation of ketoprofen—An anti-inflammatory drug using boron doped diamond and platinum electrodes. *J. Hazard. Mater.* **2010**, *180*, 753–758. [CrossRef]

83. Brocenschi, R.F.; Rocha-Filho, R.C.; Bocchi, N.; Biaggio, S.R. Electrochemical degradation of estrone using a boron-doped diamond anode in a filter-press reactor. *Electrochim. Acta* **2016**, *197*, 186–193. [CrossRef]

84. Coledam, D.A.; Pupo, M.M.; Silva, B.F.; Silva, A.J.; Eguiluz, K.I.; Salazar-Banda, G.R.; Aquino, J.M. Electrochemical mineralization of cephalexin using a conductive diamond anode: A mechanistic and toxicity investigation. *Chemosphere* **2017**, *168*, 638–647. [CrossRef] [PubMed]

85. Barışçı, S.; Turkay, O.; Ulusoy, E.; Şeker, M.G.; Yüksel, E.; Dimoglo, A. Electro-oxidation of cytostatic drugs: Experimental and theoretical identification of by-products and evaluation of ecotoxicological effects. *Chem. Eng. J.* **2018**, *334*, 1820–1827. [CrossRef]

86. Hussain, S.; Gul, S.; Steter, J.R.; Miwa, D.W.; Motheo, A.J. Route of electrochemical oxidation of the antibiotic sulfamethoxazole on a mixed oxide anode. *Environ. Sci. Pollut. Res.* **2015**, *22*, 15004–15015. [CrossRef] [PubMed]

87. Guo, X.; Li, D.; Wan, J.; Yu, X. Preparation and electrochemical property of TiO$_2$/Nano-graphite composite anode for electro-catalytic degradation of ceftriaxone sodium. *Electrochim. Acta* **2015**, *180*, 957–964. [CrossRef]

88. Lin, H.; Wu, J.; Zhang, H. Degradation of clofibric acid in aqueous solution by an EC/Fe^{3+}/PMS process. *Chem. Eng. J.* **2014**, *244*, 514–521. [CrossRef]

89. Sifuna, F.W.; Orata, F.; Okello, V.; Jemutai-Kimosop, S. Comparative studies in electrochemical degradation of sulfamethoxazole and diclofenac in water by using various electrodes and phosphate and sulfate supporting electrolytes. *J. Environ. Sci. Health Part A* **2016**, *51*, 954–961. [CrossRef]

90. Tu, X.; Xiao, S.; Song, Y.; Zhang, D.; Zeng, P. Treatment of simulated berberine wastewater by electrochemical process with Pt/Ti anode. *Environ. Earth Sci.* **2015**, *73*, 4957–4966. [CrossRef]

91. Duan, P.; Yang, X.; Huang, G.; Wei, J.; Sun, Z.; Hu, X. La$_2$O$_3$-CuO$_2$/CNTs electrode with excellent electrocatalytic oxidation ability for ceftazidime removal from aqueous solution. *Colloids Surf. A Physicochem. Eng. Asp.* **2019**, *569*, 119–128. [CrossRef]

92. Duan, P.; Gao, S.; Li, X.; Sun, Z.; Hu, X. Preparation of CeO$_2$-ZrO$_2$ and titanium dioxide coated carbon nanotube electrode for electrochemical degradation of ceftazidime from aqueous solution. *J. Electroanal. Chem.* **2019**, *841*, 10–20. [CrossRef]

93. Díaz, E.; Stożek, S.; Patiño, Y.; Ordóñez, S. Electrochemical degradation of naproxen from water by anodic oxidation with multiwall carbon nanotubes glassy carbon electrode. *Water Sci. Technol. Water Treat.* **2019**, *79*, 480–488. [CrossRef]

94. Fabiańska, A.; Ofiarska, A.; Fiszka-Borzyszkowska, A.; Stepnowski, P.; Siedlecka, E.M. Electrodegradation of ifosfamide and cyclophosphamide at BDD electrode: Decomposition pathway and its kinetics. *Chem. Eng. J.* **2015**, *276*, 274–282. [CrossRef]

95. Domínguez, J.R.; Muñoz-Peña, M.J.; González, T.; Palo, P.; Cuerda-Correa, E.M. Parabens abatement from surface waters by electrochemical advanced oxidation with boron doped diamond anodes. *Environ. Sci. Pollut. Res.* **2016**, *23*, 20315–20330. [CrossRef]

96. Rabaaoui, N.; Allagui, M.S. Anodic oxidation of salicylic acid on BDD electrode: Variable effects and mechanisms of degradation. *J. Hazard. Mater.* **2012**, *243*, 187–192. [CrossRef] [PubMed]

97. Indermuhle, C.; Martin de Vidales, M.J.; Saez, C.; Robles, J.; Canizares, P.; Garcia-Reyes, J.F.; Molina-Diaz, A.; Comninellis, C.; Rodrigo, M.A. Degradation of caffeine by conductive diamond electrochemical oxidation. *Chemosphere* **2013**, *93*, 1720–1725. [CrossRef]

98. Ambuludi, S.L.; Panizza, M.; Oturan, N.; Ozcan, A.; Oturan, M.A. Kinetic behavior of anti-inflammatory drug ibuprofen in aqueous medium during its degradation by electrochemical advanced oxidation. *Environ. Sci. Pollut. Res. Int.* **2013**, *20*, 2381–2389. [CrossRef] [PubMed]

99. Da Silva Duarte, J.L.; Solano, A.M.S.; Arguelho, M.L.; Tonholo, J.; Martínez-Huitle, C.A.; e Silva, C.L.d.P. Evaluation of treatment of effluents contaminated with rifampicin by Fenton, electrochemical and associated processes. *J. Water Process Eng.* **2018**, *22*, 250–257. [CrossRef]

100. Coledam, D.A.; Aquino, J.M.; Silva, B.F.; Silva, A.J.; Rocha-Filho, R.C. Electrochemical mineralization of norfloxacin using distinct boron-doped diamond anodes in a filter-press reactor, with investigations of toxicity and oxidation by-products. *Electrochim. Acta* **2016**, *213*, 856–864. [CrossRef]

101. Steter, J.R.; Rocha, R.S.; Dionísio, D.; Lanza, M.R.; Motheo, A.J. Electrochemical oxidation route of methyl paraben on a boron-doped diamond anode. *Electrochim. Acta* **2014**, *117*, 127–133. [CrossRef]

102. Fabiańska, A.; Białk-Bielińska, A.; Stepnowski, P.; Stolte, S.; Siedlecka, E.M. Electrochemical degradation of sulfonamides at BDD electrode: Kinetics, reaction pathway and eco-toxicity evaluation. *J. Hazard. Mater.* **2014**, *280*, 579–587. [CrossRef]

103. Brinzila, C.; Monteiro, N.; Pacheco, M.; Ciríaco, L.; Siminiceanu, I.; Lopes, A. Degradation of tetracycline at a boron-doped diamond anode: Influence of initial pH, applied current intensity and electrolyte. *Environ. Sci. Pollut. Res.* **2014**, *21*, 8457–8465. [CrossRef]

104. Haidar, M.; Dirany, A.; Sirés, I.; Oturan, N.; Oturan, M.A. Electrochemical degradation of the antibiotic sulfachloropyridazine by hydroxyl radicals generated at a BDD anode. *Chemosphere* **2013**, *91*, 1304–1309. [CrossRef]

105. González, T.; Domínguez, J.R.; Palo, P.; Sánchez-Martín, J. Conductive-diamond electrochemical advanced oxidation of naproxen in aqueous solution: Optimizing the process. *J. Chem. Technol. Biotechnol. Adv.* **2011**, *86*, 121–127. [CrossRef]

106. Domínguez, J.R.; González, T.; Palo, P.; Sánchez-Martín, J. Electrochemical advanced oxidation of carbamazepine on boron-doped diamond anodes. Influence of operating variables. *Ind. Eng. Chem. Res.* **2010**, *49*, 8353–8359. [CrossRef]

107. Sun, Y.; Li, P.; Zheng, H.; Zhao, C.; Xiao, X.; Xu, Y.; Sun, W.; Wu, H.; Ren, M. Electrochemical treatment of

108. Hu, X.; Yu, Y.; Sun, Z. Preparation and characterization of cerium-doped multiwalled carbon nanotubes electrode for the electrochemical degradation of low-concentration ceftazidime in aqueous solutions. *Electrochim. Acta* **2016**, *199*, 80–91. [CrossRef]

109. Yang, W.; Zhou, M.; Oturan, N.; Li, Y.; Su, P.; Oturan, M.A. Enhanced activation of hydrogen peroxide using nitrogen doped graphene for effective removal of herbicide 2, 4-D from water by iron-free electrochemical advanced oxidation. *Electrochim. Acta* **2019**, *297*, 582–592. [CrossRef]

110. Al-Qaim, F.F.; Mussa, Z.H.; Othman, M.R.; Abdullah, M.P. Removal of caffeine from aqueous solution by indirect electrochemical oxidation using a graphite-PVC composite electrode: A role of hypochlorite ion as an oxidising agent. *J. Hazard. Mater.* **2015**, *300*, 387–397. [CrossRef] [PubMed]

111. Frontistis, Z.; Antonopoulou, M.; Yazirdagi, M.; Kilinc, Z.; Konstantinou, I.; Katsaounis, A.; Mantzavinos, D. Boron-doped diamond electrooxidation of ethyl paraben: The effect of electrolyte on by-products distribution and mechanisms. *J. Environ. Manag.* **2017**, *195*, 148–156. [CrossRef] [PubMed]

112. Palo, P.; Domínguez, J.R.; González, T.; Sánchez-Martin, J.; Cuerda-Correa, E.M. Feasibility of electrochemical degradation of pharmaceutical pollutants in different aqueous matrices: Optimization through design of experiments. *J. Environ. Sci. Health Part A* **2014**, *49*, 843–850. [CrossRef]

113. De Vidales, M.J.M.; Millán, M.; Sáez, C.; Pérez, J.F.; Rodrigo, M.A.; Cañizares, P. Conductive diamond electrochemical oxidation of caffeine-intensified biologically treated urban wastewater. *Chemosphere* **2015**, *136*, 281–288. [CrossRef]

114. Zaghdoudi, M.; Fourcade, F.; Soutrel, I.; Floner, D.; Amrane, A.; Maghraoui-Meherzi, H.; Geneste, F. Direct and indirect electrochemical reduction prior to a biological treatment for dimetridazole removal. *J. Hazard. Mater.* **2017**, *335*, 10–17. [CrossRef]

115. Belkheiri, D.; Fourcade, F.; Geneste, F.; Floner, D.; Aït-Amar, H.; Amrane, A. Feasibility of an electrochemical pre-treatment prior to a biological treatment for tetracycline removal. *Sep. Purif. Technol.* **2011**, *83*, 151–156. [CrossRef]

116. Yahiaoui, I.; Aissani-Benissad, F.; Fourcade, F.; Amrane, A. Removal of tetracycline hydrochloride from water based on direct anodic oxidation (Pb/PbO$_2$ electrode) coupled to activated sludge culture. *Chem. Eng. J.* **2013**, *221*, 418–425. [CrossRef]

117. Rodríguez-Nava, O.; Ramírez-Saad, H.; Loera, O.; González, I. Evaluation of the simultaneous removal of recalcitrant drugs (bezafibrate, gemfibrozil, indomethacin and sulfamethoxazole) and biodegradable organic matter from synthetic wastewater by electro-oxidation coupled with a biological system. *Environ. Technol.* **2016**, *37*, 2964–2974. [CrossRef] [PubMed]

118. García-Gómez, C.; Drogui, P.; Seyhi, B.; Gortáres-Moroyoqui, P.; Buelna, G.; Estrada-Alvgarado, M.I.; Álvarez, L.H. Combined membrane bioreactor and electrochemical oxidation using Ti/PbO$_2$ anode for the removal of carbamazepine. *J. Taiwan Inst. Chem. Eng.* **2016**, *64*, 211–219. [CrossRef]

119. Ouarda, Y.; Tiwari, B.; Azais, A.; Vaudreuil, M.A.; Ndiaye, S.D.; Drogui, P.; Tyagi, R.D.; Sauve, S.; Desrosiers, M.; Buelna, G.; et al. Synthetic hospital wastewater treatment by coupling submerged membrane bioreactor and electrochemical advanced oxidation process: Kinetic study and toxicity assessment. *Chemosphere* **2018**, *193*, 160–169. [CrossRef]

120. Ensano, B.M.B.; Borea, L.; Naddeo, V.; Belgiorno, V.; de Luna, M.D.G.; Balakrishnan, M.; Ballesteros, F.C., Jr. Applicability of the electrocoagulation process in treating real municipal wastewater containing pharmaceutical active compounds. *J. Hazard. Mater.* **2019**, *361*, 367–373. [CrossRef]

121. Borea, L.; Ensano, B.M.B.; Hasan, S.W.; Balakrishnan, M.; Belgiorno, V.; de Luna, M.D.G.; Ballesteros, F.C.; Naddeo, V. Are pharmaceuticals removal and membrane fouling in electromembrane bioreactor affected by current density? *Sci. Total Environ.* **2019**, *692*, 732–740. [CrossRef]

5

Insights into the Photocatalytic Bacterial Inactivation by Flower-Like Bi_2WO_6 under Solar or Visible Light, through In Situ Monitoring and Determination of Reactive Oxygen Species (ROS)

Minoo Karbasi [1,2], Fathallah Karimzadeh [1,*], Keyvan Raeissi [1], Sami Rtimi [2], John Kiwi [2], Stefanos Giannakis [3,*] and Cesar Pulgarin [2]

[1]   Department of Materials Engineering, Isfahan University of Technology, Isfahan 84156-83111, Iran
[2]   School of Basic Sciences (SB), Institute of Chemical Science and Engineering (ISIC), Group of Advanced Oxidation Processes (GPAO), École Polytechnique Fédérale de Lausanne (EPFL), Station 6, CH-1015 Lausanne, Switzerland
[3]   Departamento de Ingeniería Civil: Hidráulica, Universidad Politécnica de Madrid (UPM), E.T.S. Ingenieros de Caminos, Canales y Puertos, Energía y Medio Ambiente, Unidad docente Ingeniería Sanitaria, c/ Profesor Aranguren, s/n, ES-28040 Madrid, Spain
*   Correspondence: karimzadeh_f@cc.iut.ac.ir (F.K.); stefanos.giannakis@upm.es (S.G.)

Abstract: This study addresses the visible light-induced bacterial inactivation kinetics over a Bi_2WO_6 synthesized catalyst. The systematic investigation was undertaken with Bi_2WO_6 prepared by the complexation of Bi with acetic acid (carboxylate) leading to a flower-like morphology. The characterization of the as-prepared Bi_2WO_6 was carried out by X-ray diffraction (XRD), scanning electron microscopy (SEM), X-ray photoelectron spectroscopy (XPS), specific surface area (SSA), and photoluminescence (PL). Under low intensity solar light (<48 mW/cm^2), complete bacterial inactivation was achieved within two hours in the presence of the flower-like Bi_2WO_6, while under visible light, the synthesized catalyst performed better than commercial TiO_2. The in situ interfacial charge transfer and local pH changes between Bi_2WO_6 and bacteria were monitored during the bacterial inactivation. Furthermore, the reactive oxygen species (ROS) were identified during *Escherichia coli* inactivation mediated by appropriate scavengers. The ROS tests alongside the morphological characteristics allowed the proposition of the mechanism for bacterial inactivation. Finally, recycling of the catalyst confirmed the stable nature of the catalyst presented in this study.

Keywords: flower-like Bi_2WO_6; *E.coli* inactivation; reactive oxygen species (ROS); photocatalysis; solar disinfection; water treatment; pollution

1. Introduction

Over the last few decades, environmental contamination has shifted from the exclusive focus of organic and inorganic pollutants [1], towards the inclusion of bacteria and other organisms [2–4]. Therefore, well-organized methods are urgently required to control the spread [5] or eradicate microorganism-related issues [6]. In recent times, beside the traditional bacterial inactivation methods such as UV disinfection and chlorination, a green, efficient, and cost-effective semiconductor photocatalysis has appeared to be a more promising technique [7,8]. TiO_2 has been extensively reported as an effective bactericidal semiconductor photocatalyst due to its high stability, strong redox potential, low cost, and non-toxic nature, but its band-gap of 3.2 eV allows light absorption up to 387 nm which makes up just over 4% of the total solar spectrum [9–11].

Since solar radiation contains more visible light (~47%) than UV, the appropriate use of this fraction becomes necessary through the employment of efficient visible-light photocatalysts [12]. As a promising visible-light-driven photocatalyst with good chemical and thermal stability, Bi_2WO_6, beside its non-toxic and environmentally friendly nature, is a typical n-type semiconductor composed of accumulated layers of alternating $(Bi_2O_2)^{2+}$ layers and $(WO_4)^{2-}$ octahedral sheets [13,14]. The valence band of Bi_2WO_6 consists of O 2p and Bi 6s hybrid orbitals, its narrowed band gap increases visible light absorption capacity, and photoactivity [15,16], while its photocatalytic activity greatly depends on morphology, particle size, surface area, and interface structure [17,18].

Constructing a unique micro/nano hierarchical structure usually shortens the pathways of water pollutants, absorb incidental light more efficiently, because of multiple-scattering increase, and easily separated from wastewater by filtration or sedimentation methods [13,19,20].

However, despite the long presence of this catalyst as a possible solution, most studies on Bi_2WO_6 have focused on the photocatalytic degradation of organic pollutants, with only a few studies investigating the photocatalytic inactivation of microorganisms. Ren et al. [21] reported *Escherichia coli* degradation in a few hours on Bi_2WO_6 nest-like structures in a pseudo-first order process. Helali et al. [22] prepared a 20 m^2/g SSA Bi_2WO_6 leading to *E. coli* inactivation within four to five hours under solar light on a hydrothermally grown mixture of Bi-nitrate and Na-tungstate in a 65–35% ratio while a similar study has been reported by Amano et al. [23]. However, there is a relatively wide gap in literature on effective preparation of robust structures with high specific surface areas in order to promote efficient disinfection, and a gap in interpreting the pathways to bacterial inactivation by this catalyst.

This study aims to assess a facile preparation method for flower-like Bi_2WO_6 photocatalysts destined for disinfection applications. As such, we assess the preparation parameters (aging, temperature, pH) in order to modify the structural (crystalline) and morphological characteristics (flower-like, nanoparticles). These modifications are envisioned to create a series of catalysts, and their activity under low-intensity solar or visible light will be assessed. Furthermore, the robustness of the catalyst in serial reuse cycles will be evaluated for its stability. Last but not least, special focus will be given to the identification of the pathways that lead to bacterial inactivation in an effort to decrypt the mechanistic action mode of the flower-like Bi_2WO_6.

2. Materials and Methods

2.1. Synthesis of Flower-Like Bi_2WO_6 Samples

All chemicals were of analytical grade. They were used as received without any further purification and were purchased from Merck, Germany. All solutions were prepared with Milli-Q water (18.2 $M\Omega$ cm). In a typical hydrothermal procedure for the synthesis of flower-like Bi_2WO_6, 0.5 mmol of $Na_2WO_4 \cdot 2H_2O$ was dissolved in an 80 mL solvent containing 16 mL acetic acid and 64 mL Milli-Q water until attaining a clear solution. Then, 1 mmol of $Bi(NO_3)_3 \cdot 5H_2O$ solid was added to the solution, and a white precipitate immediately emerged.

Next, the reaction mixture was stirred for 1 h, transferred into a 120 mL Teflon-lined stainless-steel reactor, and heated at 160 °C for 12 h. The as-formed yellow precipitates were collected, washed with distilled water, and dried in vacuum at 70 °C for 10 h. A schematic representation of the synthesis is illustrated in Scheme 1. The influence of the hydrothermal reaction time and temperature has been explored as shown in Table 1. In order to investigate the effect of morphology on photocatalysis, Bi_2WO_6 nanoparticles (BWO6) were prepared applying the same hydrothermal method at 200 °C for 24 h by the regulation of pH to 10.

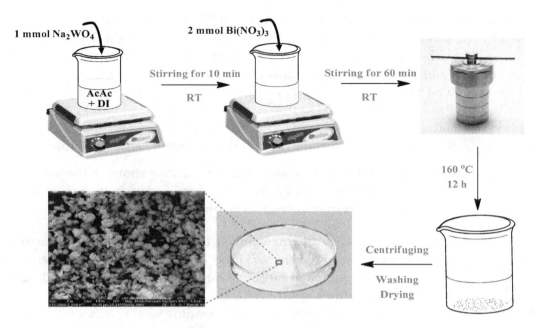

Scheme 1. Schematic illustration of the preparation of flower-like Bi_2WO_6 by hydrothermal method.

Table 1. Bi_2WO_6 obtained at the different synthetic conditions.

Samples	Reaction Time (h)	Reaction Temperature (°C)
BWO1	12	160
BWO2	18	160
BWO3	24	160
BWO4	24	180
BWO5	24	200
BWO6	24	200 (pH = 10)

2.2. Physical Characterization of the Bi_2WO_6 Flakes

The crystallinity and phase identification of the as-prepared samples were determined by powder X-ray diffraction (XRD) using an X'Pert MPD PRO (Panalytical) analyzer, equipped with a ceramic tube (Cu anode, $\lambda = 1.54060$ Å), and with a continuous scanning rate in the range of $5° < 2\theta < 80°$. The results were studied with Rietvield refinement by the FullProf program. The morphology developments of the samples were characterized using scanning electron microscopy (SEM, FEI Quanta 200). Before SEM imaging, the samples were coated with a thin layer of gold. The specific surface area and porosity size were obtained using Brunauer–Emmett–Teller (BET) analysis, performed with a BELSORP-mini II analyzer, Japan. The photoluminescence (PL) measurement was carried out using a fluorescence spectrophotometer (Perkin Elmer LS55) equipped with a xenon lamp at an excitation wavelength of $\lambda = 340$ nm. The surface atomic percentage of the element in the as-synthesized sample was analyzed using an AXIS NOVA photoelectron spectrometer with a mono-chromatic Al Ka X-ray (h$\nu = 1486.6$ eV) source (Kratos Analytical, Manchester, UK). The interfacial in situ voltage and pH variation during the bacterial inactivation was monitored in a pH/mV/Temp meter (Jenco 6230N) equipped with a microprocessor and a RS-232-C IBM interface for data recording.

2.3. Photocatalytic Antibacterial Activity on Bi_2WO_6 and Light Sources

The bacterial strain used was a wild type *E. coli* K12, supplied by the German Collection of Microorganisms and Cell Cultures, DSMZ (No. 498). The master plate and stock solution were prepared according to previous research reported by our laboratory [24,25]. The bacterial concentration of the samples was measured in Colony Forming Units (CFU/mL) and was determined by plating on a non-selective cultivation media, namely, Plate Count Agar (PCA). A total of 1 mL of the sample was withdrawn after each interval and then serial dilutions were made in a sterile 0.8% NaCl/KCl solution. A 100 µL aliquot was pipetted onto a nutrient agar plate and processed using the standard plate count method. The plates were incubated at 37 °C followed by the bacterial evaluation. Experimental results were carried in triplicate runs applying statistical analysis for the calculation of mean and standard deviation (reported in the graphs). Samples were irradiated in the cavity of a SUNTEST solar simulator CPS (Atlas GmbH, Hanau, Germany) with an overall light irradiance of 48 mW/cm^2 (~0.8 × 10^{16} photons/s, Supplementary Figure S1). A cut-off filter was used in the SUNTEST cavity to filter the light <310 nm. A second cut-off filter was also used during bacterial inactivation under visible light with a cut-off blocking the wavelength < 405 nm rendering (Supplementary Figure S2). Finally, after the two filters, the visible light irradiance reaching the sample was 38 mW/cm^2.

3. Results

3.1. Synthesis and Characterization of Bi$_2$WO$_6$: X-Ray Diffraction (XRD), Scanning Electron Microscopy (SEM), X-Ray Photoelectron Spectroscopy (XPS), and SSA Determination

Figure 1 depicts the XRD patterns of the as-synthesized Bi$_2$WO$_6$ via the hydrothermal method at different reaction times and temperatures. All of the XRD patterns illustrated that characteristic peaks were in good agreement with the orthorhombic phased Bi$_2$WO$_6$ in the standard JCPDS card (39-0256) [26]. No other diffraction peaks arising from possible impurities were detected. With the holding time increasing to 24 h, the characteristic peaks became much sharper due to an increase in crystallinity. Understandably, the increment of the temperature with the constant reaction time for 24 h resulted in the same trend because of grain growth. Table 2 illustrates the crystallite size of the samples (using the Scherrer formula based on the half-width of their (113) peak) calculated by the Rietveld method using the FullProf program.

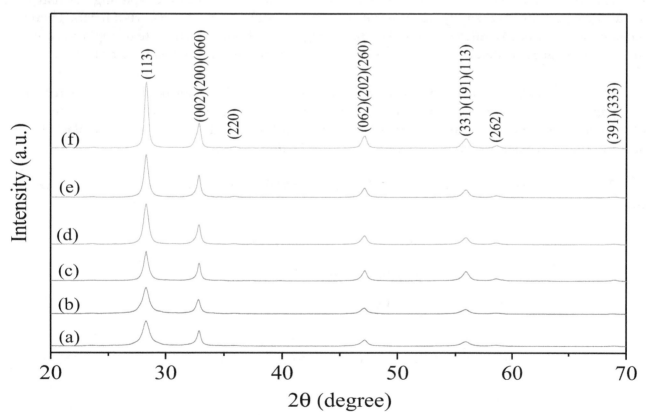

Figure 1. X-ray diffraction patterns of the Bi$_2$WO$_6$ samples: (**a**) BWO1, (**b**) BWO2, (**c**) BWO3, (**d**) BWO4, (**e**) BWO5, and (**f**) BWO6. Profiles are shifted in y-scale for clarity.

Table 2. Rietveld structural parameters of the samples.

Samples	Crystallite Size (nm)
BWO1	9
BWO2	10
BWO3	17
BWO4	20
BWO5	22
BWO6	31

The scanning electron microscopy (SEM) images of the Bi_2WO_6 samples prepared under different experimental conditions are shown in Figure 2. Heating at 160 °C for 12 h and 18 h (Figure 2a,b) led to aggregated irregular small Bi_2WO_6 nanoparticles and flower-like microspheres. However, when the heating time was prolonged to 24 h (see Figure 2c), organized hierarchical flower-like Bi_2WO_6 microspheres composed of nanoplates were obtained and the aggregated nanoparticles totally disappeared (Figure 2e,f). The SEM images of as-prepared Bi_2WO_6 nanoparticles are also shown in Figure 2f. The joint effect of nanoparticles assembly followed by the localized ripening mechanism as well as the hierarchical assembly of nanoplates have been also previously reported for the formation mechanism of flower-like microspheres [27,28]. Owing to the absence of discrete nanoplates according to the SEM images at different reaction times (Figure 2), the former mechanism seems to predominate.

Scheme 2 illustrates the proposed formation mechanism of flower-like Bi_2WO_6 microspheres. Nanoparticles initially aggregated, then the self-assembled nanoparticles preferentially grew along <010>. Longer reaction times and higher temperatures result in dissolution of some nanoplates leading concomitantly to re-deposition by Ostwald ripening [27,28].

The relevant reactions leading to the Bi_2WO_6 synthesis in aqueous solutions when working in acetic acid media can be suggested as follows:

$$Bi(NO_3)_3 \rightarrow Bi^{3+} + 3NO_3^-, \tag{1}$$

$$Na_2WO_4 \rightarrow 2Na^+ + WO_4^{2-}, \tag{2}$$

$$CH_3-COOH + H_2O \rightarrow CH_3-COO^- + H_3O^+ \quad pK_a 4.75, \tag{3}$$

$$Bi_3^+ + WO_4^{2-} + CH_3-COO^- \rightarrow CH_3COO^- + \left| \begin{smallmatrix} CH_3-COO^- \\ \\ CH_3-COO^- \end{smallmatrix} \right\rangle M, \tag{4}$$

$$\left| \begin{smallmatrix} CH_3-COO^- \\ \\ CH_3-COO^- \end{smallmatrix} \right\rangle M \rightarrow Bi_2WO_6 \quad flower-like\ microspheres\ (see\ Figure\ 2)? . \tag{5}$$

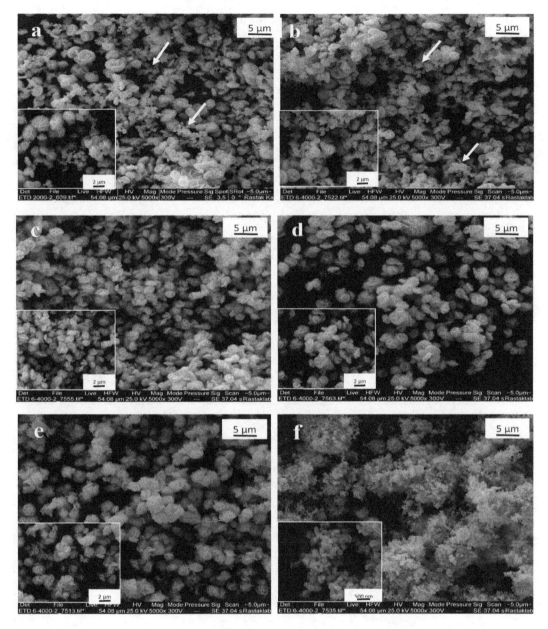

Figure 2. Scanning electron microscopy (SEM) images of Bi_2WO_6 samples prepared under different conditions: (**a**) BWO1, (**b**) BWO2, (**c**) BWO3, (**d**) BWO4, (**e**) BWO5, and (**f**) BWO6.

The initial complex between Bi and acetic acid presents a stability constant of $10^{2.6-2.7}$ [29], which is not in the range found for insoluble complexes/precipitates $>10^{11-12}$ [30–32]. This coordination complex is suggested in Equation (4) (Bi = M). The complex formation which is the precursor of Bi_2WO_6 does not lead to precipitate formation and gradually decomposes releasing Bi^{3+} which reacts with WO_4^{2-}. Therefore, the nanoplate formation leads to aggregates which present inner pores/voids and provide the required contact area for the photocatalytic bacterial inactivation.

In addition to the crystal structure and morphology, the surface chemical composition of the as-synthesized flower-like sample at 200 °C for 24 h was examined by XPS. As shown in the survey XPS spectrum in Figure 3, the Bi, O, W, and C elements were present in the pure Bi_2WO_6. The C element peak can be attributed to adventitious carbon from the sample preparation and/or the XPS instrument itself [33]. The surface atomic concentration ratio of Bi:W:O estimating from XPS peak areas is around 2.0:0.8:5.4, which further confirms its composition of Bi_2WO_6. Furthermore, the peaks centering at 164.7 and 159.4 eV are attributed to the binding energies of Bi $4f_{5/2}$ and Bi $4f_{7/2}$, respectively (inset of Figure 3), confirming Bi^{3+} ions in the crystalline structure [34–36]. The W4f energy region can be designated to be the +6 oxidation state of tungsten in accordance with previous reports [33,36].

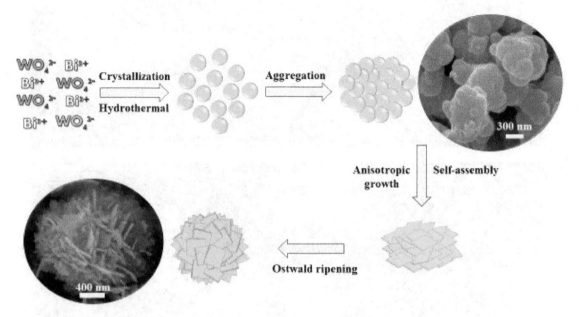

Scheme 2. Schematic illustration of the growth process of the flower-like Bi_2WO_6 microspheres.

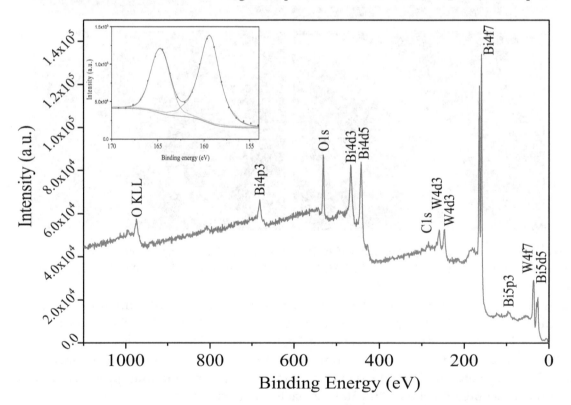

Figure 3. XPS survey spectra of the hydrothermally prepared Bi_2WO_6 sample at 200 °C for 24 h. Inset is the zoom of XPS scans over the $Bi_{4f7/2}$ peak in the 154–170 eV region.

The N_2 adsorption–desorption isotherms of the well-organized flower-like (BWO5) and nanoparticles (BWO6) Bi_2WO_6 are presented in Figure 4. According to IUPAC classification, it can be seen that the isotherm shape for both samples exhibited a typical type IV isotherm with a clear hysteresis loop H3, suggesting the presence of mesopores in the size range of 2–50 nm [37].The insets show the Barrett–Joyner–Halenda (BJH) pore-size distributions and present the evidence for the existence of mesopores (2–50 nm). Table 3 summarizes the BET specific surface areas (SSA) and the pore volumes of BWO5 and BWO6.

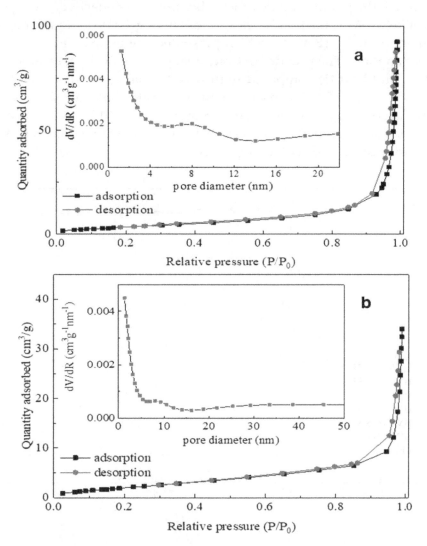

Figure 4. N_2 adsorption–desorption isotherm of the samples: (**a**) flower-like Bi_2WO_6, (**b**) nanoparticle Bi_2WO_6. The insert shows the pore size distribution.

Table 3. Brunauer–Emmett–Teller (BET) parameters of the Bi_2WO_6 samples at various temperatures.

Samples	Surface Areas ($m^2 \ g^{-1}$)	Total Pore Volumes ($cm^3 \ g^{-1}$)
BWO5	14.475	0.142
BWO6	6.87	0.0532

3.2. E. Coli Inactivation Kinetics: Effect of the Bacterial Concentration, Amount of Catalyst, Light Dose, and Applied Light Wavelength

Figure 5 shows the complete bacterial inactivation mediated by the BWO5 being faster under low-intensity simulated solar light, compared to the other samples. The *E. coli* inactivation was 95% after 2 h. The effectiveness of a disinfection process resides in the time necessary to inactivate a determined percentage of bacteria. In the Chick–Watson model [38,39], the simplest inactivation model, the inactivation rate shown in Figure 5 is seen to be dependent on the residual bacteria after each specific time during the inactivation process and this allows comparing the effect of the different Bi_2WO_6 samples. Neither irradiation in the absence of Bi_2WO_6 (photolysis) nor runs in the presence of this catalyst in the dark lead to bacterial inactivation of up to 4 h. The latter provides the proof that Bi_2WO_6 is not toxic to *E. coli* and a photocatalytic process is required for their inactivation. As the treatment time increased, the photocatalytic process became more effective, owing to the formation of hierarchical flower-like Bi_2WO_6 microspheres and loss of aggregates and the higher crystallite size. Nevertheless, the nanoparticles (BWO6), which presented lower specific surface area than BWO5, led to lower inactivation rates. The pseudo first-order rates of the Bi_2WO_6 samples during flower-like development (BWO1 and BWO5) compared with Bi_2WO_6 nanoparticles (BWO6) are given in the supplementary material, Figure S3. The pseudo first-order rate constants (k_{app}) of the BWO1, BWO5, and BWO6 were estimated to be 0.0331 min^{-1}, 0.0488 min^{-1}, and 0.0195 min^{-1}, respectively. As can be seen, the photocatalytic inactivation of bacteria mediated by as-developed flower-like Bi_2WO_6 (BWO5) is around 2.5 times faster compared with nanoparticles.

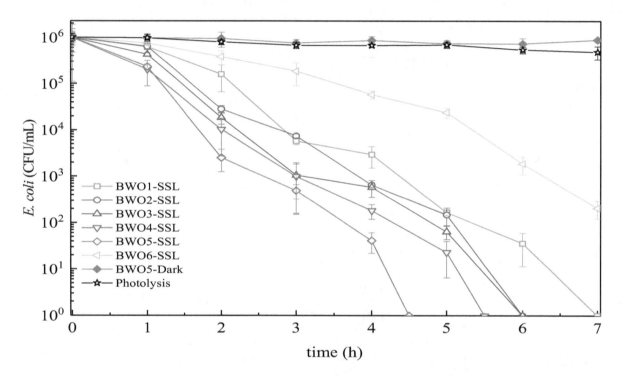

Figure 5. Photocatalytic inactivation of *Escherichia coli* in aqueous dispersions on different Bi_2WO_6 samples in the dark and under simulated solar light (SSL). Experimental conditions: [Catalyst]$_0$ = 0.2 g/L, [bacteria]$_0$ = 2 × 10^6 Colony Forming Units (CFU)/mL and light intensity: 48 mW/cm^2.

The photoluminescence spectrum of the prepared catalysts was used as a practical method to verify the separation efficiency of photo-generated electron–hole pairs in the semiconductors. Generally, a lower photoluminescence (PL) intensity represents a lower recombination rate of photo-generated charge carriers. The photoluminescence (PL) spectra of the Bi_2WO_6 samples during the flower-like development (BWO1 and BWO5) in comparison with Bi_2WO_6 nanoparticles (BWO6) is shown in Figure 6. The wide absorption-band was observed between 350 nm and 600 nm which is due to the Bi_2WO_6 electron-hole recombination giving rise to the free and bound-exciton luminescence [40]. The PL spectra of the as-synthesized samples through flower-like development (BWO1 and BWO5) exhibited significantly decreased PL intensity related to that of the Bi_2WO_6 nanoparticles. It could be ascribed that the recombination of photo-generated charge carriers is greatly inhibited in the hierarchically flower-like composed of nanosheets. Hence, the efficient separation of photo-generated electron–hole pairs and rapid transfer of electrons to the surface of crystal would be obtained. Moreover, the lower PL-intensity bands shown in BWO5 reflected a higher crystallinity in comparison with BWO1, allowing a lower amount of crystal defects, leading to a higher electron-hole separation and an increased photocatalytic activity [41], a fact that corroborates with the faster inactivation of bacteria (Figure S3).

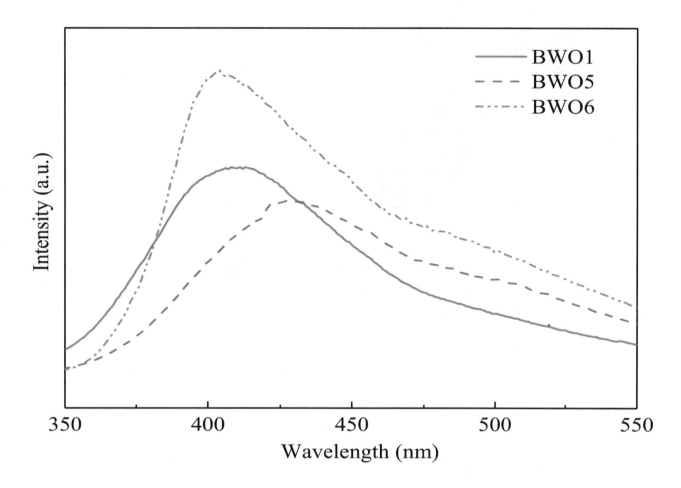

Figure 6. Photoluminescence (PL) spectroscopy of the synthesized samples at different conditions. BWO1: 12 h, 160 °C. BWO5, 24 h, 200 °C. BWO6: 24 h, 200 °C. pH = 10.

Following, the effects of initial catalyst or bacterial concentration were studied, and the results are summarized in Figure 7. The effect of the Bi_2WO_6 concentration on *E. coli* inactivation is shown in Figure 7a. Although increasing Bi_2WO_6 concentration of up to 0.2 mg/mL resulted in higher inactivation rates, increasing the catalyst concentration to 0.4 mg/mL resulted in a slower bacterial inactivation kinetics, most possibly due to a loss in surface area by catalyst agglomeration (particle–particle interactions), as well as a decrease in the penetration of the photon flux by the solution opacity, thereby decreasing the photocatalytic inactivation rate [42].

The effect of the initial concentration on the *E. coli* kinetics mediated by Bi_2WO_6 catalysts is presented in Figure 7b, showing a delay in the time necessary for bacterial inactivation at higher bacterial concentrations. Although this effect can be ascribed to the exhaustion of surface active sites due to opacity in solution [43], we note here that in absolute numbers, the higher the amount of bacteria in solution, the higher the number of available bacteria (for inactivation). Hence, by calculating the amount of cells inactivated in 4 h per mg of catalyst and per minute, we get 2075, 208, and 21 cells min^{-1} mg^{-1} for 10^8, 10^7, and 10^6, respectively. As a result, we report that this catalyst can effectively disinfect higher amounts of microorganisms, albeit in a higher residence time.

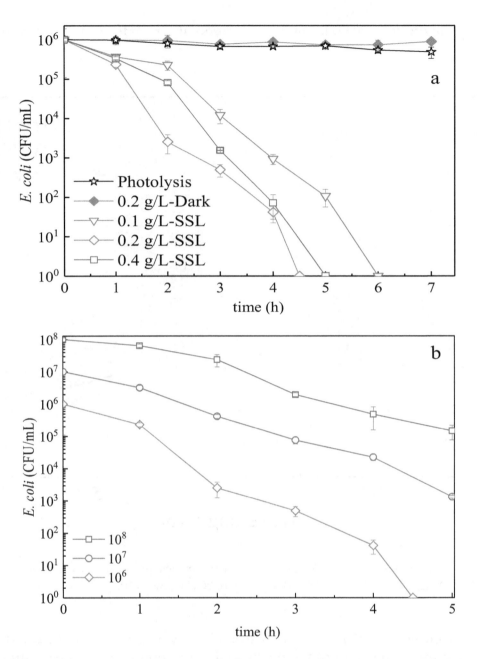

Figure 7. Effect of catalyst and bacterial concentration on inactivation kinetics. (**a**) *E. coli* survival on Bi_2WO_6 samples in the dark and under low intensity solar simulated light. Experimental conditions: $(bacteria)_0 = 2 \times 10^6$ CFU/mL and light intensity: 48 mW/cm². (**b**) Initial concentration of *E. coli* (CFU/mL) effects on the bacterial inactivation kinetics mediated by Bi_2WO_6 (200 °C for 4 h) under low intensity solar simulated light. Experimental conditions: $(Catalyst)_0 = 0.2$ g/L and light intensity: 48 mW/cm².

Next up in the operational parameters investigation, we assessed the possibility of photonic limitation or saturation of the system. As such, Figure 8a,b shows the effects of the light intensity and composition (UVA–vis or Vis only) on the bacterial degradation kinetics. A higher light dose accelerated the bacterial inactivation because of a higher amount of charges generated in the semiconductor during bacterial disinfection under band-gap irradiation (Figure 8a), since the direct inactivation by light was previously excluded. Figure 8 b illustrates that under visible light, a solution containing 0.2 g/L of Bi_2WO_6 was still efficiently inactivating bacteria and was more effective compared to commercial TiO_2 P25 Degussa (used as reference). These results come from the optical absorption of up to ~450 nm in the visible region by Bi_2WO_6, which is significantly wider than that of TiO_2 P25 Degussa with an absorption of up to 387 nm for the 20 nm particles, making up the bulk of this mixed TiO_2 P25 Degussa rutile–anatase [44].

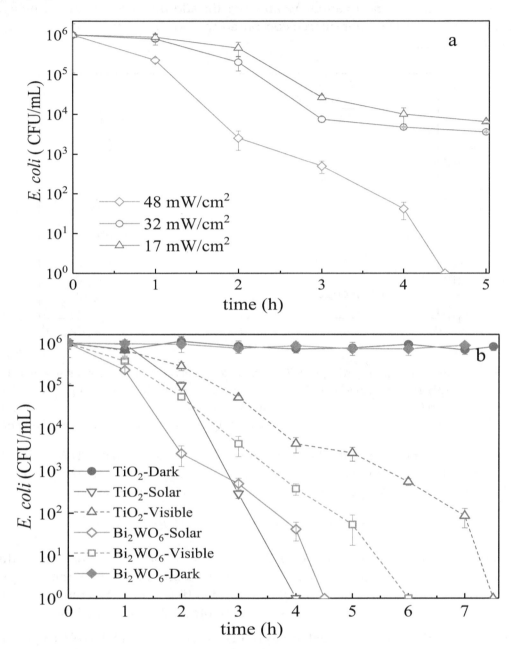

Figure 8. Effect of light irradiance and composition on inactivation kinetics. (**a**) *E. coli* inactivation on Bi_2WO_6 (200 °C for 24 h) under different solar light irradiation intensities. (**b**) *E. coli* inactivation mediated by Bi_2WO_6 (200 °C for 24 h) and TiO_2 under low intensity solar simulated (48 mW/cm^2) and visible light (38 mW/cm^2). Experimental conditions: $(Catalyst)_0 = 0.2$ g/L and $(bacteria)_0 = 2 \times 10^6$ CFU/mL.

3.3. Mechanistic Interpretation: ROS-Species Involvement, Interfacial Charge Transfer, and Catalyst Reuse During Bacterial Inactivation

The reactive oxygen species (ROS) such as $^{\cdot}OH$, $O_2{^{\cdot-}}$, and vb (h$^+$) play a pivotal role in the photo-degradation of organic pollutants and bacterial inactivation [22,45–47]. To determine the main ROS followed by the photodegradation mechanism, appropriate radical-scavengers such as isopropanol ($^{\cdot}OH$ scavenger), sodium oxalate (a vbh$^+$ hole scavenger), and superoxide dismutase ($O_2{^{\cdot-}}$ scavenger) were used in the present study. Figure 9 depicts the results of scavenging experiments mediated by the optimized flower-like Bi_2WO_6 (BWO5). The photocatalytic bacterial inactivation could be remarkably suppressed by the addition of isopropanol and sodium oxalate. It is very likely that ·OH and h^+ intervene jointly in the bacterial inactivation. Meanwhile, the addition of SOD ($O_2^{·-}$ scavenger) inhibits the bacterial inactivation to a smaller degree compared to vb(h^+) and the ·OH-radical as shown in Figure 9, traces (a) and (b).

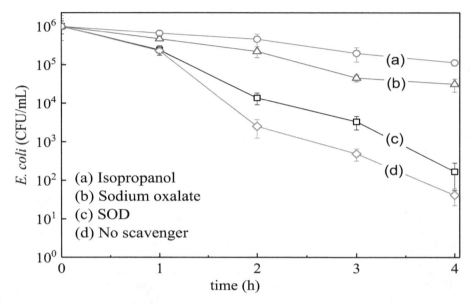

Figure 9. Effect of the scavengers during *E. coli* inactivation on Bi_2WO_6 under solar simulated light for (a) isopropanol as OH-radical scavenger, (b) sodium oxalate a hole vb(h^+) scavenger, (c) superoxide dismutase (SOD) as an $O_2^{·-}$ scavenger, (d) no scavenger. Runs under low intensity solar simulated light (48 mW/cm^2). The solutions contained Bi_2WO_6 (0.2 g/L) and scavenger concentration of 0.1 mM.

The possible reaction mechanism for the inactivation of *E. coli* mediated by Bi_2WO_6 can be proposed as the following, which is shown in Scheme 3. Under visible-light irradiation, the photo-excitation of Bi_2WO_6 implies the transfer of an electron from the valence band (Equation (6)).

$$Bi_2WO_6 + hv \rightarrow e^-(CB) + h^+(VB). \tag{6}$$

As mentioned before, the valence band of Bi_2WO_6 is a hybrid band made up by the O2p and Bi6s orbitals. Under light irradiation, the O2p and Bi6s hybrid orbitals increase the charge transfer in the W5d orbitals of Bi_2WO_6. This moves the valence band (VB) potential to a more positive potential energy narrowing the band-gap and inducing a higher photocatalytic activity [48].

Based on the references, CB and VB potentials of Bi_2WO_6 are 3.08 and 0.36 eV, respectively [49,50]. The redox potential for the dissolved oxygen/superoxide couple (E^0 ($O_2/O_2^{·-}$)), $O_2/HO_2^·$, and $OH^-/·OH$ are −0.33 eV, −0.046 eV, and 1.98 eV vs NHE [49], respectively. Comparing the band edge energy level of Bi_2WO_6 with the redox potentials of ROS, it is obvious that the excited holes in the valence band of Bi_2WO_6 were sufficiently more positive than that of $OH^-/·OH$, suggesting that the photogenerated holes on the surface of Bi_2WO_6 could react with OH^-/H_2O to form "non-selective" ·OH radicals (Equation (7)). However, the conduction band edge potential of Bi_2WO_6, which is more positive than the standard redox potential of $O_2/O_2^{·-}$ and $O_2/HO_2^·$, cannot directly reduce O_2 to $O_2^{·-}$ or $HO_2^·$. As shown in Figure 9, the bacterial inactivation is reduced in the presence of SOD-scavengers, which confirms the presence of the $HO_2^·$ radicals. Considering the redox potential of O_2/H_2O_2 = +0.682 eV vs NHE [51], H_2O_2 seems to be generated initially (Equation (8)) which is followed by the formation of different species according to the relations 9–10 in the photocatalytic reaction. It is worth noting that the powerful hole can directly attack bacteria cells in the photocatalytic oxidation process, which was also confirmed by the hole scavenger [45,52].

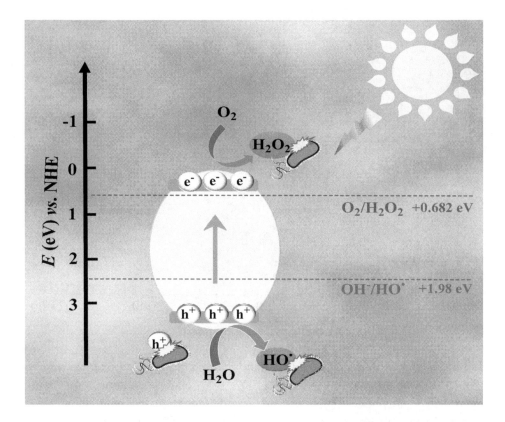

Scheme 3. Schematic diagram showing the photocatalytic inactivation of bacteria on the Bi_2WO_6.

$$h^+(VB) + H_2O \rightarrow HO\bullet + H^+, \tag{7}$$

$$O_2 + 2H^+ + 2e^- \rightarrow H_2O_2, \tag{8}$$

$$H_2O_2 + e^- \rightarrow OH\bullet + OH^-, \tag{9}$$

$$H_2O_2 + h^+ \rightarrow O_2^-\bullet + 2H^+ \ \ or \ \ HO_2^-\bullet + H^+. \tag{10}$$

Figure 10 shows the variation of the interfacial potential and the local pH shift under simulated solar light. At pH ~6, the bacterial inactivation preferentially proceeds via the $O_2^{\cdot-}$ species over $HO_2^{-\cdot}$ as shown in Equation (11) and Figure 9, trace (c).

$$HO_2^{-\cdot} \Leftrightarrow O_2^{\cdot-} + H^+ \qquad pK_a \quad 4.8. \tag{11}$$

The initial pH at time zero in Figure 10 was observed to decrease slightly from 6.0 to 5.9 within four hours of irradiation. The initial pH of 6.0 in this figure is seen to decrease drastically to 5.4 after 8000 s due to the concomitant production of long-lived intermediates carboxylic acids, owing to the degradation of the bacterial membrane. The interface potential is shown to drastically drop within 8000 s (2.2 h) when the bacterial reduction is reduced by 99.90%, which is equivalent to 3 logs as shown in Figure 5. The interface potential recovers to its initial value as shown in Figure 10 after the inactivation of bacteria [52]. The recovery to the initial pH-level occurs when the intermediate acids are mineralized to CO_2 by the photo-Kolbe reaction according to Equation (12) [53,54].

$$RCOO^- + solar\ light \rightarrow R + CO^2. \tag{12}$$

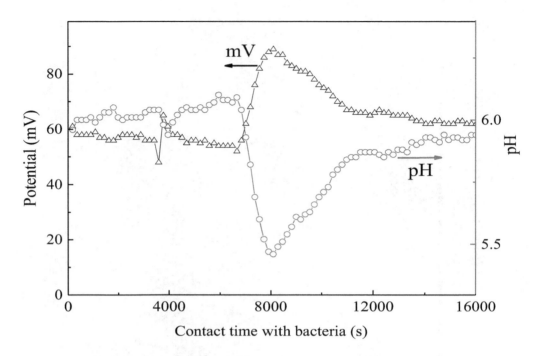

Figure 10. Evolution of the interfacial potential and local pH of an *E. coli* suspension in contact with Bi_2WO_6 under low intensity light irradiation (48 mW/cm^2). Catalyst concentration 0.2 g/L.

Finally, we provide the evidence for synthesizing a stable Bi_2WO_6 flower-like photocatalyst by a repetitive inactivation of a *E. coli* test, which results are shown in Figure 11. In order to evaluate the bacterial inactivation after each cycle, the pseudo first-order rate constants (*kapp*) were calculated and are reported in Table 4. The recycled sample used in Figure 11 was thoroughly washed after each cycle. Practically, no loss of bacterial inactivation was observed. These results show the stable repetitive bacterial inactivation mediated by flower-like Bi_2WO_6 up to five cycles and confirm the potential for the practical application of this photocatalyst in *E. coli* inactivation.

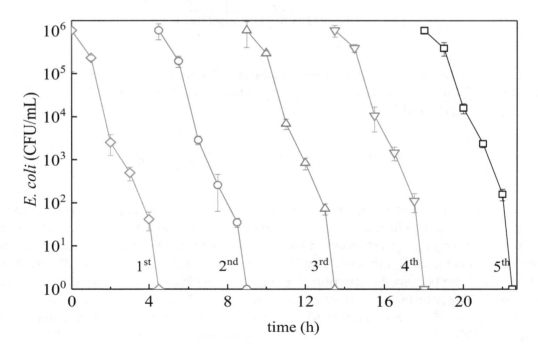

Figure 11. Reusability of flower-like Bi_2WO_6 under low intensity solar simulated light (48 mW/cm^2). Solution parameters: (Catalyst)$_0$ = 0.2 g/L and (bacteria)$_0$ = 2 × 10^6 CFU/mL.

Table 4. Pseudo first-order rate constants (k_{app}) for *E. coli* inactivation under different conditions consistent with Figure 11.

Cycle Number	k_{app} (min^{-1})
First	0.0488 ± 0.005
Second	0.0494 ± 0.004
Third	0.0484 ± 0.005
Fourth	0.0480 ± 0.006
Fifth	$0.0471 + 0.007$

4. Conclusions

In the present study, Bi_2WO_6 flower-like samples were prepared at 200 °C attaining a high crystallinity and led a low amount of crystal by hydrothermal growth in acetic acid media. By SEM, XRD, XPS, and PL analysis, the properties of the flower-like Bi_2WO_6 samples and nanoparticles were investigated. These catalysts resulted in effective bacterial inactivation even under visible light and were faster than TiO_2. In addition to higher SSA of flower-like Bi_2WO_6, its lower PL intensity leads to lower recombination of photo-generated electron–hole pairs as a consequence of more efficient photocatalytic activity. The photocatalytic inactivation of bacteria mediated by as-developed flower-like Bi_2WO_6 (BWO5) is around 2.5 times faster when compared with nanoparticles. The samples under light lead to effective Bi_2WO_6 charge separation and the generation of ROS inducing bacterial inactivation. The intermediate ROS species produced by Bi_2WO_6 were identified by the use of the appropriate scavengers, and the ˙OH-radical was identified to be the dominant inactivation mechanism. Finally, the stable performance of the synthesized catalyst during recycling indicates its robustness and may suggest practical application potential.

Author Contributions: Conceptualization, M.K., S.G., C.P., S.R., and J.K.; methodology, M.K., S.G., S.R., J.K., and C.P.; software, M.K.; validation, M.K., S.G., S.R., and J.K.; investigation, M.K.; resources, C.P., F.K., and K.R.; writing—original draft preparation, M.K., F.K., K.R., S.R., J.K., S.G., and C.P.; writing—review & editing, M.K. and S.G.; visualization, M.K.; supervision, F.K., K.R., S.R., J.K., and C.P.; project administration, C.P., F.K., and K.R.; funding acquisition, S.G., C.P., F.K., and K.R. All authors have read and agreed to the published version of the manuscript.

References

1. Giannakis, S.; Jovic, M.; Gasilova, N.; Pastor Gelabert, M.; Schindelholz, S.; Furbringer, J.M.; Girault, H.; Pulgarin, C. Iohexol degradation in wastewater and urine by UV-based Advanced Oxidation Processes (AOPs): Process modeling and by-products identification. *J. Environ. Manag.* **2017**, *195*, 174–185. [CrossRef] [PubMed]

2. Chong, M.N.; Jin, B.; Chow, C.W.K.; Saint, C. Recent developments in photocatalytic water treatment technology: A review. *Water Res.* **2010**, *44*, 2997–3027. [CrossRef] [PubMed]

3. Pelaez, M.; Nolan, N.T.; Pillai, S.C.; Seery, M.K.; Falaras, P.; Kontos, A.G.; Dunlop, P.S.M.; Hamilton, J.W.J.; Byrne, J.A.; O'Shea, K.; et al. A review on the visible light active titanium dioxide photocatalysts for environmental applications. *Appl. Catal. B Environ.* **2012**, *125*, 331–349. [CrossRef]

4. Campoccia, D.; Montanaro, L.; Arciola, C.R. A review of the biomaterials technologies for infection-resistant surfaces. *Biomaterials* **2013**, *34*, 8533–8554. [CrossRef]

5. Giannakis, S.; Merino Gamo, A.I.; Darakas, E.; Escalas-Cañellas, A.; Pulgarin, C. Impact of different light intermittence regimes on bacteria during simulated solar treatment of secondary effluent: Implications of the inserted dark periods. *Sol. Energy* **2013**, *98*, 572–581. [CrossRef]

6.   Giannakis, S.; Darakas, E.; Escalas-Cañellas, A.; Pulgarin, C. Environmental considerations on solar disinfection of wastewater and the subsequent bacterial (re)growth. *Photochem. Photobiol. Sci.* **2015**, *14*, 618–625. [CrossRef]

7.   Byrne, J.A.; Dunlop, P.S.M.; Hamilton, J.W.J.; Fernández-Ibáñez, P.; Polo-López, I.; Sharma, P.K.; Vennard, A.S.M. A Review of Heterogeneous Photocatalysis for Water and Surface Disinfection. *Molecules* **2015**, *20*, 5574–5615. [CrossRef]

8.   Rizzo, L.; Della Sala, A.; Fiorentino, A.; Li Puma, G. Disinfection of urban wastewater by solar driven and UV lamp–TiO_2 photocatalysis: Effect on a multi drug resistant Escherichia coli strain. *Water Res.* **2014**, *53*, 145–152. [CrossRef]

9.   Carré, G.; Hamon, E.; Ennahar, S.; Estner, M.; Lett, M.C.; Horvatovich, P.; Gies, J.P.; Keller, V.; Keller, N.; Andre, P. TiO_2 photocatalysis damages lipids and proteins in Escherichia coli. *Appl. Environ. Microbiol.* **2014**, *80*, 2573–2581. [CrossRef]

10. Ouyang, K.; Dai, K.; Walker, S.L.; Huang, Q.; Yin, X.; Cai, P. Efficient photocatalytic disinfection of Escherichia coli O157:H7 using C70-TiO_2hybrid under visible light irradiation. *Sci. Rep.* **2016**, *6*, 1–8. [CrossRef]

11. Nakata, K.; Fujishima, A. TiO_2 photocatalysis: Design and applications. *J. Photochem. Photobiol. C Photochem. Rev.* **2012**, *13*, 169. [CrossRef]

12. Zhang, L.; Wang, W.; Chen, Z.; Zhou, L.; Xu, H.; Zhu, W. Fabrication of flower-like Bi_2WO_6superstructures as high performance visible-light driven photocatalysts. *J. Mater. Chem.* **2007**, *17*, 2526–2532. [CrossRef]

13. Zhang, L.; Wang, H.; Chen, Z.; Wong, P.K.; Liu, J. Bi_2WO_6 micro/nano-structures: Synthesis, modifications and visible-light-driven photocatalytic applications. *Appl. Catal. B Environ.* **2011**, *106*, 1–13. [CrossRef]

14. Saison, T.; Gras, P.; Chemin, N.; Chanéac, C.; Durupthy, O.; Brezová, V.; Colbeau-Justin, C.; Jolivet, J.P. New insights into Bi_2WO_6 properties as a visible-light photocatalyst. *J. Phys. Chem. C* **2013**, *117*, 22656–22666. [CrossRef]

15. Zhang, L.; Zhu, Y. A review of controllable synthesis and enhancement of performances of bismuth tungstate visible-light-driven photocatalysts. *Catal. Sci. Technol.* **2012**, *2*, 694–706. [CrossRef]

16. Chen, M. Degradation of Antibiotic Norfloxacin by Solar Light/Visible Light-Assisted Oxidation Processes in Aqueous Phase. Ph.D. Thesis, The Hong Kong Polytechnic University, Hong Kong, China, 2013.

17. Shang, Y.; Cui, Y.; Shi, R.; Yang, P. Effect of acetic acid on morphology of $Bi_2WO_6$with enhanced photocatalytic activity. *Mater. Sci. Semicond. Process.* **2019**, *89*, 240–249. [CrossRef]

18. Zhang, Q.; Chen, J.; Xie, Y.; Wang, M.; Ge, X. Inductive effect of poly (vinyl pyrrolidone) on morphology and photocatalytic performance of Bi_2WO_6. *Appl. Surf. Sci.* **2016**, *368*, 332–340. [CrossRef]

19. Shen, R.; Jiang, C.; Xiang, Q.; Xie, J.; Li, X. Surface and interface engineering of hierarchical photocatalysts. *Appl. Surf. Sci.* **2018**, *471*, 43–87. [CrossRef]

20. Li, C.; Chen, G.; Sun, J.; Rao, J.; Han, Z.; Hu, Y.; Zhou, Y. A Novel Mesoporous Single-Crystal-Like $Bi_2WO_6$with Enhanced Photocatalytic Activity for Pollutants Degradation and Oxygen Production. *ACS Appl. Mater. Interfaces* **2015**, *7*, 25716–25724. [CrossRef]

21. Ren, J.; Wang, W.; Zhang, L.; Chang, J.; Hu, S. Photocatalytic inactivation of bacteria by photocatalyst Bi_2WO_6 under visible light. *Catal. Commun.* **2009**, *10*, 1940–1943. [CrossRef]

22. Helali, S.; Polo-López, M.I.; Fernández-Ibáñez, P.; Ohtani, B.; Amano, F.; Malato, S.; Guillard, C. Solar photocatalysis: A green technology for E. coli contaminated water disinfection. Effect of concentration and different types of suspended catalyst. *J. Photochem. Photobiol. A Chem.* **2014**, *276*, 31–40. [CrossRef]

23. Amano, F.; Nogami, K.; Ohtani, B. Visible Light-Responsive Bismuth Tungstate Photocatalysts: Effects of Hierarchical Architecture on Photocatalytic Activity. *J. Phys. Chem* **2009**, *113*, 1536–1542. [CrossRef]

24. Porras, J.; Giannakis, S.; Torres-Palma, R.A.; Fernandez, J.J.; Bensimon, M.; Pulgarin, C. Fe and Cu in humic acid extracts modify bacterial inactivation pathways during solar disinfection and photo-Fenton processes in water. *Appl. Catal. B Environ.* **2018**, *235*. [CrossRef]

25. Marjanovic, M.; Giannakis, S.; Grandjean, D.; de Alencastro, L.F.; Pulgarin, C. Effect of MM Fe addition, mild heat and solar UV on sulfate radical-mediated inactivation of bacteria, viruses, and micropollutant degradation in water. *Water Res.* **2018**, *140*, 220–231. [CrossRef]

26. Zhang, C.; Zhu, Y. Synthesis of Square Bi2WO6 Nanoplates as High-Activity Visible-Light-Driven Photocatalysts, Chem. *Chem. Mater.* **2005**, *17*, 3537–3545. [CrossRef]

27. Liang, Y.; Shi, J.; Fang, B. Synthesis and electrochemical performance of bismuth tungsten oxides with different composition and morphology. *Chem. Phys. Lett.* **2019**, *716*, 112–118. [CrossRef]

28. Guan, J.; Liu, L.; Xu, L.; Sun, Z.; Zhang, Y. Nickel flower-like nanostructures composed of nanoplates: One-pot synthesis, stepwise growth mechanism and enhanced ferromagnetic properties. *CrystEngComm* **2011**, *13*, 2636. [CrossRef]

29. Donaldson, J.D.; Knifton, J.F.; Ross, S.D. The fundamental vibrational spectra of the formates of the main group elements. *Spectrochim. Acta* **1964**, *20*, 847–851. [CrossRef]

30. Anderegg, G.; Arnaud-Neu, F.; Delgado, R.; Felcman, J.; Popov, K. Critical evaluation of stability constants of metal complexes of complexones for biomedical and environmental applications (IUPAC Technical Report). *Pure Appl. Chem.* **2005**, *77*, 1445–1495. [CrossRef]

31. Portanova, R.; Lajunen, L.H.J.; Tolazzi, M.; Piispanen, J. Critical evaluation of stability constants for alpha-hydroxycarboxylic acid complexes with protons and metal ions and the accompanying enthalpy changes. Part II. Aliphatic 2-hydroxycarboxylic acids (IUPAC Technical Report). *Pure Appl. Chem.* **2003**, *75*, 495–540. [CrossRef]

32. Briand, G.G.; Burford, N. Coordination Complexes of Bismuth(III) Involving Organic Ligands with Pnictogen or Chalcogen Donors. *Adv. Inorg. Chem.* **2000**, *50*, 285–357.

33. Tang, R.; Su, H.; Sun, Y.; Zhang, X.; Li, L.; Liu, C.; Wang, B.; Zeng, S.; Sun, D. Facile Fabrication of Bi_2WO_6/Ag2S Heterostructure with Enhanced Visible-Light-Driven Photocatalytic Performances. *Nanoscale Res. Lett.* **2016**, *11*, 126. [CrossRef] [PubMed]

34. Xu, X.; Ge, Y.; Wang, H.; Li, B.; Yu, L.; Liang, Y.; Chen, K.; Wang, F. Sol–gel synthesis and enhanced photocatalytic activity of doped bismuth tungsten oxide composite. *Mater. Res. Bull.* **2016**, *73*, 385–393. [CrossRef]

35. Li, W.; Wang, Q.; Huang, L.; Li, Y.; Xu, Y.; Song, Y.; Zhang, Q.; Xu, H.; Li, H. Synthesis and characterization of BN/Bi_2WO_6 composite photocatalysts with enhanced visible-light photocatalytic activity. *RSC Adv.* **2015**, *5*, 88832–88840. [CrossRef]

36. Tang, R.; Su, H.; Sun, Y.; Zhang, X.; Li, L.; Liu, C.; Zeng, S.; Sun, D. Journal of Colloid and Interface Science Enhanced photocatalytic performance in Bi_2WO_6/SnS heterostructures: Facile synthesis, influencing factors and mechanism of the photocatalytic process. *J. Colloid Interface Sci.* **2016**, *466*, 388–399. [CrossRef]

37. Guo, Y.; Zhang, G.; Gan, H.; Zhang, Y. Micro/nano-structured CaWO4/Bi_2WO_6 composite: Synthesis, characterization and photocatalytic properties for degradation of organic contaminants. *Dalton Trans.* **2012**, *41*, 12697. [CrossRef]

38. Chick, H. An Investigation of the Laws of Disinfection. *J. Hyg. (Lond)* **1908**, *8*, 92–158. [CrossRef]

39. Watson, H.E. A Note on the Variation of the Rate of Disinfection with Change in the Concentration of the Disinfectant. *J. Hyg. (Lond)* **1908**, *8*, 536–542. [CrossRef]

40. Sun, X.; Zhang, H.; Wei, J.; Yu, Q.; Yang, P.; Zhang, F. Preparation of point-line Bi_2WO_6@TiO_2 nanowires composite photocatalysts with enhanced UV/visible-light-driven photocatalytic activity. *Mater. Sci. Semicond. Process.* **2016**, *45*, 51–56. [CrossRef]

41. Docampo, P.; Guldin, S.; Steiner, U.; Snaith, H.J. Charge Transport Limitations in Self-Assembled TiO_2 Photoanodes for Dye-Sensitized Solar Cells. *J. Phys. Chem. Lett.* **2013**, *4*, 698–703. [CrossRef]

42. Kh, M.; Reza, A.; Kurny, F.G. Parameters affecting the photocatalytic degradation of dyes using TiO_2: A review. *Appl. Water Sci.* **2017**, *7*, 1569–1578.

43. Adhikari, S.; Banerjee, A.; Eswar, N.K.R.; Sarkar, D.; Madras, G. Photocatalytic inactivation of E. Coli by ZnO-Ag nanoparticles under solar radiation. *RSC Adv.* **2015**, *5*, 51067–51077. [CrossRef]

44. Winkler, J. *Titanium Dioxide: Production, Properties and Effective Usage*; Vincentz Network: Hanover, Germany, 2013; ISBN 9783866308121.

45. Nosaka, Y.; Nosaka, A.Y. Generation and Detection of Reactive Oxygen Species in Photocatalysis. *Chem. Rev.* **2017**, *117*, 11302–11336. [CrossRef] [PubMed]

46. Benabbou, A.K.; Derriche, Z.; Felix, C.; Lejeune, P.; Guillard, C. Photocatalytic inactivation of *Escherischia coli*: Effect of concentration of TiO2 and microorganism, nature, and intensity of UV irradiation. *Appl. Catal. B Environ.* **2007**, *76*, 257–263.

47. Regmi, C.; Joshi, B.; Ray, S.K.; Gyawali, G.; Pandey, R.P. Understanding Mechanism of Photocatalytic Microbial Decontamination of Environmental Wastewater. *Front. Chem.* **2018**, *6*, 33. [CrossRef]

48. Wei, Z.; Zhu, Y.; Kudo, A. *Nanostructured Photocatalysts*; Springer International Publishing: Basel, Switzerland, 2016; ISBN 978-3-319-26077-8.

49. Xu, J.; Wang, W.; Sun, S.; Wang, L. Enhancing visible-light-induced photocatalytic activity by coupling with wide-band-gap semiconductor: A case study on Bi_2WO_6/TiO_2. *Appl. Catal. B Environ.* **2012**, *111–112*, 126–132. [CrossRef]

50. Schneider, J.; Bahnemann, D.; Ye, J.; Puma, G.L.; Dionysiou, D.D. *Photocatalysis*; Royal Society of Chemistry: Cambridge, UK, 2016; ISBN 9781782620419.

51. Wu, D.; Wang, W.; Ng, T.W.; Huang, G.; Xia, D.; Yip, H.Y.; Lee, H.K.; Li, G.; An, T.; Wong, P.K. Visible-light-driven photocatalytic bacterial inactivation and the mechanism of zinc oxysulfide under LED light irradiation. *J. Mater. Chem. A* **2016**, *4*, 1052–1059. [CrossRef]

52. Sherman, I.; Gerchman, Y.; Sasson, Y.; Gnayem, H.; Mamane, H. Disinfection and Mechanistic Insights of Escherichia coli in Water by Bismuth Oxyhalide Photocatalysis. *Photochem. Photobiol.* **2016**, *92*, 826–834. [CrossRef]

53. Kiwi, J.; Nadtochenko, V. New Evidence for TiO_2 Photocatalysis during Bilayer Lipid Peroxidation. *J. Phys. Chem. B* **2004**, *108*, 17675–17684. [CrossRef]

54. Kraeutler, B.; Bard, A.J. Heterogeneous photocatalytic decomposition of saturated carboxylic acids on titanium dioxide powder. Decarboxylative route to alkanes. *J. Am. Chem. Soc.* **1978**, *100*, 5985–5992. [CrossRef]

6

Comparing the Effects of Types of Electrode on the Removal of Multiple Pharmaceuticals from Water by Electrochemical Methods

Yu-Jung Liu [1] , Yung-Ling Huang [2], Shang-Lien Lo [1] and Ching-Yao Hu [2],*

[1]   Graduate Institute of Environmental Engineering, National Taiwan University, Taipei 106, Taiwan;
       yujungliu77@gmail.com (Y.-J.L.); sllo@ntu.edu.tw (S.-L.L.)
[2]   School of Public Health, Taipei Medical University, Taipei 110, Taiwan; m508101008@tmu.edu.tw
*   Correspondence: cyhu@tmu.edu.tw

Abstract: Considering the lack of information on simultaneously removing multiple pharmaceuticals from water or wastewater by electrochemical methods, this study aimed to investigate the removal of multiple pharmaceuticals by electro-coagulation and electro-oxidation based on two types of electrodes (aluminum and graphite). The synthetic wastewater contained a nonsteroidal anti-inflammatory drug (diclofenac), a sulfonamide antibiotic (sulfamethoxazole) and a β-blocker (atenolol). The pharmaceutical removal with electro-oxidation was much higher than those with the electro-coagulation process, which was obtained from a five-cell graphite electrode system, while the removal of pharmaceuticals with aluminum electrodes was about 20% (20 μM). In the electro-coagulation system, pharmaceutical removal was mainly influenced by the solubility or hydrophilicity of the compound. In the electro-oxidation system, the removal mechanism was influenced by the dissociation status of the compounds, which are attracted to the anode due to electrostatic forces and have a higher mass transformation rate with the electro-oxidation process. Therefore, atenolol, which was undissociated, cannot adequately be eliminated by electro-oxidation, unless the electrode's surface is large enough to increase the mass diffusion rate.

Keywords: β-blockers; electro-coagulation; electro-oxidation; nonsteroidal anti-inflammatory drugs (NSAIDs); sulfonamide antibiotics

1. Introduction

The presence of emerging pharmaceutical contaminants has drawn much attention in recent years; however, these substances are inefficiently removed by conventional unit operations utilized by most wastewater treatment plants (WWTPs) due to their intricate properties like high water solubility and poor biodegradability [1–6]. Among such pharmaceuticals, nonsteroidal anti-inflammatory drugs (NSAIDs), beta-blockers, and antibiotics are widely used groups, and thus these groups are most often found in wastewater, which may co-exist in water bodies and soil environment [1,6,7]. They may induce adverse effects on aquatic systems [6,8–11].

An electrochemical process was applied to broad applicability like textile, cellulose, paper factories, laundry, and various kinds of different characteristic wastewater [12–18]. This technology can be operated by separation and degradation, called electro-coagulation and electrochemical oxidation, which depends on the characteristics of the electrodes [18–20]. In a separation system, electro-coagulation involves applying electric current to sacrificial electrodes where coagulants and gas bubbles are generated in situ by the current, and destabilizes, suspends, emulsifies or dissolves pollutants in an aqueous medium, or floats pollutants to the surface by tiny bubbles of hydrogen and oxygen gases generated from water electrolysis [21]. Additionally, the stable complexation between pollutants and flocs might be due to the structure of the pollutants, which might be the assumption for high selectivity by the metal adsorption capacity of the ligand-based-like flocs [22–24]. In a degradation system, electro-oxidation occurs through two routes, (1) direct oxidation, where the pollutants eliminated at the anode surface; (2) indirect oxidation, where mediators (reactive oxygen species or active chlorine species) are anodically generated to carry out the reaction [20,25]. The electrochemical methods have the advantages of simple equipment, flexible operation, and being chemical free [26,27]. Although it requires wastewater under high conductivity and energy consumption, it still has high treatment efficiency with chloride without any secondary pollutants, and the reusability of chloride with relatively low costs, which is the concept of green chemistry [21,26–30].

Previous studies have focused on single pharmaceutical removal efficiencies by using electro-coagulation and electro-oxidation processes [19,20,31]. However, many pharmaceuticals are simultaneously present in the environment. To the best of our knowledge from reviewing the literature, less reports are available to date on removing multiple pharmaceuticals from water or wastewater using an electrochemical process. Accordingly, this study aimed to evaluate the effect of laboratory-scale electrochemical treatments of wastewater on removing multiple pharmaceuticals from synthetic water and spiked wastewater. We evaluated the effectiveness of aluminum and graphite electrodes in monopolar and bipolar arrangements under consideration for electrochemical utilization methods to eliminate a beta-blocker (atenolol; ATE), an NSAID (diclofenac; DIC), and a sulfonamide antibiotic (sulfamethoxazole; SMX).

2. Materials and Methods

2.1. Materials

Characteristics of the target compounds are shown in Table 1. All the chemicals were an analytical grade ($\geq$98%). DIC ($C_{14}H_{11}Cl_2NO_2$) and SMX ($C_{10}H_{11}N_3O_3S$) were purchased from Sigma-Aldrich; ATE ($C_{14}H_{22}N_2O_3$) was from TCI; and sodium chloride (NaCl) (99.5%; Wako) was used as the supporting electrolyte to prevent passive film generated during the reaction, which may increase the resistance and diminish the release of the coagulants from the electrode [21,32,33]. Monopotassium phosphate (KH_2PO_4) (99%; Showa), methanol (high-performance liquid chromatography (HPLC) grade; Scharlau), and acetonitrile (HPLC grade; J.T. Baker) were used in the HPLC analyses. Stock solutions were prepared with deionized Milli-Q water obtained from a Merck-Millipore system and were stored in amber glass bottles at 4 °C. The standard solutions of target compounds at various concentrations were prepared by diluting the stock solutions prior to use. The anodes and cathodes used were 99% pure aluminum as the sacrificial electrodes, and graphite as the non-sacrificial electrodes.

Table 1. Characteristics of the target compounds.

Categories	Compound (CAS Number)	Structure	MW (g mol^{-1})	Solubility (mg/L)	pK$_a$	Log K$_{ow}$
β-blocker	Atenolol (ATE) (29122-68-7)		266.34	300	9.6	0.16
Nonsteroidal anti-inflammatory drug (NSAID)	Diclofenac (DIC) (15307-86-5)		296.13	2.37	4.15	4.51
Sulfonamide antibiotic	Sulfamethoxazole (SMX) (723-46-6)		253.28	370	5.7	0.89

2.2. Electrochemical (EC) Experiments and Analyses

EC experiments were carried out in a 1 L reactor, which is a double layered cylindrical glass container, with a diameter of 10 cm and a height of 20 cm, as schematically shown in Figure 1. The cell was operated with 2 and 6 electrodes held vertically, respectively; each electrode was 1 mm thick and had an effective area of 125 cm^2; the distance between the electrodes was 100 and 24 mm, respectively. All electrochemical experiments were carried out under potent stirring with a magnetic bar at 260 rpm and a constant current intensity (I = 0.5 A) provided by direct current (DC) (GPR-30H10D, Good Will Instrument). Desired concentrations of the pharmaceutical stock solutions (10 mM) were made by adding proper amounts of pharmaceuticals to distilled water. The volume (V) of the solution of each batch was 1 L with 100 µM of each pharmaceutical (ATE, DIC, and SMX) and 0.01 M NaCl as the supporting electrolyte. Solutions were maintained at 25 ± 1 °C in a water-bath. Samples were taken at assigned time intervals following the achievement of the EC process. The total reaction time was set to 20 or 40 min, which depended on what was required for the electrode cells to achieve a suitable removal efficiency. Pharmaceutical concentrations were investigated by high-performance liquid chromatography (HPLC; Hitachi, L-7200, Japan) with a diode array detector (DAD; Hitachi, L-7455, Japan), equipped with a C$_{18}$ column (RP-18 GP 150 × 4.6 mm, 5 µm, Mightysil). HPLC–DAD is a relatively cheap and simple operation technique for screening and analyzing purposes on gradient elution [34]. Mobile phase A was KH$_2$PO$_4$, B was acetonitrile, and the specific conditions are shown in Table 2. The flow rate of mobile phases A and B were both set to 1 mL min^{-1}. The injected volume of each sample was 20 µl. The deviations of all analyses were within 5%. Quality assurance/quality control (QA/QC) standards were prepared by diluting and combining the 0.01 M pharmaceutical stock solution accordingly, and to monitor HPLC–DAD performances.

Table 2. Conditions of HPLC–diode array detector (DAD).

Time (min)	Mobile Phase	
	A KH$_2$PO$_4$ (%)	B Acetonitrile (%)
0	80	20
3	80	20
4	60	40
10.5	60	40
11.5	40	60
18.5	40	60
19.5	80	20

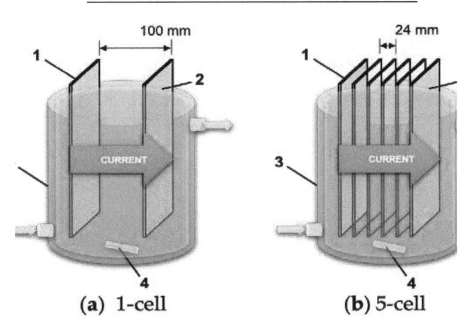

(a) 1-cell **(b) 5-cell**

Figure 1. Profile of the batch reactor: (**a**) 1 cell and (**b**) 5 cell. 1: anode (aluminum, graphite). 2: cathode (aluminum, graphite). 3: Double layered cylindrical glass container (V = 1 L) with a water bath. 4: magnetic stirring bar.

3. Results and Discussion

3.1. Removal of ATE, DIC, and SMX with Different Electrode Systems

Three electrode arrangements were investigated for single pharmaceutical removal in the batch electrochemical method with sacrificial and non-sacrificial anodes, under conditions of the present study, and two mechanisms were responsible for the removal of the pollutants: electro-coagulation, and electro-oxidation. We choose aluminum electrodes used as model sacrificial anode; graphite electrodes used as non-sacrificial anode. Figure 2a shows that removal rates of DIC, SMX, and ATE by the aluminum electrode system were 17.9%, 4.8%, and 2.3%, respectively. These results are similar to those of our previous study [19]; the substance with lipophilic characteristic could be removed either by precipitation of the flocs or by flotation with hydrogen gas during the electro-coagulation reaction, which depended on adsorption [35,36] and charge neutralization on the surface of hydrogen bubbles and aluminum hydroxide [37]. Additionally, it might be due to the soft atoms that are favorable for stable complexation mechanism, and this might be the main assumption for the high selectivity by the Al (II) adsorption capacity of the ligand-based flocs [22].

The main mechanism of pharmaceutical removal in the graphite electrode system should be electrochemical oxidation, which involves two procedures when an electric current is passed through a non-sacrificial electrode. The first one is the direct electrolysis of the pharmaceuticals on the surface of the anodes. The other one is indirect oxidation which involves a strong oxidant ($\bullet$OH or HOCl) produced by the electrolysis of chloride or other compounds [20]. The oxidant is then transferred to the bulk solution to degrade the pharmaceuticals, and this process may cause the structure of the initial pollutant to be converted into byproducts [18,20]. DIC, SMX, and ATE, as shown in Figure 2b, were reduced by 100%, 99.8%, and 85%, respectively. The higher removal of DIC and SMX may have been due to the dissociation of the two compounds in the reaction conditions. As shown in Table 1, pKa values of both DIC and SMX are relatively smaller than that of ATE, which means the two compounds are dissociated in a neutral condition but ATE is not. The mass transfer rate of anions should be much higher than that of neutral compounds because of electrostatic forces. The removal efficiencies of ionic compounds such as DIC and SMX by electro-oxidation were higher than that of the neutral compound of ATE. The schemes of the mechanism of electro-coagulation and electro-oxidation are shown in Figure 3.

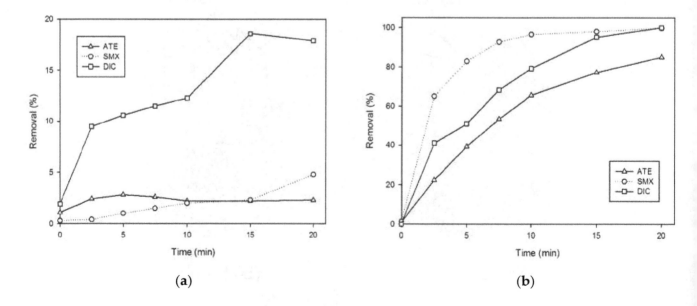

(a)   (b)

Figure 2. Removal efficiencies of single pharmaceuticals in a 1 L batch reactor with a 5-cell system of (a) aluminum or (b) graphite ([Phar.]$_0$ = 100 μM, [NaCl] = 0.01 M, I = 0.5 A).

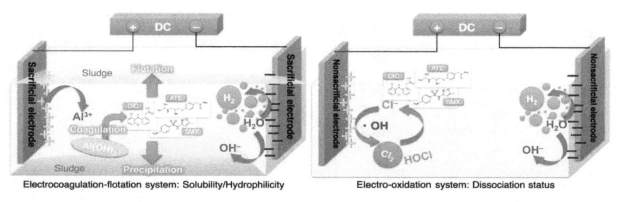

Figure 3. Schemes of the mechanism of electro-coagulation and electro-oxidation.

3.2. Removal Efficiencies of Multiple Pharmaceuticals

The above observations reveal that the removal efficiencies of single pharmaceuticals changed with different electrodes, especially in the graphite electrode system. To utilize this system in a practical way, as shown in Figure 4, the influences of the electrode cell size in the triple-pharmaceutical solutions should be considered. The removal of multiple pharmaceuticals in a 1-cell system was inefficient, especially for ATE. Compared to the one-cell electrode, five-cell electrodes were significantly more effective at removing multiple pharmaceuticals from the water, with removal rates of SMX, DIC, and ATE of 75%, 68%, and 55%, respectively. The reason could also be due to the dissociation of the selected pharmaceuticals. On the other hand, the synergistic effects of direct and mediated oxidation may be another possibility causing this circumstance. Hydroxyl radicals ($\bullet$OH) produced by the oxidation of water are known to be a very active reagent and lead to the formation of higher-state oxides or superoxides (Equations (1) and (2)). Using sodium chloride (NaCl) as the electrolyte provided the formation of hypochlorous acid (HOCl) (Equation (3)). Accordingly, the oxidation of pharmaceuticals occurred more effectively since $\bullet$OH and HOCl were generated in the five-cell electrode system:

$$M + H_2O \rightarrow M(\bullet OH) + H^+ + e^- \tag{1}$$

$$M(\bullet OH) \rightarrow MO + H^+ + e^- \tag{2}$$

$$Cl^- + H_2O \rightarrow HOCl + H^+ + 2e^- \tag{3}$$

To clarify the interactions between multiple pharmaceuticals in the electro-oxidation reaction, variations in removal rates of double pharmaceuticals with the reaction times in the graphite electrode system are shown in Figure 4. When ATE and SMX existed simultaneously (Figure 5a), only SMX could effectively be removed, at 69.9%. The same tendency is shown in Figure 5b of DIC and ATE concurrently. Elimination rates of DIC and ATE were 94.8% and 63.4%. Figure 5c shows that elimination rates of SMX and DIC were 75.2% and 73.7%.

As a result of the behavior described above, the structure and characteristics of the pharmaceuticals were the most important factors. Difficulty with cleavage affected the removal of multiple pharmaceuticals, which is related to competitive oxidation [38]. The S–N bond, isoxazole ring, and benzene ring of SMX and DIC were easily cleaved during oxidation [39]. Moreover, it was found that DIC and SMX were more effectively removed than ATE. After the reaction, the pH of the solution was about 7 ± 0.8, but the pKa of ATE was 9.6. This observation indicated that ATE was still in a molecular state and had not reached the anode; DIC and SMX were dissociated into ionic compounds and were attracted to the anode, and thus were rapidly degraded through the electrochemical oxidation system.

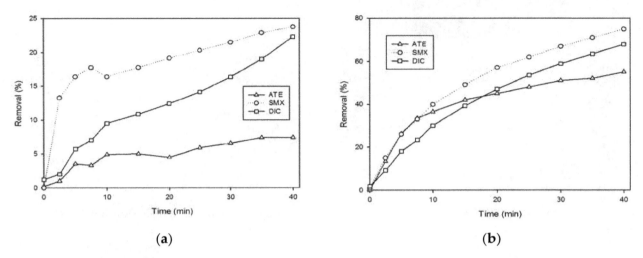

Figure 4. Removal efficiencies of triple pharmaceuticals in a 1 L batch reactor with a graphite electrode and a (**a**) 1-cell or (**b**) 5-cell system ([Phar.]$_0$ = 100 μM, [NaCl] = 0.01 M, I = 0.5 A).

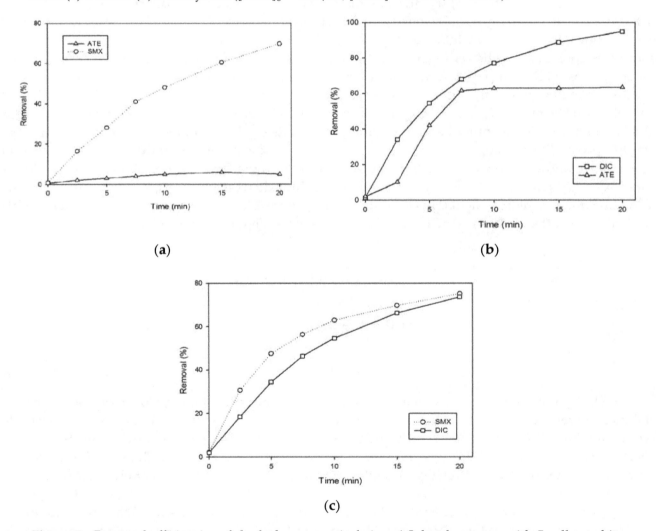

Figure 5. Removal efficiencies of dual pharmaceuticals in a 1 L batch reactor with 5-cell graphite electrodes ([Phar.]$_0$ = 100 μM, [NaCl] = 0.01 M, I = 0.5 A). (**a**) ATE + SMX; (**b**) DIC + ATE; (**c**) SMX + DIC.

3.3. Influences of Multiple Pharmaceuticals on Hospital Wastewater Matrixes

To simulate pharmaceutical interactions in real wastewater matrixes, we spiked our target compounds into an actual hospital effluent. Table 3 shows the characteristics of the hospital effluent. Figure 6 shows the performances of multiple pharmaceuticals in the electro-oxidation system with graphite electrodes. Martins et al. indicated that if a solution has large amounts of organic compounds or complex matrixes, pharmaceuticals may be adsorbed or attached onto other species or structures, which decreases the removal efficiency [40]. These results reflect the same circumstances of our experiment outcomes. Compared to laboratory outcomes, the declines in ATE, DIC, and SMX in an actual hospital effluent were only 11%, 47.8%, and 43%, respectively.

Table 3. Characteristics of hospital effluent.

Parameters	Hospital Effluent
Temperature (°C)	25.0
Dissolved oxygen (DO) (ppm)	3.35
pH	7.20
Turbidity (NTU)	10.9
Suspended solids (SS) (mg/L)	0.018
Chemical oxygen demand (COD) (mg/L)	41.5
Biochemical oxygen demand (BOD) (mg/L)	4.21

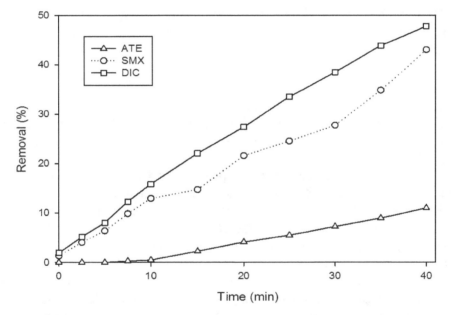

Figure 6. Removal efficiencies of spiked pharmaceuticals in actual hospital effluent matrices ([Phar.]$_s$ = 100 µM, [NaCl] = 0.01 M, I = 0.5 A).

4. Conclusions

To develop promising treatments for removing pharmaceuticals from sewage, the ability of a physicochemical process, electrochemical treatment, to remove selected pharmaceuticals was assessed. Aluminum was used for the active electrodes with the electrocoagulation–flotation process; graphite was used for the inert electrodes with electro-oxidation. The electrochemical process with the graphite electrode system removed the pharmaceuticals in water more effectively with multiple electrodes by the electro-oxidation process. Removal efficiencies were affected by the hydrophobic character of the pharmaceuticals under the electrocoagulation–flotation system, and the uneven electron structural characteristics of the pharmaceuticals under the electro-oxidation system. According to these results, the degradation of pharmaceuticals was improved and more efficient than the separation of medicines.

Author Contributions: Y.-L.H. carried out the experiments. Y.-J.L. processed the experiments, analyzed the data, and wrote the manuscript in consultation with S.-L.L. and C.-Y.H. S.-L.L. devised the conceptual ideas, discussed the results, and commented on the manuscript. C.-Y.H. conceived the original idea, helped supervise the project, and final approval of the version to be published. All authors have read and agreed to the published version of the manuscript.

Abbreviations

ATE	Atenolol
DAD	Diode Array Detector
DC	Direct Current
DIC	Diclofenac
EC	Electrochemical
HOCl	Hypochlorous Acid
HPLC	High-Performance Liquid Chromatography
I	Current
KH_2PO_4	Monopotassium Phosphate
NaCl	Sodium Chloride
NSAIDs	Nonsteroidal Anti-Inflammatory Drugs
•OH	Hydroxyl Radicals
$[Phar]_0$	Initial Concentration of Pharmaceutical
$[Phar]_s$	Initial Concentration of Spiked Pharmaceuticals
QA/QC	Quality Assurance/Quality Control
SMX	Sulfamethoxazole

References

1. Rivera-Utrilla, J.; Sánchez-Polo, M.; Ferro-García, M.Á.; Prados-Joya, G.; Ocampo-Pérez, R. Pharmaceuticals as emerging contaminants and their removal from water. A review. *Chemosphere* **2013**, *93*, 1268–1287. [CrossRef] [PubMed]

2. Crouse, B.A.; Ghoshdastidar, A.J.; Tong, A.Z. The presence of acidic and neutral drugs in treated sewage effluents and receiving waters in the Cornwallis and Annapolis River watersheds and the Mill CoveSewage Treatment Plant in Nova Scotia, Canada. *Environ. Res.* **2012**, *112*, 92–99. [CrossRef] [PubMed]

3. Li, Y.; Zhang, S.; Zhang, W.; Xiong, W.; Ye, Q.; Hou, X.; Wang, C.; Wang, P. Life cycle assessment of advanced wastewater treatment processes: Involving 126 pharmaceuticals and personal care products in life cycle inventory. *J. Environ. Manag.* **2019**, *238*, 442–450. [CrossRef] [PubMed]

4. Mestre, A.S.; Pires, R.A.; Aroso, I.; Fernandes, E.M.; Pinto, M.L.; Reis, R.L.; Andrade, M.A.; Pires, J.; Silva, S.P.; Carvalho, A.P. Activated carbons prepared from industrial pre-treated cork: Sustainable adsorbents for pharmaceutical compounds removal. *Chem. Eng. J.* **2014**, *253*, 408–417. [CrossRef]

5. Phoon, B.L.; Ong, C.C.; Mohamed Saheed, M.S.; Show, P.-L.; Chang, J.-S.; Ling, T.C.; Lam, S.S.; Juan, J.C. Conventional and emerging technologies for removal of antibiotics from wastewater. *J. Hazard. Mater.* **2020**, *400*, 122961. [CrossRef]

6. Khan, A.H.; Khan, N.A.; Ahmed, S.; Dhingra, A.; Singh, C.P.; Khan, S.U.; Mohammadi, A.A.; Changani, F.; Yousefi, M.; Alam, S.; et al. Application of advanced oxidation processes followed by different treatment technologies for hospital wastewater treatment. *J. Clean. Prod.* **2020**, *269*, 122411. [CrossRef]

7. Petrovic, M.; Hernando, M.D.; Diaz-Cruz, M.S.; Barcelo, D. Liquid chromatography-tandem mass spectrometry for the analysis of pharmaceutical residues in environmental samples: A review. *J. Chromatogr. A* **2005**, *1067*, 1–14. [CrossRef]

8. Verlicchi, P.; Al Aukidy, M.; Zambello, E. Occurrence of pharmaceutical compounds in urban wastewater: Removal, mass load and environmental risk after a secondary treatment—A review. *Sci. Total Environ.* **2012**, *429*, 123–155. [CrossRef]

9. Berninger, J.P.; Brooks, B.W. Leveraging mammalian pharmaceutical toxicology and pharmacology data to predict chronic fish responses to pharmaceuticals. *Toxicol. Lett.* **2010**, *193*, 69–78. [CrossRef]

10. Camacho-Muñoz, D.; Martín, J.; Santos, J.L.; Aparicio, I.; Alonso, E. Occurrence, temporal evolution and risk assessment of pharmaceutically active compounds in Donana Park (Spain). *J. Hazard. Mater.* **2010**, *183*, 602–608. [CrossRef]

11. Xu, M.; Huang, H.; Li, N.; Li, F.; Wang, D.; Luo, Q. Occurrence and ecological risk of pharmaceuticals and personal care products (PPCPs) and pesticides in typical surface watersheds, China. *Ecotox. Environ. Safe* **2019**, *175*, 289–298. [CrossRef] [PubMed]

12. Essadki, A.H.; Bennajah, M.; Gourich, B.; Vial, C.; Azzi, M.; Delmas, H. Electrocoagulation/electroflotation in an external-loop airlift reactor—Application to the decolorization of textile dye wastewater: A case study. *Chem. Eng. Process. Process Intensif.* **2008**, *47*, 1211–1223. [CrossRef]

13. Boroski, M.; Rodrigues, A.C.; Garcia, J.C.; Gerola, A.P.; Nozaki, J.; Hioka, N. The effect of operational parameters on electrocoagulation–flotation process followed by photocatalysis applied to the decontamination of water effluents from cellulose and paper factories. *J. Hazard. Mater.* **2008**, *160*, 135–141. [CrossRef] [PubMed]

14. Chou, W.-L.; Wang, C.-T.; Chang, S.-Y. Study of COD and turbidity removal from real oxide-CMP wastewater by iron electrocoagulation and the evaluation of specific energy consumption. *J. Hazard. Mater.* **2009**, *168*, 1200–1207. [CrossRef]

15. Ge, J.; Qu, J.; Lei, P.; Liu, H. New bipolar electrocoagulation–electroflotation process for the treatment of laundry wastewater. *Sep. Purif. Technol.* **2004**, *36*, 33–39. [CrossRef]

16. Emamjomeh, M.M.; Sivakumar, M. Fluoride removal by a continuous flow electrocoagulation reactor. *J. Environ. Manag.* **2009**, *90*, 1204–1212. [CrossRef]

17. Bansal, S.; Kushwaha, J.P.; Sangal, V.K. Electrochemical Treatment of Reactive Black 5 Textile Wastewater: Optimization, Kinetics, and Disposal Study. *Water Environ. Res* **2013**, *85*, 2294–2306. [CrossRef]

18. Sirés, I.; Brillas, E. Remediation of water pollution caused by pharmaceutical residues based on electrochemical separation and degradation technologies: A review. *Environ. Int.* **2012**, *40*, 212–229. [CrossRef]

19. Liu, Y.-J.; Lo, S.-L.; Liou, Y.-H.; Hu, C.-Y. Removal of nonsteroidal anti-inflammatory drugs (NSAIDs) by electrocoagulation–flotation with a cationic surfactant. *Sep. Purif. Technol.* **2015**, *152*, 148–154. [CrossRef]

20. Liu, Y.-J.; Hu, C.-Y.; Lo, S.-L. Direct and indirect electrochemical oxidation of amine-containing pharmaceuticals using graphite electrodes. *J. Hazard. Mater.* **2019**, *366*, 592–605. [CrossRef]

21. Mollah, M.Y.A.; Schennach, R.; Parga, J.R.; Cocke, D.L. Electrocoagulation (EC)—Science and applications. *J. Hazard. Mater.* **2001**, *84*, 29–41. [CrossRef]

22. Awual, M.R.; Hasan, M.M. A ligand based innovative composite material for selective lead(II) capturing from wastewater. *J. Mol. Liq.* **2019**, *294*, 111679. [CrossRef]

23. Awual, M.R.; Hasan, M.M.; Rahman, M.M.; Asiri, A.M. Novel composite material for selective copper(II) detection and removal from aqueous media. *J. Mol. Liq.* **2019**, *283*, 772–780. [CrossRef]

24. Awual, M.R. A novel facial composite adsorbent for enhanced copper(II) detection and removal from wastewater. *Chem. Eng. J.* **2015**, *266*, 368–375. [CrossRef]

25. Anglada, Á.; Urtiaga, A.; Ortiz, I. Contributions of electrochemical oxidation to waste-water treatment: Fundamentals and review of applications. *J. Chem. Technol. Biot.* **2009**, *84*, 1747–1755. [CrossRef]

26. Liu, H.; Zhao, X.; Qu, J. Electrocoagulation in Water Treatment. In *Electrochemistry for the Environment*; Comninellis, C., Chen, G., Eds.; Springer: New York, NY, USA, 2010; pp. 245–262. [CrossRef]

27. Rychen, P.; Provent, C.; Pupunat, L.; Hermant, N. Domestic and Industrial Water Disinfection Using Boron-Doped Diamond Electrodes. In *Electrochemistry for the Environment*; Comninellis, C., Chen, G., Eds.; Springer: New York, NY, USA, 2010; pp. 143–161. [CrossRef]

28. Pulkka, S.; Martikainen, M.; Bhatnagar, A.; Sillanpää, M. Electrochemical methods for the removal of anionic contaminants from water—A review. *Sep. Purif. Technol.* **2014**, *132*, 252–271. [CrossRef]

29. Griesbach, U.; Malkowsky, I.M.; Waldvogel, S.R. Green Electroorganic Synthesis Using BDD Electrodes. In *Electrochemistry for the Environment*; Comninellis, C., Chen, G., Eds.; Springer: New York, NY, USA, 2010; pp. 125–141. [CrossRef]

30. Steckhan, E.; Arns, T.; Heineman, W.R.; Hilt, G.; Hoormann, D.; Jörissen, J.; Kröner, L.; Lewall, B.; Pütter, H. Environmental protection and economization of resources by electroorganic and electroenzymatic syntheses. *Chemosphere* **2001**, *43*, 63–73. [CrossRef]

31. Dos Santos, A.J.; Cabot, P.L.; Brillas, E.; Sirés, I. A comprehensive study on the electrochemical advanced oxidation of antihypertensive captopril in different cells and aqueous matrices. *Appl. Catal. B Environ.* **2020**, *277*, 119240. [CrossRef]

32. Chen, X.; Chen, G.; Yue, P.L. Investigation on the electrolysis voltage of electrocoagulation. *Chem. Eng. Sci.* **2002**, *57*, 2449–2455. [CrossRef]

33. He, C.-C.; Hu, C.-Y.; Lo, S.-L. Integrating chloride addition and ultrasonic processing with electrocoagulation to remove passivation layers and enhance phosphate removal. *Sep. Purif. Technol.* **2018**, *201*, 148–155. [CrossRef]

34. Teixeira, S.; Delerue-Matos, C.; Alves, A.; Santos, L. Fast screening procedure for antibiotics in wastewaters by direct HPLC-DAD analysis. *J. Sep. Sci.* **2008**, *31*, 2924–2931. [CrossRef] [PubMed]

35. Chou, W.-L.; Wang, C.-T.; Huang, K.-Y.; Liu, T.-C. Electrochemical removal of salicylic acid from aqueous solutions using aluminum electrodes. *Desalination* **2011**, *271*, 55–61. [CrossRef]

36. Zhang, G.H.; Yin, L.L.; Zhang, S.T.; Li, X. Adsorption Behavior of Sulfamethoxazole as Inhibitor for Mild Steel in 3% HCl Solution. *Adv. Mater. Res.* **2011**, *194–196*, 8–15. [CrossRef]

37. Ren, M.; Song, Y.; Xiao, S.; Zeng, P.; Peng, J. Treatment of berberine hydrochloride wastewater by using pulse electro-coagulation process with Fe electrode. *Chem. Eng. J.* **2011**, *169*, 84–90. [CrossRef]

38. Indermuhle, C.; Martín de Vidales, M.J.; Sáez, C.; Robles, J.; Cañizares, P.; García-Reyes, J.F.; Molina-Díaz, A.; Comninellis, C.; Rodrigo, M.A. Degradation of caffeine by conductive diamond electrochemical oxidation. *Chemosphere* **2013**, *93*, 1720–1725. [CrossRef]

39. Gao, S.; Zhao, Z.; Xu, Y.; Tian, J.; Qi, H.; Lin, W.; Cui, F. Oxidation of sulfamethoxazole (SMX) by chlorine, ozone and permanganate—A comparative study. *J. Hazard. Mater.* **2014**, *274*, 258–269. [CrossRef]

40. Martins, A.F.; Mallmann, C.A.; Arsand, D.R.; Mayer, F.M.; Brenner, C.G.B. Occurrence of the Antimicrobials Sulfamethoxazole and Trimethoprim in Hospital Effluent and Study of Their Degradation Products after Electrocoagulation. *CLEAN Soil Air Water* **2011**, *39*, 21–27. [CrossRef]

7

Towards the Removal of Antibiotics Detected in Wastewaters in the POCTEFA Territory: Occurrence and TiO$_2$ Photocatalytic Pilot-Scale Plant Performance

Samuel Moles [1,*], Rosa Mosteo [1], Jairo Gómez [2], Joanna Szpunar [3], Sebastiano Gozzo [3], Juan R. Castillo [4] and María P. Ormad [1]

[1] Water and Environmental Health Research Group, c/María de Luna 3, 50018 Zaragoza, Spain; mosteo@unizar.es (R.M.); mpormad@unizar.es (M.P.O.)
[2] Navarra de Infraestructuras Locales SA, av. Barañain 22, 31008 Pamplona, Spain; jgomez@nilsa.com
[3] Institute of Analytical Sciences and Physico-Chemistry for Environment and Materials (IPREM), Centre National de la Recherche Scientifique (CNRS), Universite de Pau et des Pays de l'Adour, CEDEX 9 Pau, France; joanna.szpunar@univ-pau.fr (J.S.); sebastiano.gozzo@univ-pau.fr (S.G.)
[4] Analytical Spectroscopy and Sensors Group Analytic Chemistry Department, Science Faculty, Environmental Science Institute, University of Zaragoza, 50009 Zaragoza, Spain; jcastilo@unizar.es
* Correspondence: sma@unizar.es

Abstract: This research aims to assess the presence of four antibiotic compounds detected in the influent and effluent of wastewater treatment plants (WWTPs) in the POCTEFA territory (north of Spain and south of France) during the period of 2018–2019, and to relate the removal of antibiotic compounds with the processes used in the WWTPs. The performance of a photocatalytic TiO$_2$/UV-VIS pilot-scale plant was then evaluated for the degradation of selected antibiotics previously detected in urban treated effluent. The main results reflect that azithromycin had the highest mass loadings (11.3 g/day per 1000 inhabitants) in the influent of one of the selected WWTPs. The results also show considerable differences in the extent of antibiotics removal in WWTPs ranging from 100% for sulfadiazine to practically 0% for trimethoprim. Finally, the photocatalytic TiO$_2$/UV-VIS pilot-scale plant achieved the removal of the four antibiotics after 240 min of treatment from 78%–80% for trimethoprim and enrofloxacin, up to 100% for amoxicillin, sulfadiazine and azithromycin. The catalyst recovery via mechanical coagulation–flocculation–decantation was almost total. The Ti concentration in the effluent of the TiO$_2$/UV-VIS pilot-scale plant was lower than 0.1% (w/w), and its release into the environment was subsequently minimized.

Keywords: antibiotics; wastewater; removal efficiency; photocatalysis; slurry reactor

Highlights:

- Antibiotics mass loadings range from 11,332 mg/day·1000 inhabitants to undetectable levels.
- Sulfadiazine, amoxicillin and azithromycin can be removed from wastewaters, while 80% of trimethoprim and enrofloxacin removal can be achieved after the photocatalytic treatment.
- The facilities provided with trickling filters proved to be more effective in removing antibiotics from wastewaters.

1. Introduction

The problem of the presence of pharmaceutical compounds in wastewater has recently become a matter for concern, not only in terms of human health, but also for the preservation of the environment [1]. Antibiotics are an important group of medicines suitable for the treatment of human infections and in veterinary medicine. Many of them are not completely metabolized by the body so between 30%–90% are excreted and, as a result, they end up in wastewater [2,3]. The main difference with other organic pollutants is that antibiotics represent a potential risk if they are released into the environment because they a direct biological action on microorganisms, generating antimicrobial-resistant bacteria (ARB). As suggested by other authors [4], ARB of animal origin can also be transmitted to humans.

Several studies have pointed out that conventional wastewater treatment plants (WWTPs) are not designed to remove pharmaceuticals, metabolites or drugs [3–5]. Besides urban plants and hospitals, slaughterhouses also generate wastewaters which are not usually incorporated into sewage systems. As a result, they represent a significant source of antibiotics released into the environment. The European Surveillance of Veterinary Antimicrobial Consumption (ESVAC) collects information on how antimicrobial medicines are used in animals across the European Union (EU). According to their latest report [6], Spain is known to be one of the main consumers of veterinary drugs in the EU. As a result, many studies have monitored the occurrence of the most commonly administered pharmaceuticals in urban wastewater, groundwaters and surface water in Spain. The literature informs that concentrations of antibiotics from ng/L to µg/L [7–10]. sulfonamides [11,12], trimethoprim [13,14], β-lactams [15,16], fluoroquinolones [17,18] and macrolides [19–21] all represent a potential risk for the environment. Consequently, a representative antibiotic from each one these groups was analyzed in this research work: sulfadiazine (veterinary use, sulfonamide), trimethoprim (human and veterinary use, trimethoprim), amoxicillin (human and veterinary use, β-lactam), enrofloxacin (veterinary use, fluoroquinolone) and azithromycin (human use, macrolide).

Among the various water treatment techniques used to eliminate these drugs, advanced oxidation processes (AOPs) are suitable for antibiotic degradation [22–24]. Other techniques, such as activated carbon or reverse osmosis, only transfer the contaminants from one phase to another without degrading them. Nevertheless, photocatalysis has been demonstrated to be effective for wastewater treatment as it is cost-effective and simultaneously oxidizes various organic contaminants into inorganic compounds, water and carbon dioxide, and pathogenic microorganisms [24]. Several semiconductors are used in photocatalysis, such as TiO_2, ZnO, and CdS. Among these, TiO_2 has been widely used because of its strong oxidizing power, availability, nontoxicity and price. The catalyst can be employed either in a colloidal or in an immobilized form. Although immobilizing the catalyst might improve the catalyst recovery, immobilized systems show lower degradation efficiencies compared to the suspended counterpart because of a reduction in the surface area [25,26]. Whenever the nanoparticles are dispersed in an aqueous medium, the depth of penetration of the radiation is limited because of absorption/scattering by the catalyst nanoparticles and the dissolved organic species. These systems also require an additional separation process to prevent Ti emission to the environment, and this stage induces further costs [27].

Pilot-scale plants represent the previous step to industrial scale plant. Literature suggest how to operate at lab-scale photocatalytic systems [28–30]. However, design and operation with a pilot-scale plant are necessary to determine how to deal with possible operational problems and establish the optimal operational parameters for real scale operation. Pilot-scale plants also allow one to determine if the real scale process would be economically feasible. Some studies about the application of TiO_2 photocatalysis in wastewater have been reported [13,14,28–31]. However, these research works do not focus on the simultaneous antibiotic removal by a TiO_2 photocatalysis pilot-scale plant applied to real wastewater.

The aim of this research work is to evaluate the presence of selected antibiotic compounds in the inlet and outlet of four WWTPs for the period of 2018–2019. Another objective is to treat selected antibiotics present in real wastewater in a photocatalytic plant by applying TiO_2 in suspension. Finally, the Ti concentration in the final effluent was controlled, to prevent Ti emission to the environment.

2. Materials and Methods

2.1. Site Description and Sample Collection

This research is focused on four WWTPs located in the POCTEFA territory (north of Spain and south of France). These WWTPs are designed to treat urban wastewater of domestic and industrial origins. Table 1 shows the main characteristics of selected WWTPs.

Samples were collected in four sampling campaigns for two years (in the spring and autumn of 2018 and 2019). The inlet and outlet of each WWTP were selected as sampling points to estimate the current removal performance of selected antibiotics in the four WWTPs, aiming to compare the different treatment lines.

Table 1. Main characteristics of each wastewater treatment plants (WWTP).

#WWTP	Population Equivalent	Total Inlet Flow (m³/day)	Water Treatment Line
1	695,232	129,600	Grit and grease separator/Activated Sludge/Decanter
2	82,500	22,150	Grit and grease separator/Decanter/Trickling filter (first stage) /Decanter/Trickling filter (second stage)/Decanter
3	10,470	10,995	Decanter/Trickling filter (first stage) /Decanter
4	51,336	7500	Grit and grease separator/Decanter/ Moving bed biofilm reactor/Decanter

The sampling was carried out following the EPA method 1694 for the analysis of pharmaceuticals and personal care products in water, soil, sediment, and biosolids by liquid chromatography tandem mass spectrometry (LC/MS/MS) [32]. Amber glass bottles were used to collect 1000 mL samples which were stored under refrigeration at 4 °C. The bottles were fully filled to avoid the presence of air and properly sealed by means of a PTFE seal. According to EPA 1694, the filtration of the samples is necessary in order to remove suspension solids. Two filtration steps were carried out prior to analysis using glass fiber filters (1.6 μm, supplied by GVS) for the first filtration stage and nylon filters (0.45 μm, supplied by GVS) for the second stage, as suggested in other research works [33–35].

Mass loadings of the antibiotics were calculated in each sampling period as the product of the individual concentration of each antibiotic in the samples and the daily flow rate of each WWTP.

Removal efficiencies of the target compounds were determined as the difference between the inlet mass loading and the outlet mass loading divided by the inlet mass loading and expressed as a percentage (Equation (1)).

$$Removal\,efficiency\,(\%): \frac{(m_{inf} - m_{eff})}{m_{inf}} \times 100 \tag{1}$$

2.2. Antibiotic Characterization

The quantification of the concentration of antibiotics was carried out via HPLC/MS/MS. Samples were centrifuged for 10 min at 13,000 rpm in Eppendorf tubes and then diluted 40-fold with 0.1% formic acid/MeOH/ACN (80%/10%/10%) before LC-MS/MS analysis. Chromatographic separations were carried out on an Ultra Performance Liquid Chromatography (UPLC) Ultimate 3000 RSLC system (Thermo Fisher Scientific). The column used was an Accucore C18 100 × 2.1 mm, 2.6 μm (Thermo Fisher Scientific). The mobile phases were A H$_2$O 0.4% formic acid + 5 mM ammonium formate and B MeOH/ACN 1:1 (v/v). A 20μL sample aliquot was injected. Detection was performed on a Q Exactive Plus (Thermo Fisher Scientific) mass spectrometer operated in the targeted single ion monitoring

(SIM) positive mode with a resolution of 70,000. External calibration was used for quantification and validated by standard additions for selected samples; the samples were prepared and analyzed in triplicate. The limits of detection and quantification of each antibiotic are featured in Table 2.

Table 2. Limits of detection and quantification of selected antibiotics.

Antibiotic	LOD (ng/L)	LOQ (ng/L)
Sulfadiazine	0.8	2.5
Trimethoprim	0.8	2.5
Amoxicillin	10	30
Enrofloxacin	1.2	3.7
Azithromycin	2.0	6.5

The four antibiotics investigated in this research work were selected according to their potential risk for the environment and reported occurrence [11–21]. All of them are representative human-use and veterinary-use antibiotics belonging to the main antibiotic groups. The standards were supplied by Sigma-Aldrich. Some characteristics of the selected antibiotic compounds are included in Table 3.

Table 3. Selected antibiotics and chemical information.

Antibiotic	Group	Chemical Abstracts Service Registry Number (CAS Nr.)	MW (g/mol)
Sulfadiazine	Sulfonamide	68-35-9	250
Trimethoprim	Trimethoprim	738-70-5	290
Amoxicillin	β-lactam	26787-78-0	365
Enrofloxacin	Fluoroquinolone	93106-60-6	359
Azithromycin	Macrolide	83905-01-5	749

2.3. Total Ti Assessment in the Effluent

The effluent from the pilot-scale plant might contain some Ti which would then be emitted to the environment. The Ti concentration was quantified by an Inductively Coupled Plasma Mass Spectrometry (ICP/MS) ELAN DRC-e, PerkinElmer, Toronto, Canada. A discrete volume sample (100 µL) was injected through a six-way valve, and the carrier was delivered directly to the nebulizer of the spectrometer. A glass concentric slurry nebulizer with a cyclonic spray chamber (Glass Expansion, Melbourne, Australia) was used. Default values were used for the rest of the instrumental parameters. The quantification of TiO_2 was based on monitoring the ICP-MS signal of the isotope ^{49}Ti, using ^{74}Ge as an internal standard. From an on-line calibration with an ionic titanium standard diluted in nitric acid (1%), intensity signals from the ICP-MS for samples were transformed into mass values by integrating the area of the transient signals obtained. All samples were injected in triplicate. The limit of detection of the method was established at 0.81 µg/L and the limit of quantification at 2.70 µg/L.

2.4. Photocatalytic Oxidation Experiment

The oxidation assays were carried out in the facility detailed in Figures 1 and 2. First, a 1 m³ storage tank provided with a stirrer was filled with the water sample. The solution was then pumped (8–16 L/min) to a 0.1 m³ mixer decanter where the catalyst was stored. The mixer decanter was provided with a stirrer to mix the influent with the catalyst. Subsequently, the mixture was placed in four identical slurry reactors. These reactors are made of aluminium because this material is known to have a high degree of light reflection. The reactors had a volume of 17 L and were provided with a UVA lamp (330–390 nm) of 40W. When the reactors were completely full, the UV/vis lamp in each reactor was turned on and stirring by means of compressed air took place. After treatment in the reactors, the treated water was pumped again to the decanter where a mechanical coagulation–flocculation–decantation treatment (CFD) was applied. Two steps take place in the separation process, coagulation 200 rpm during 5 min, flocculation 40 rpm during 25 min. and 90 min of decantation, resulting in a 120 min total process. Coagulant was not added to the mechanical CFD separation process.

The effluent (clarified phase) was generated and the catalyst remained in the decanter for the next cycle.

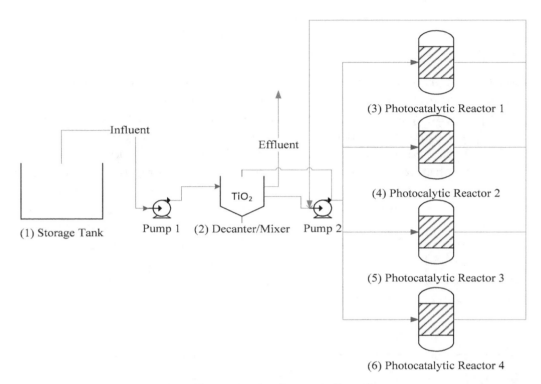

Figure 1. Pilot-scale plant process flow-diagram.

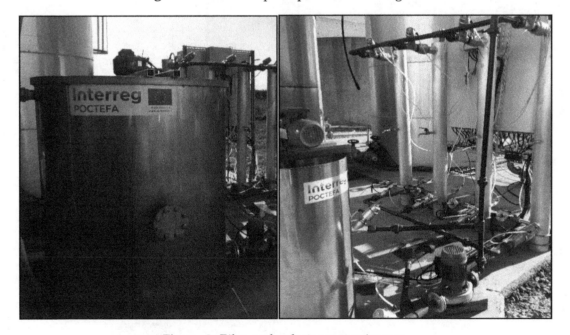

Figure 2. Pilot-scale plant process images.

The sample was prepared by the addition of individual concentrations of 1 mg/L of each antibiotic (amoxicillin, azithromycin, enrofloxacin, trimethoprim and sulfadiazine) simultaneously in the effluent from WWTP2. The physicochemical characteristics of this wastewater were pH = 7.6, DQO = 90 mg/L, Turbidity =11 NTU. The catalyst was applied in suspension in a concentration of 1 g/L of TiO_2 FN2 (supplied by Levenger S.L.). Radiation per unity of volume was 0.3 W/L in each reactor. The temperature ranged from 14 °C to 20 °C during the experiment. The experiments were conducted twice, and their average is represented in the results. The individual antibiotic concentration was quantified via HPLC/MS/MS during 240 min of photocatalytic treatment following the procedure described in Section 2.2.

3. Results

3.1. Occurrence of Target Antibiotics in Urban Wastewaters

Figure 3 shows the mass loading of the selected antibiotics at the four WWTPs during 2018 and 2019. It should be noted that atypical points are not represented. However, they are all available in Tables S1 and S2 of the Supplementary Material. The tables show that the highest mass loading corresponds to azithromycin. This human-use antibiotic presented an average load of around 925 mg/day per 1000 inhabitants and reached a maximum load of 11,332 mg/day per 1000 inhabitants in WWTP4. The azithromycin concentration increased over the four sampling campaigns in all the WWTPs. This fact might be attributed to a major increase in the use of azithromycin [36]. Enrofloxacin (fluoroquinolone group) also followed the same trend: higher loads were detected over the campaigns analyzed in this study. This veterinary-use antibiotic showed median mass loadings of around 200 mg/day per 1000 inhabitants and a maximum of 4329 mg/day per 1000 inhabitants. The enrofloxacin mass load was higher in the spring of both years, while it was rarely found in autumn, and always in lower loads, reflecting a seasonal use of this fluoroquinolone. The seasonal appearance of enrofloxacin has also been recently reported in another research work [37].

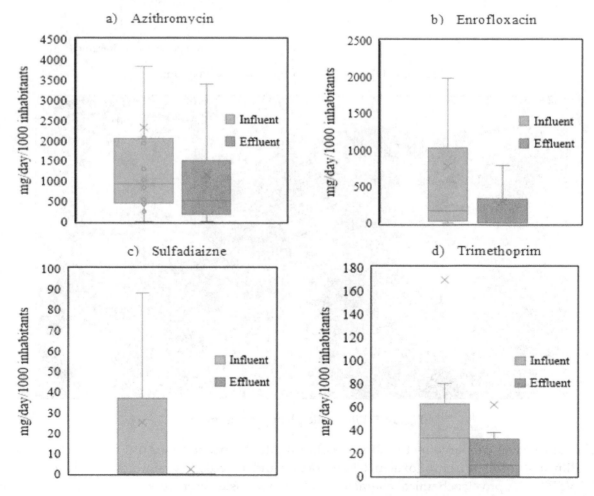

Figure 3. Boxplots of each antibiotic indicating total mass loading values: (**a**) azithromycin, (**b**) enrofloxacin, (**c**) sulfadiazine (**d**) trimethoprim.

By contrast, the average mass loadings of the other two antibiotics (trimethoprim and sulfadiazine) ranged from 35 mg/day per 1000 inhabitants for trimethoprim, to undetectable levels for sulfadiazine. Comparing the median mass loadings of trimethoprim and sulfadiazine with enrofloxacin and azithromycin reveals a difference greater than one order of magnitude.

Amoxicillin was not found in any sample (influent and effluent of the WWTPs). This fact could be attributed to the low stability of amoxicillin and the subsequent generation of degradation products such as amoxicillin penicilloic acid or amoxicillin-diketopiperazine-2', 5', as is suggested in the literature [38–40]. These degradation products were found in subsequent campaigns carried out in the same sampling points.

Finally, Tables S1 and S2 show that the reported load of each antibiotic varies significantly, by more than one order of magnitude, between the years and seasons. This trend could be related to the differences in rainfall patterns between the two years: high flows in particular were reported in spring 2018 and in autumn 2019 in both the Ebro River basin and the Cantabrico Occidental River basin. However, the total mass loading of the four antibiotics was relatively higher in the spring: $\sum_{\text{SPRING}}$ = 69.4 g/day per 1000 inhabitants versus $\sum_{\text{AUTUMN}}$ = 13.6 g/day per 1000 inhabitants.

3.2. Removal Efficiency of Selected WWTPs

The results of the removal efficiency of the antibiotics in each WWTP is shown in Figure 2. These results are also fully detailed in Table S3. Figure 4 shows that the removal efficiencies range from 2%–100%, demonstrating the fact that WWTPs can partially or almost totally remove the target antibiotics.

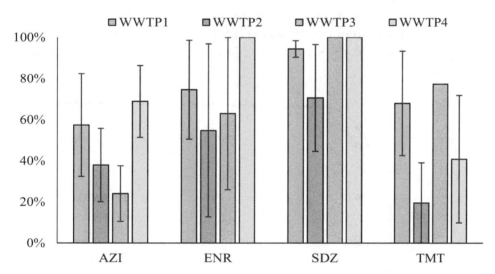

Figure 4. Wastewater treatment plants' removal efficiency for each antibiotic in 2018–2019 (AZI = azithromycin, ENR = enrofloxacin, SDZ = sulfadiazine, TMT = trimethoprim).

The results also suggest a significant variation for each antibiotic in each WWTP, indicating that the removal efficiency strongly depends not only on the specific matrix but also on the season and associated flow. This variation might also be attributed to the different physicochemical characteristics of the antibiotics, such as the degradation rates in the water, organic carbon–water partition coefficients or acid dissociation constants and water solubilities. More precisely, the results suggest that sulfadiazine is the antibiotic with the highest removal efficiency in every case. This trend may be due to the low mass loadings of this antibiotic, which make its removal easier [41]. Enrofloxacin also presents a high removal efficiency after the wastewater treatment, reaching 100% in several samples. By contrast, azithromycin and trimethoprim showed lower removal efficiencies in all the WWTPs. It should be noted that azithromycin had the highest mass loadings, more than one order of magnitude greater than the other antibiotics. It might be also attributed to the fact that azithromycin and trimethoprim have similar carbon–water partition coefficients [42]. Some studies suggest that biological processes, which are present in the four WWTPs, can remove fluoroquinolones effectively, while trimethoprim is more difficult to remove by means of biological treatments [43].

Comparing the different treatments of each WWTP, it can be observed that the facilities provided with a trickling filter (WWTP2 and WWTP3) showed higher removal efficiencies of the target antibiotics. This is consistent with some studies which demonstrate that trickling filters can remove antibiotics and other pharmaceuticals as well as personal care products [44–46]. WWTP4 showed the highest average antibiotic removal in most samples. However, it is important to note that this WWTP is in the Cantabrico River basin where there is higher rainfall than in the Ebro River Basin. As a result, the selected antibiotics had lower mass loadings in WWTP4.

3.3. Photocatalytic Oxidation of Antibiotics

Figure 5 shows the performance of the photocatalytic assays for the simultaneous oxidation of amoxicillin, enrofloxacin, sulfadiazine, trimethoprim and azithromycin during 4 h of treatment. The results show that in only 30 min of treatment, azithromycin and amoxicillin reached a degradation rate of 85% and 75%, respectively. Moreover, after 120 min of treatment, both antibiotics were completely removed from the wastewater. Amoxicillin was previously reported to be easily removed from waters by TiO_2 photocatalysis applied to the isolated compound at lab scale [47]. However, sulfadiazine shows a slower degradation rate, achieving degradation yields of 25% and 100% in 30 min and 240 min, respectively. In contrast, enrofloxacin and trimethoprim were not completely removed from the wastewater after the treatment. The degradation rate of enrofloxacin was relatively high at the beginning of the process (degradation yield of 50% in 30 min of treatment), but complete degradation was not achieved by the end of the treatment. A similar trend has been reported in other research work at lab scale [31]. Finally, trimethoprim showed the slowest initial degradation rate during the first 120 min, as reported in other studies at lab scale [45], while its degradation yield at the end of the treatment was close to 70%.

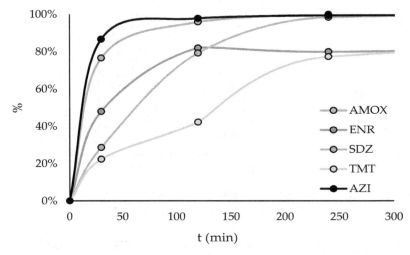

Figure 5. TiO_2 photocatalysis degradation yields of each antibiotic in WWTPE during 240 min of treatment (individual antibiotic initial concentration = 1 mg/L, TiO_2 initial concentration = 1 g/L, radiation per unity of volume = 0.3 W/L). AMX =amoxicillin, AZI = azithromycin, ENR = enrofloxacin, SDZ = sulfadiazine, TMT = trimethoprim.

3.4. Ti Assessment in the Effluent of the Photocatalytic Treatment Plant

In order to determine whether Ti is released into the environment, a quantification of the Ti concentration in the effluent was carried out applying a coagulation–flocculation–decantation treatment. TiO_2 has been demonstrated to be effective after several cycles of photocatalytic treatment for the degradation of pharmaceuticals [48], so it is important to recover it effectively. Fortunately, the results of the Ti assessment suggest that no more than 0.1% of the initial Ti concentration remained in the effluent. This reflects the fact that recovery by means of mechanical coagulation–flocculation–decantation treatment is quite efficient for recovering TiO_2 when used in suspension [48].

4. Conclusions

This research work evaluates the behavior of four antibiotic compounds in four different WWTPs located in the north of Spain. The mass loadings of amoxicillin, enrofloxacin, sulfadiazine, trimethoprim and azithromycin were analyzed in the influent and effluent of the WWTPs. The performance of a TiO_2 photocatalytic treatment plant applied to the simultaneous removal of the antibiotics from real urban treated water was evaluated. This showed that the technology can be used to totally remove some of the selected antibiotics at slightly higher concentrations than those commonly found in wastewaters. The conclusions can be summarized as follows:

1. The mass loadings of the antibiotics ranged from 11,332 mg/day·1000 inhabitants to undetectable levels. Azithromycin had the highest mass loadings, followed by enrofloxacin, trimethoprim, sulfadiazine and amoxicillin.

2. The use of enrofloxacin and azithromycin increased in the locations of the WWTPs during the period of this study.

3. Sulfadiazine, amoxicillin and azithromycin were totally removed from wastewaters in the TiO_2 photocatalytic pilot-scale plant, while 80% removal of trimethoprim and enrofloxacin was achieved by the treatment. Moreover, the facility was able to recover the catalyst after the treatment, minimizing the Ti released into the environment and allowing catalyst reuse.

4. Although WWTPs are not designed to remove antibiotics, they do reduce them. This research shows that biological treatments have a significant influence on antibiotic removal. In particular, the presence of a trickling filter in the water treatment line of the WWTPs has been demonstrated to lead to a higher degree of antibiotic removal. However, the efficiency of the antibiotic removal depends on the physicochemical properties of the antibiotics and on the characteristics of the wastewater.

Author Contributions: Planning photocatalysis experimentation, R.M. and S.M.; general experimental work coordinator, M.P.O.; LC/MS/MS analysis, S.G. (technician) and J.S. (responsible); design and construction of pilot-scale plant, J.G. and S.M.; responsible of nanoparticles detection in the effluent, J.R.C.; experimental photocatalysis technician and principal author, S.M. All authors have read and agreed to the published version of the manuscript.

Acknowledgments: This work was financed by DGA_FSE Research Team "Water and Environmental Health" Ref: B43-20R in the framework of the project EFA 183/16/OUTBIOTICS, Program Interreg-POCTEFA 2014-2020, funded by FEDER.

References

1. Klein, E.Y.; Van Boeckel, T.P.; Martinez, E.M.; Pant, S.; Gandra, S.; evin, S.A.; Goossens, H.; Laxminarayan, R. Global increase and geographic convergence in antibiotic consumption between 2000 and 2015. *Proc. Natl. Acad. Sci. USA* **2018**, *115*, E3463–E3470. [CrossRef] [PubMed]

2. Santos, L.H.; Gros, M.; Rodriguez-Mozaz, S.; Delerue-Matos, C.; Pena, A.; Barceló, D.; Montenegro, M.C. Contribution of hospital effluents to the load of pharmaceuticals in urban wastewaters: Identification of ecologically relevant pharmaceuticals. *Sci. Total Environ.* **2013**, *461–462*, 302–316. [CrossRef] [PubMed]

3. Mceneff, G.; Barron, L.; Kelleher, B.; Paull, B.; Quinn, B. A year-long study of the spatial occurrence and relative distribution of pharmaceutical residues in sewage effluent, receiving marine waters and marine bivalves. *Sci. Total Environ.* **2014**, *476–477*, 317–326. [CrossRef] [PubMed]

4. Kuehn, B.M. Antibiotic-Resistant "Superbugs" May Be Transmitted From Animals to Humans. *Med. News Perspect Futur.* **2007**, *298*, 2125–2126. [CrossRef] [PubMed]

5.   Watkinson, A.J.; Murby, E.J.; Costanzo, S.D. Removal of antibiotics in conventional and advanced wastewater treatment: Implications for environmental discharge and wastewater recycling. *Water Res.* **2007**, *41*, 4164–4176. [CrossRef] [PubMed]

6.   EMA. Sales of Veterinary Antimicrobial Agents in 31 European Countries in 2017 Trends from 2010 to 2017. Ema/294674/2019. 2019. Available online: https://www.ema.europa.eu/en/documents/report/sales-veterinary-antimicrobial-agents-31-european-countries-2017_en.pdf (accessed on 18 February 2020).

7.   García-Galán, M.J.; Garrido, T.; Fraile, J.; Ginebreda, A.; Díaz-Cruz, M.S.; Barceló, D. Simultaneous occurrence of nitrates and sulfonamide antibiotics in two ground water bodies of Catalonia (Spain). *J. Hydrol.* **2010**, *383*, 93–101. [CrossRef]

8.   Jurado, A.; Walther, M.; Díaz-Cruz, M.S. Occurrence, fate and environmental risk assessment of the organic microcontaminants included in the Watch Lists set by EU Decisions 2015/495 and 2018/840 in the groundwater of Spain. *Sci. Total Environ.* **2019**, *663*, 285–296. [CrossRef]

9.   Boy-Roura, M.; Mas-Pla, J.; Petrovic, M.; Gros, M.; Soler, D.; Brusi, D.; Menció, A. Towards the understanding of antibiotic occurrence and transport in groundwater: Findings from the Baix Fluvià alluvial aquifer (NE Catalonia, Spain). *Sci. Total Environ.* **2018**, *612*, 1387–1406. [CrossRef]

10. García-Gil, A.; Garrido Schneider, E.; Mejías, M.; Barceló, D.; Vázquez-Suñé, E.; Díaz-Cruz, S. Occurrence of pharmaceuticals and personal care products in the urban aquifer of Zaragoza (Spain) and its relationship with intensive shallow geothermal energy exploitation. *J. Hydrol.* **2018**, *566*, 629–642. [CrossRef]

11. Senta, I.; Terzic, S.; Ahel, M. Occurrence and fate of dissolved and particulate antimicrobials in municipal wastewater treatment. *Water Res.* **2013**, *47*, 705–714. [CrossRef]

12. Babić, S.; Ašperger, D.; Mutavdžić, D.; Horvat, A.J.M.; Kaštelan-Macan, M. Solid phase extraction and HPLC determination of veterinary pharmaceuticals in wastewater. *Talanta* **2006**, *70*, 732–738. [CrossRef] [PubMed]

13. Golovko, O.; Kumar, V.; Fedorova, G.; Randak, T.; Grabic, R. Seasonal changes in antibiotics, antidepressants/psychiatric drugs, antihistamines and lipid regulators in a wastewater treatment plant. *Chemosphere* **2014**, *111*, 418–426. [CrossRef] [PubMed]

14. Al Aukidy, M.; Verlicchi, P.; Jelic, A.; Petrovic, M.; Barcelò, D. Monitoring release of pharmaceutical compounds: Occurrence and environmental risk assessment of two WWTP effluents and their receiving bodies in the Po Valley, Italy. *Sci. Total Environ.* **2012**, *438*, 15–25. [CrossRef] [PubMed]

15. Tuc Dinh, Q.; Alliot, F.; Moreau-Guigon, E.; Eurin, J.; Chevreuil, M.; Labadie, P. Measurement of trace levels of antibiotics in river water using on-line enrichment and triple-quadrupole LC-MS/MS. *Talanta* **2011**, *85*, 1238–1245. [CrossRef]

16. Rossmann, J.; Schubert, S.; Gurke, R.; Oertel, R.; Kirch, W. Simultaneous determination of most prescribed antibiotics in multiple urban wastewater by SPE-LC-MS/MS. *J. Chromatogr. B Anal. Technol. Biomed. Life Sci.* **2014**, *969*, 162–170. [CrossRef]

17. Tamtam, F.; Mercier, F.; Le Bot, B.; Eurin, J.; Dinh, Q.T.; Clément, M.; Chevreuil, M. Occurrence and fate of antibiotics in the Seine River in various hydrological conditions. *Sci. Total Environ.* **2008**, *393*, 84–95. [CrossRef]

18. Wagil, M.; Kumirska, J.; Stolte, S.; Puckowski, A.; Maszkowska, J.; Stepnowski, P.; Białk-Bielińska, A. Development of sensitive and reliable LC-MS/MS methods for the determination of three fluoroquinolones in water and fish tissue samples and preliminary environmental risk assessment of their presence in two rivers in northern Poland. *Sci. Total Environ.* **2014**, *493*, 1006–1013. [CrossRef]

19. Dan, A.; Zhang, X.; Dai, Y.; Chen, C.; Yang, Y. Occurrence and removal of quinolone, tetracycline, and macrolide antibiotics from urban wastewater in constructed wetlands. *J. Clean Prod.* **2020**, *252*, 119677. [CrossRef]

20. Samir, A.; Abdel-Moein, K.A.; Zaher, H.M. Emergence of penicillin-macrolide-resistant Streptococcus pyogenes among pet animals: An ongoing public health threat. *Comp. Immunol. Microbiol. Infect. Dis.* **2020**, *68*, 101390. [CrossRef]

21. Milaković, M.; Vestergaard, G.; González-Plaza, J.J.; Petrić, I.; Šimatović, A.; Senta, I.; Kublik, S.; Schloter, M.; Smalla, K.; Udiković-Kolić, N. Pollution from azithromycin-manufacturing promotes macrolide-resistance gene propagation and induces spatial and seasonal bacterial community shifts in receiving river sediments. *Environ. Int.* **2019**, *123*, 501–511. [CrossRef]

22. Yao, N.; Li, C.; Yu, J.; Xu, Q.; Wei, S.; Tian, Z.; Yang, Z.; Yang, W.; Shen, J. Insight into adsorption of combined

antibiotic-heavy metal contaminants on graphene oxide in water. *Sep. Purif. Technol.* **2019**, *236*, 116278. [CrossRef]

23. Pan, S.F.; Zhu, M.P.; Chen, J.P.; Yuan, Z.H.; Zhong, L.B.; Zheng, Y.M. Separation of tetracycline from wastewater using forward osmosis process with thin film composite membrane—Implications for antibiotics recovery. *Sep. Purif. Technol.* **2015**, *153*, 76–83. [CrossRef]

24. Moles, S.; Valero, P.; Escuadra, S.; Mosteo, R.; Gómez, J.; Ormad, M.P. Performance comparison of commercial TiO2: Separation and reuse for bacterial photo-inactivation and emerging pollutants photo-degradation. *Environ. Sci. Pollut. Res.* **2020**, 1–15. [CrossRef] [PubMed]

25. van Grieken, R.; Marugán, J.; Pablos, C.; Furones, L.; López, A. Comparison between the photocatalytic inactivation of Gram-positive E. faecalis and Gram-negative E. coli faecal contamination indicator microorganisms. *Appl. Catal. B Environ.* **2010**, *100*, 212–220. [CrossRef]

26. Gumy, D.; Rincon, A.G.; Hajdu, R.; Pulgarin, C. Solar photocatalysis for detoxification and disinfection of water: Different types of suspended and fixed TiO2 catalysts study. *Sol. Energy* **2006**, *80*, 1376–1381. [CrossRef]

27. Malato, S.; Fernández-Ibáñez, P.; Maldonado, M.I.; Blanco, J.; Gernjak, W. Decontamination and disinfection of water by solar photocatalysis: Recent overview and trends. *Catal. Today* **2009**, *147*, 1–59. [CrossRef]

28. Bernabeu, A.; Vercher, R.F.; Santos-Juanes, L.; Simón, P.J.; Lardín, C.; Martínez, M.A.; Vicente, J.A.; González, R.; Llosá, C.; Arques, A.; et al. Solar photocatalysis as a tertiary treatment to remove emerging pollutants from wastewater treatment plant effluents. *Catal. Today* **2011**, *161*, 235–240. [CrossRef]

29. Malesic-Eleftheriadou, N.; Evgenidou, E.; Kyzas, G.Z.; Bikiaris, D.N.; Lambropoulou, D.A. Removal of antibiotics in aqueous media by using new synthesized bio-based poly(ethylene terephthalate)-TiO2 photocatalysts. *Chemosphere* **2019**, *234*, 746–755. [CrossRef]

30. Biancullo, F.; Moreira, N.F.; Ribeiro, A.R.; Manaia, C.M.; Faria, J.L.; Nunes, O.C.; Castro-Silva, S.M.; Silva, A.M. Heterogeneous photocatalysis using UVA-LEDs for the removal of antibiotics and antibiotic resistant bacteria from urban wastewater treatment plant effluents. *Chem. Eng. J.* **2019**, *36*, 304–313. [CrossRef]

31. Cai, Q.; Hu, J. Decomposition of sulfamethoxazole and trimethoprim by continuous UVA/LED/TiO2 photocatalysis: Decomposition pathways, residual antibacterial activity and toxicity. *J. Hazard. Mater.* **2017**, *323*, 527–536. [CrossRef]

32. Englert, B. *Method 1694: Pharmaceuticals and Personal Care Products in Water, Soil, Sediment, and Biosolids by HPLC/MS/MS*; US Environmental Protection Agency (EPA): Washington, DC, USA, 2007.

33. Mirzaei, R.; Yunesian, M.; Nasseri, S.; Gholami, M.; Jalilzadeh, E.; Shoeibi, S.; Bidshahi, H.S.; Mesdaghinia, A. An optimized SPE-LC-MS/MS method for antibiotics residue analysis in ground, surface and treated water samples by response surface methodology- central composite design. *J. Environ. Health Sci. Eng.* **2017**, *15*, 1–16. [CrossRef] [PubMed]

34. Díaz-Bao, M.; Barreiro, R.; Miranda, J.M.; Cepeda, A.; Regal, P. Fast HPLC-MS/MS method for determining penicillin antibiotics in infant formulas using molecularly imprinted solid-phase extraction. *J. Anal. Methods Chem.* **2015**, *2015*, 959675. [CrossRef] [PubMed]

35. Gros, M.; Petrovié, M.; Barceló, D. Multi-residue analytical methods using LC-tandem MS for the determination of pharmaceuticals in environmental and wastewater samples: A review. *Anal. Bioanal. Chem.* **2006**, *386*, 941–952. [CrossRef]

36. European Centre for Disease Prevention and Control. *Antimicrobial Consumption. ECDC. Annual Epidemiological Report for 2017*; ECDC: Stockholm, Sweden, 2018.

37. Lei, K.; Zhu, Y.; Chen, W.; Pan, H.Y.; Cao, Y.X.; Zhang, X.; Guo, B.B.; Sweetman, A.; Lin, C.Y.; Ouyang, W.; et al. Spatial and seasonal variations of antibiotics in river waters in the Haihe River Catchment in China and ecotoxicological risk assessment. *Environ. Int.* **2019**, *130*, 104919. [CrossRef]

38. Lamm, A.; Gozlan, I.; Rotstein, A.; Avisar, D. Detection of amoxicillin-diketopiperazine-2′, 5′ in wastewater samples. *J. Environ. Sci. Health Part A Toxic/Hazardous Subst Environ. Eng.* **2009**, *44*, 1512–1517. [CrossRef]

39. Gozlan, I.; Rotstein, A.; Avisar, D. Amoxicillin-degradation products formed under controlled environmental conditions: Identification and determination in the aquatic environment. *Chemosphere* **2013**, *91*, 985–992. [CrossRef]

40. Li, L.; Guo, C.; Ai, L.; Dou, C.; Wang, G.; Sun, H. Research on degradation of penicillins in milk by β-lactamase using ultra-performance liquid chromatography coupled with time-of-flight mass spectrometry. *J. Dairy Sci.* **2014**, *97*, 4052–4061. [CrossRef]

41. Xia, S.; Jia, R.; Feng, F.; Xie, K.; Li, H.; Jing, D.; Xu, X. Effect of solids retention time on antibiotics removal performance and microbial communities in an A/O-MBR process. *Bioresour. Technol.* **2012**, *106*, 36–43. [CrossRef]

42. Abegglen, C.; Joss, A.; McArdell, C.S.; Fink, G.; Schlüsener, M.P.; Ternes, T.A.; Siegrist, H. The fate of selected micropollutants in a single-house MBR. *Water Res.* **2009**, *43*, 2036–2046. [CrossRef]

43. Tran, N.H.; Chen, H.; Reinhard, M.; Mao, F.; Gin, K.Y.H. Occurrence and removal of multiple classes of antibiotics and antimicrobial agents in biological wastewater treatment processes. *Water Res.* **2016**, *104*, 461–472. [CrossRef]

44. Thompson, A.; Griffin, P.; Stuetz, R.; Cartmell, E. The Fate and Removal of Triclosan during Wastewater Treatment. *Water Environ Res.* **2005**, *77*, 63–67. [CrossRef] [PubMed]

45. Drewes, J.E.; Heberer, T.; Reddersen, K. Fate of pharmaceuticals during indirect potable reuse. *Water Sci. Technol.* **2002**, *46*, 73–80. [CrossRef] [PubMed]

46. Kasprzyk-Hordern, B.; Dinsdale, R.M.; Guwy, A.J. The removal of pharmaceuticals, personal care products, endocrine disruptors and illicit drugs during wastewater treatment and its impact on the quality of receiving waters. *Water Res.* **2009**, *43*, 363–380. [CrossRef] [PubMed]

47. Awfa, D.; Ateia, M.; Fujii, M.; Johnson, M.S.; Yoshimura, C. Photodegradation of pharmaceuticals and personal care products in water treatment using carbonaceous-TiO2 composites: A critical review of recent literature. *Water Res.* **2018**, *142*, 26–45. [CrossRef] [PubMed]

48. Conde-Cid, M.; Álvarez-Esmorís, C.; Paradelo-Núñez, R.; Nóvoa-Muñoz, J.C.; Arias-Estévez, M.; Álvarez-Rodríguez, E.; Fernández-Sanjurjo, M.J.; Núñez-Delgado, A. Occurrence of tetracyclines and sulfonamides in manures, agricultural soils and crops from different areas in Galicia (NW Spain). *J. Clean Prod.* **2018**, *197*, 491–500. [CrossRef]

8

Degradation of Ketamine and Methamphetamine by the UV/H$_2$O$_2$ System: Kinetics, Mechanisms and Comparison

De-Ming Gu [1,2], Chang-Sheng Guo [1] 🅾, Qi-Yan Feng [2], Heng Zhang [1] and Jian Xu [1,*]

[1]   Center for Environmental Health Risk Assessment and Research, Chinese Research Academy of Environmental Sciences, Beijing 100012, China; goodmingaust@163.com (D.-M.G.); guocs@craes.org.cn (C.-S.G.); zhangheng_craes@163.com (H.Z.)

[2]   School of Environment Science and Spatial Informatics, China University of Mining and Technology, Xuzhou 221116, China; fqycumt@126.com

*   Correspondence: xujian@craes.org.cn

Abstract: The illegal use and low biodegradability of psychoactive substances has led to their introduction to the natural water environment, causing potential harm to ecosystems and human health. This paper compared the reaction kinetics and degradation mechanisms of ketamine (KET) and methamphetamine (METH) by UV/H$_2$O$_2$. Results indicated that the degradation of KET and METH using UV or H$_2$O$_2$ alone was negligible. UV/H$_2$O$_2$ had a strong synergizing effect, which could effectively remove 99% of KET and METH (100 µg/L) within 120 and 60 min, respectively. Their degradation was fully consistent with pseudo-first-order reaction kinetics ($R^2 > 0.99$). Based on competition kinetics, the rate constants of the hydroxyl radical with KET and METH were calculated to be 4.43×10^9 and 7.91×10^9 M$^{-1} \cdot$s^{-1}, respectively. The apparent rate constants of KET and METH increased respectively from 0.001 to 0.027 and 0.049 min^{-1} with the initial H$_2$O$_2$ dosage ranging from 0 to 1000 µM at pH 7. Their degradation was significantly inhibited by HCO$_3^-$, Cl$^-$, NO$_3^-$ and humic acid, with Cl$^-$ having relatively little effect on the degradation of KET. Ultraperformance liquid chromatography with tandem mass spectrometry was used to identify the reaction intermediates, based on which the possible degradation pathways were proposed. These promising results clearly demonstrated the potential of the UV/H$_2$O$_2$ process for the effective removal of KET and METH from contaminated wastewater.

Keywords: ketamine; methamphetamine; UV/H$_2$O$_2$; degradation kinetics; reaction intermediates

1. Introduction

Illicit drugs are nonprescribed or psychostimulant substances which cannot be completely removed by conventional wastewater treatment, resulting in their widespread occurrence in aquatic environments [1,2]. Ketamine (KET) and methamphetamine (METH) were detected most frequently, with concentration levels up to 275 ng/L for KET and 239 ng/L for METH, in surface waters in China [3]. METH removal at most wastewater treatment plants was more than 80%, while the elimination of KET was less than 50% or even negative [4]. It was confirmed that chronic environmental concentrations of METH can lead to health issues in aquatic organisms [5]. Liao et al. [6] also reported that blood circulation and incubation time in medaka fish embryos could be significantly delayed at environmental concentration levels (0.004–40 µM) of KET and METH, which altered the swimming behavior of medaka fish larvae. Thus, there is an urgent need to explore new, efficient methods for eliminating these emerging contaminants in water.

Advanced oxidation processes (AOPs) have been employed to destroy illicit drugs due to their high efficiency and lower environmental impact [7,8]. The UV/H_2O_2 process is one of the AOPs and generates the strong, oxidizing hydroxyl radical ($^{\bullet}$OH, E_0 = 2.72 V), which attacks the organic compounds with rate constants ranging from 10^8 to 10^{10} M^{-1} s^{-1} [9]. Benzoylecgonine (BE), a metabolite of cocaine, was effectively removed by UV/H_2O_2 from different matrices [10]. The degradation of KET and METH was investigated using various AOPs, but no available report, so far, has addressed $^{\bullet}$OH assisted by UV/H_2O_2 treatment. After 3 min, 100 µg/L of METH that had been added to deionized water was completely eliminated by TiO_2 photocatalysis under UV_{365nm} irradiation [11]. Wei et al. [12] studied the synthesis of a novel sonocatalyst Er^{3+}:$YAlO_3$/Nb_2O_5 and its application for METH degradation. Gu et al. [13] observed that complete removal of KET was achieved by UV/persulfate, and possible transformation pathways were proposed.

To the best of our knowledge, there is little information about the theoretical calculation of the reactivity of KET and METH by radical attack using the UV/H_2O_2 process. Water constituents in actual wastewater could affect the degradation efficacy; therefore, a comprehensive understanding of the degradation of KET and METH using the UV/H_2O_2 system is needed. The aim of this study was to investigate the degradation kinetics and mechanisms of KET and METH during the UV/H_2O_2 process. The influence of various parameters on KET and METH removal was evaluated, including initial H_2O_2 dosage, pH and water background components. The degradation products were analyzed by ultraperformance liquid chromatography with tandem mass spectrometry (UPLC-MS/MS), and possible transformation paths were proposed.

2. Materials and Methods

2.1. Materials

The KET and METH were obtained from Cerilliant Corporation (Round Rock, TX); detailed information is listed in Table 1. HPLC grade acetonitrile (ACN) and methanol (MeOH) were purchased from Fisher Scientific (Poole, UK). Formic acid (FA, ≥98%) and benzoic acid (BA) were purchased from Sigma-Aldrich (Bellefonte, USA). Analytical grade H_2O_2 (30%, v/v), $NaHCO_3$ (≥99.7%), NaCl (≥99.0%), $NaNO_3$ (≥99.5%), NaOH (≥99.5%), humic acid (HA) and H_2SO_4 (≥98%) were obtained from Sinopharm Chemical Reagent Co., Ltd. (Beijing, China). All reaction solutions were configured with Milli-Q water produced by an ultrapure water system (Millipore, MA, USA).

Table 1. Chemical structures and properties of ketamine and methamphetamine.

Compound	Chemical Formula	Structure	CAS Number	pKa	Log K_{ow}
Ketamine	$C_{13}H_{16}ClNO$		6740-88-1	7.5	2.18
Methamphetamine	$C_{10}H_{15}N$		4846-07-5	9.9	2.07

2.2. Experimental Section

The experiments were operated in the quartz tubes (25 mm in diameter and 175 mm in length), which were placed in a photochemical reactor (Figure 1, XPA-7, Xujiang Machinery Factory, Nanjing, China). A low-pressure mercury lamp (11 W, emission at 254 nm, Philips Co., Zhuhai, China) was placed in the quartz sleeve. The UV lamp was preheated for 30 min to ensure irradiation stability. The UV fluence rate of 0.1 mW cm^{-2} was determined using three different methods [14]. The newly configured KET/METH and H_2O_2 stock solutions were supplemented with appropriate volumes to achieve a

50 mL reaction solution, which was then stirred thoroughly at 300 rpm with electromagnetic stirrers. Upon UV irradiation, the reaction started at pH 7.0 and room temperature. Specific samples were immediately quenched using a catalase and passed through 0.22 µm nylon filter before further analysis.

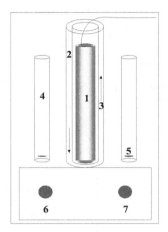

Figure 1. The schematic diagram of the experiment setup: (1) low-pressure Hg UV lamp, (2) quartz tube, (3) cooling water, (4) photoreactor, (5) magnetic stirrer, (6) magnetic stirrer apparatus, (7) thermostat.

2.3. Analytical Methods

The concentrations of KET and METH were quantified by UPLC-MS/MS equipped with a Waters Acquity liquid chromatography system and an Xevo T_QS triple quadrupole mass spectrometer (Waters Co., Milford, MA, USA). The analytes were separated by a reverse phase column (Acquity UPLC BEH C18, 1.7 µm, 50 × 2.1 mm, Waters, MA, USA). The mobile phases A and B, with a flow rate of 450 µL min^{-1}, were 0.1% FA in Milli-Q water and ACN, respectively. Ten percent of phase B was kept for 0.5 min at the initial proportion, linearly increased to 45% at 1.8 min, then increased to 95% within 0.1 min, held for 1.0 min, reverted to 10% at 3.0 min and held for 1.5 min. The injection volume was 5 µL with the column temperature at 40 °C. The chromatograms were recorded in the positive ion multiple reaction monitoring (MRM) mode. Nitrogen was used as the desolvation and nebulizing gas. The capillary voltage was set at 0.5 kV, and the desolvation temperature was 400 °C. Optimized UPLC-MS/MS parameters are given in Table 2.

Table 2. Detailed ultraperformance liquid chromatography with tandem mass spectrometry (UPLC-MS/MS) parameters for ketamine and methamphetamine.

Compound	Parent Ion (m/z)	Retention Time (min)	Production (m/z)	Cone Voltage (V)	Collision Voltage (V)
Ketamine	238	1.31	125	16	24
			179	16	16
Methamphetamine	150	1.11	91	22	16
			119	22	10

3. Results and Discussion

3.1. Degradation Kinetics of KET and METH

Figure 2 shows the degradation of KET and METH under different treatment processes. UV or H_2O_2 alone exhibited negligible effects on their degradation, suggesting that treatment by UV or H_2O_2 alone was unable to destroy KET and METH. However, nearly complete removal of KET and METH was achieved within 120 and 60 min, respectively, when treated with the combination of UV/H_2O_2. Similar results were reported regarding ofloxacin degradation, which was drastically increased due to the large amount of hydroxyl radicals ($^\bullet$OH) generated via the breakage of the H_2O_2

bond (Equation (1)) [15]. The degradation of KET and METH was consistent with the pseudo-first-order reaction kinetics. The apparent degradation rate constants (k_{obs}) of KET and METH by UV/H$_2$O$_2$ were 0.027 and 0.049 min^{-1}, respectively.

$$H - O - O - H + hv \rightarrow 2^{\bullet}OH \tag{1}$$

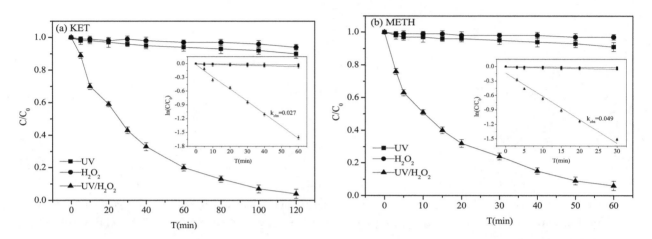

Figure 2. Degradation kinetics of ketamine (KET) (**a**) and methamphetamine (METH) (**b**) by different treatments. Conditions: Initial concentrations of KET and METH = 100 µg/L, Initial concentration of hydrogen peroxide (H$_2$O$_2$)$_0$ = 500 µM, pH$_0$ = 7.0, Temperature (T) = 25 ± 1 °C.

3.2. Determination of Bimolecular Reaction Rate

The generation of $^{\bullet}$OH in the UV/H$_2$O$_2$ system was proved by the photoluminescence (PL) technique using a probe molecule with terephthalic acid, which tends to react with $^{\bullet}$OH to form 2-hydroxyterephthalic acid, a highly fluorescent product [16]. The PL intensity of 2-hydroxyterephthalic acid is proportional to the amount of $^{\bullet}$OH radicals produced in water [17]. Figure 3 shows the PL spectral changes in the 5×10^{-4} M terephthalic acid solution with a concentration of 2×10^{-3} M NaOH (excitation at 315 nm), as described by Yu et al. [17]. Similar fluorescence intensity was found in the reaction systems with initial concentrations of 100 and 1000 µM of H$_2$O$_2$, suggesting a constant concentration of $^{\bullet}$OH with the initial H$_2$O$_2$ dosage ranging from 100 to 1000 µM. The PL signal at 425 nm increased with the irradiation time, which was attributed to the reaction of terephthalic acid with $^{\bullet}$OH generated in the UV/H$_2$O$_2$ system.

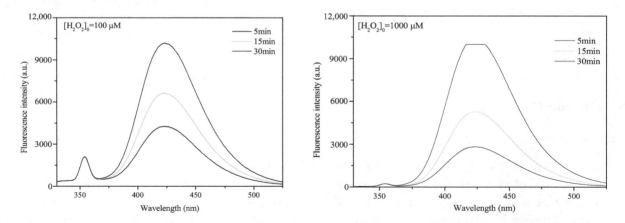

Figure 3. Photoluminescence (PL) spectral changes observed in the UV/H$_2$O$_2$ system in a 5×10^{-4} M basic solution of terephthalic acid (excitation at 315 nm).

The bimolecular reaction rates of KET and METH reacting with •OH were determined through the competition experiments at pH 7 (phosphate buffer solution, 5 mM). BA was used as the reference compound, with which the constant reaction rate of •OH is known to be 5.9×10^9 M^{-1} s^{-1} [18]. It is important to note that the degradation of KET, METH and BA using UV alone was negligible at less than 9%. Equations (2) and (3) describe the competing kinetics of KET and METH with •OH in the UV/H_2O_2 oxidation process, through which the bimolecular reaction rates of KET and METH reacting with •OH were 4.43×10^9 and 7.91×10^9 M^{-1} s^{-1}, respectively (Figure 4).

$$\ln \frac{(KET)_0}{(KET)_t} = \frac{k_{•OH-KET}}{k_{•OH-BA}} \ln \frac{(BA)_0}{(BA)_t} \tag{2}$$

$$\ln \frac{(METH)_0}{(METH)_t} = \frac{k_{•OH-METH}}{k_{•OH-BA}} \ln \frac{(BA)_0}{(BA)_t} \tag{3}$$

where $(KET)_0$, $(METH)_0$ and $(BA)_0$ are the initial concentrations (µmol/L) of target compounds. $(KET)_t$, $(METH)_t$ and $(BA)_t$ are the concentrations (µmol/L) at time t (min). $k_{•OH-KET}$, $k_{•OH-METH}$ and $k_{•OH-BA}$ are the bimolecular reaction rates of KET, METH and BA reacting with •OH, respectively.

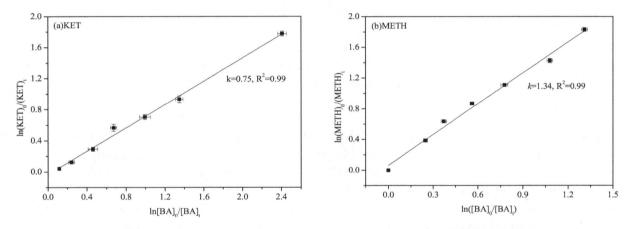

Figure 4. (a) The reaction rate constant of KET with •OH. Conditions: $(KET)_0 = (BA)_0 = 0.42$ µM, $(H_2O_2)_0 = 1$ mM, pH = 7, T = 25 ± 1 °C. (b) The reaction rate constant of METH with •OH. Conditions: $(METH)_0 = (BA)_0 = 0.67$ µM, $(H_2O_2)_0 = 1$ mM, pH = 7, temperature = 25 ± 1 °C.

3.3. Effect of H_2O_2 Dosage

The KET and METH degradation under different initial H_2O_2 dosages were consistent with the pseudo-first-order reaction model ($R^2 > 0.99$, Figure 5). The k_{obs} of KET and METH increased dramatically from 0.001 min^{-1} to 0.027 and 0.049 min^{-1} with the initial H_2O_2 dosage ranging from 0 to 1000 µM. The reason for this phenomenon is that the production of •OH increased with the initial H_2O_2 dosage ranging from 0 to 1000 µM, thus accelerating the degradation rate of target compounds [19]. However, the k_{obs} of METH decreased slightly with the initial concentration of H_2O_2 increased to 2000 µM. A similar phenomenon was observed in a previous report that indicated that the degradation rates of cyclophosphamide and 5-fluorouracil were proportional to the H_2O_2 dosage and slightly decreased with excess H_2O_2 [20]. An excessive amount of H_2O_2 would cause the self-scavenging effect of •OH to form HO_2• and O_2^-• (Equations (4) and (5)) [21], the low reactivity of which could reduce the degradation rate. Similar results were obtained concerning the degradation of ofloxacin [15] and chloramphenicol [22]. Moreover, large amounts of •OH were dimerized to H_2O_2, and the generated HO_2• and O_2^-• subsequently participated in other reactions (Equations (6)–(9)) [23]. This negative effect was not observed in this study, probably because the maximum H_2O_2 dosage (2000 µM) was not high enough to inhibit the KET degradation.

$$H_2O_2 + {}^•OH \rightarrow HO_2^• + H_2O \tag{4}$$

$$H_2O_2 + {}^{\bullet}OH \rightarrow O_2^{-\bullet} + H^+ + H_2O \qquad (5)$$

$${}^{\bullet}OH + {}^{\bullet}OH \rightarrow H_2O_2 \qquad (6)$$

$$HO_2^{\bullet} + H_2O_2 \rightarrow {}^{\bullet}OH + H_2O + O_2 \qquad (7)$$

$$HO_2^{\bullet} + {}^{\bullet}OH \rightarrow H_2O + O_2 \qquad (8)$$

$$O_2^{-\bullet} + H_2O_2 \rightarrow {}^{\bullet}OH + OH^- + O_2 \qquad (9)$$

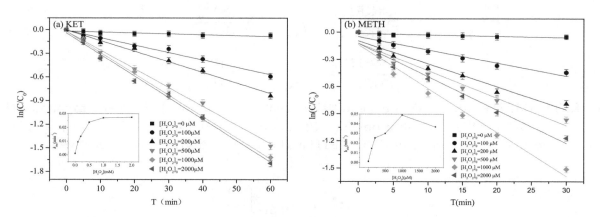

Figure 5. Effect of H_2O_2 dosage on KET (**a**) and METH (**b**) degradation in the UV/H_2O_2 system. Conditions: $(KET)_0 = (METH)_0 = 100$ µg/L, $(H_2O_2)_0 = 0$–2000 µM, $pH_0 = 7.0$, $T = 25 \pm 1$ °C.

3.4. Effect of Initial pH

Figure 6 illustrates the KET and METH destruction at different initial pHs, which were adjusted with an H_2SO_4 or NaOH solution (0.1 M). No buffer was used due to its inhibiting effect on the decomposition of organics [24]. The KET and METH degradation at different initial pHs followed the pseudo-first-order reaction model well. The k_{obs} of KET and METH reached the highest levels in a neutral environment at 0.027 and 0.085 min^{-1}, respectively. Due to the greater stability of H_2O_2 at pH 5 and 7, the degradation rates of KET and METH under acidic and neutral conditions were obviously better than those under alkaline conditions. Under alkaline conditions, $^{\bullet}OH$ could be quenched by the HO_2^- produced by H_2O_2 dissociation, thus reducing the yield of $^{\bullet}OH$ in the system.

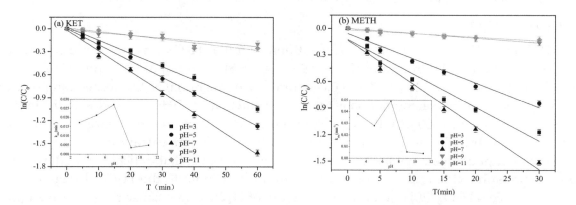

Figure 6. Effects of different initial pHs on the degradation of KET (**a**) and METH (**b**) in the UV/H_2O_2 system. Conditions: $(KET)_0 = (METH)_0 = 100$ µg/L, $(H_2O_2)_0 = 500$ µM, $pH_0 = 3$–11, $T = 25 \pm 1$ °C.

3.5. Effect of Water Background Components on Degradation Efficiency of Target Compounds

There are many different substrates in natural water, including different kinds of anions, cations and organic matter. These ions could react with free radicals in advanced oxidation processes, thus inhibiting or promoting the reaction and affecting the overall oxidation effect. Therefore, it is of great significance to study the influence of different ion types and contents on the practical application of advanced oxidation technology.

3.5.1. Effect of HCO_3^-

The decomposition of KET and METH was significantly inhibited with the addition of HCO_3^- at different initial dosages in the UV/H_2O_2 oxidation process (Figure 7). When the initial dosage of HCO_3^- ranged from 0 to 10 mM, the reaction rate of KET and METH decreased from 0.027 and 0.049 min^{-1} to 0.008 and 0.011 min^{-1}, respectively. The reason for this experimental phenomenon was that HCO_3^- was the quenching agent for $^\bullet OH$ which was also consumed by the competing reaction of ionized CO_3^{2-} (Equations (10)–(13)). Therefore, the inhibitory effect of KET and METH degradation was more obvious with the increase of the HCO_3^- concentration.

$$CO_3^{2-} + {}^\bullet OH \rightarrow CO_3^{-\bullet} + OH^- \tag{10}$$

$$HCO_3^- + {}^\bullet OH \rightarrow HCO_3^\bullet + OH^- \tag{11}$$

$$HCO_3^\bullet \rightarrow CO_3^\bullet + H^+ \tag{12}$$

$$CO_3^{-\bullet} + H_2O_2 \rightarrow HCO_3^- + HO_2^\bullet \tag{13}$$

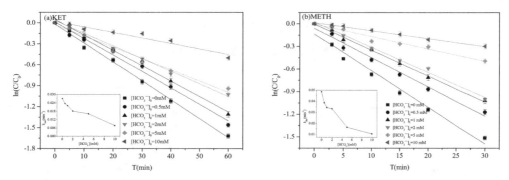

Figure 7. Effect of HCO_3^- on KET (**a**) and METH (**b**) degradation in UV/H_2O_2 system. Conditions: $(KET)_0 = (METH)_0 = 100$ μg/L, $(H_2O_2)_0 = 500$ μM, $pH_0 = 7.0$, T = 25 ± 1 °C.

3.5.2. Effect of Cl^-

With the initial concentration of Cl^- ranging from 0 to 10 mM, the destruction of KET was dramatically inhibited with the rate constant of KET decreased from 0.027 to 0.018 min^{-1} (Figure 8), which could be due to the elimination of $^\bullet OH$ by Cl^- according to Equations (14)–(16) [25]. The degradation reaction rate changed slightly as more Cl^- was added. However, the METH degradation was less affected by Cl^-, with the reaction rate remaining basically unchanged (0.0446–0.0485 min^{-1}).

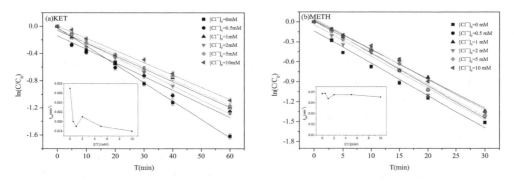

Figure 8. Effect of Cl^- on KET (**a**) and METH (**b**) degradation in the UV/H_2O_2 system. Conditions: $(KET)_0 = (METH)_0 = 100$ μg/L, $(H_2O_2)_0 = 500$ μM, $pH_0 = 7.0$, T = 25 ± 1 °C.

3.5.3. Effect of NO_3^-

$$\bullet OH + Cl^- \rightarrow Cl^\bullet + OH^- \tag{14}$$

$$Cl^\bullet + Cl^- \rightarrow Cl_2^{-\bullet} \tag{15}$$

$$Cl^\bullet + Cl^\bullet \rightarrow Cl_2 \tag{16}$$

The influence of NO_3^- on the decomposition of KET and METH is illustrated in Figure 9. With the initial concentration of NO_3^- ranging from 0 to 10 mM, the degradation of both target compounds was obviously inhibited. The reaction rate of KET and METH decreased from 0.027 and 0.049 min^{-1} to 0.007 and 0.012 min^{-1}, respectively. The above experimental phenomena were attributed to the following: First, a large amount of $\bullet OH$ could be produced from NO_3^- under UV irradiation (Equations (17)–(18)), which is an important source of $\bullet OH$ in natural water [26]. Second, as a photosensitizer, NO_3^- has a strong absorption in the ultraviolet range, which results in the formation of an internal filter that prevents the effective light transmittance and leads to the decline of $\bullet OH$ production in the UV/H_2O_2 system [27]. The latter was found to be dominant after the degradation effect of the reaction was analyzed.

$$NO_3^- + hv \rightarrow NO_2^{-\bullet} + O^{-\bullet} \tag{17}$$

$$O^{-\bullet} + H_2O \rightarrow \bullet OH + OH^- \tag{18}$$

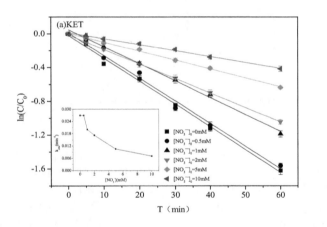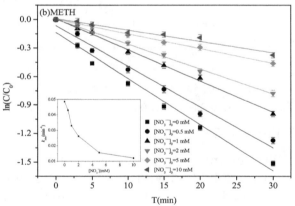

Figure 9. Effect of NO_3^- on KET (**a**) and METH (**b**) degradation in the UV/H_2O_2 system. Conditions: $(KET)_0 = (METH)_0 = 100$ μg/L, $(H_2O_2)_0 = 500$ μM, $pH_0 = 7.0$, T = 25 ± 1 °C.

3.5.4. Effect of HA

Due to its complex structure, HA may have uncontrollable effects on the destruction of target compounds. As illustrated in Figure 10, KET and METH degradation was dramatically inhibited once HA was added with different dosages in the UV/H_2O_2 system. As more HA (0–0.1 mM) was added, the reaction rate of KET and METH declined from 0.027 and 0.049 min^{-1} to 0.001 and 0.008 min^{-1}, respectively, while the degradation reaction rate changed slightly with the continued addition of the HA. UV irradiation was absorbed by HA, creating an inner filter (Figure 11) and significantly inhibiting the UV transmittance for UV photons, thus limiting the generation of $\bullet OH$ in the UV/H_2O_2 process [28]. Moreover, the degradation of target compounds can be inhibited by the competing reaction of HA with the active radicals [29].

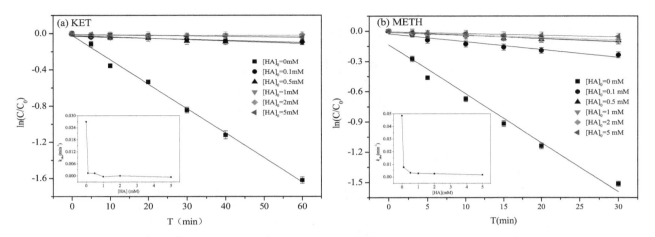

Figure 10. Effect of HA on KET (**a**) and METH (**b**) degradation in the UV/H$_2$O$_2$ system. Conditions: (KET)$_0$ = (METH)$_0$ = 100 μg/L, (H$_2$O$_2$)$_0$ = 500 μM, pH$_0$ = 7.0, T = 25 ± 1 °C.

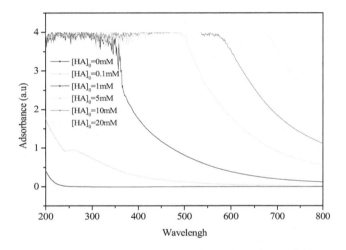

Figure 11. The ultraviolet–visible spectroscopy of reaction solutions at different concentrations of HA.

3.6. Degradation Products and Mechanism

Degradation intermediates and products of METH produced in the UV/H$_2$O$_2$ oxidation process were determined by using UPLC/MS/MS under full scans and product ion scans. During the whole METH degradation process, the mass spectra were compared to identify the intermediates. The structure of the transformation products was analyzed with the specific molecular ions and fragmentation patterns rather than direct comparison with corresponding standards. Figure 12 illustrates the mass spectra and possible structures of the degradation intermediates, based on which the possible transformation pathways of METH during UV/H$_2$O$_2$ are shown in Figure 13. The proposed degradation mechanisms of METH degradation involved in the UV/H$_2$O$_2$ system include hydrogenation, hydroxylation and electrophilic substitution.

With the molecular weight of 149, intermediate product 2 (P2, m/z = 150) was formed as a result of hydrogenation of METH. P1 (m/z = 91) with a stable structure was generated from the fracture of the C-C bond of the branched chain. Intermediates P3 (m/z = 110) and P4 (m/z = 73) were formed by electrophilic substitution of hydroxyl. METH was hydroxylated to form ephedrine (m/z = 165), of which the C-C bond of branched chain was fractured to form intermediate product P5 (m/z = 57). The hydroxylation of ephedrine induced the formation of intermediate P6 (m/z = 181) which was then achieved to form intermediate P7 (m/z = 89) after further hydroxylation. The mineralization of KET and METH was characterized by removal of total organic carbon (TOC), which achieved 41% and 57% within 60 min under UV/H$_2$O$_2$ treatment (Figure 14). The intermediate products were further degraded as the reaction continued.

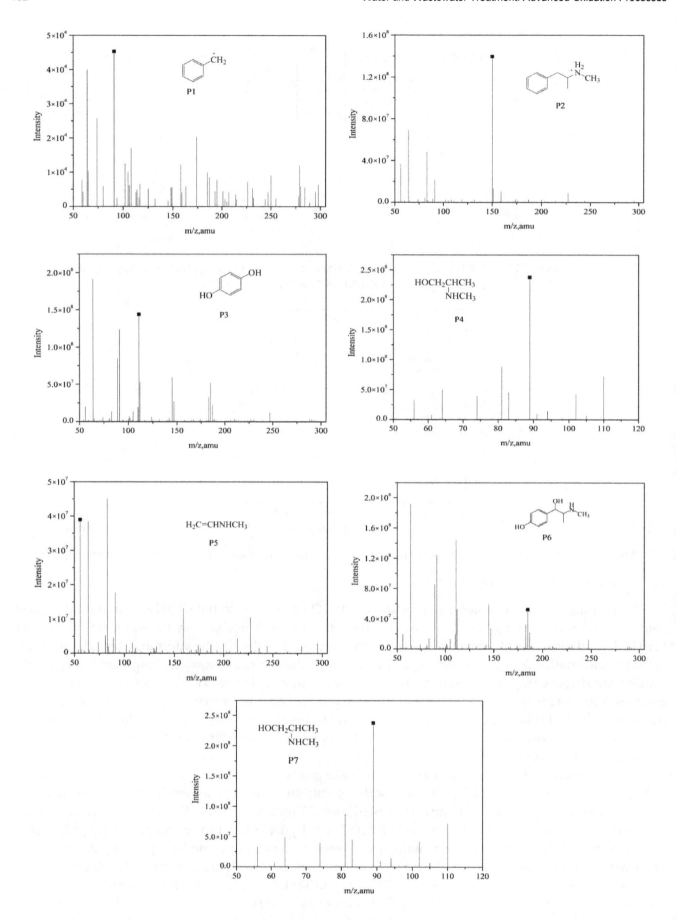

Figure 12. Mass spectra of the intermediate products of METH in the UV/PS system.

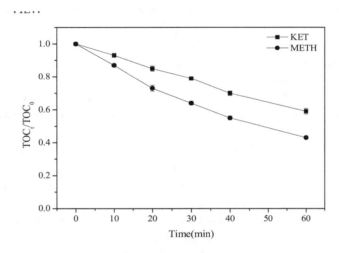

Figure 13. Tentative transformation of METH pathways in the UV/H$_2$O$_2$ system.

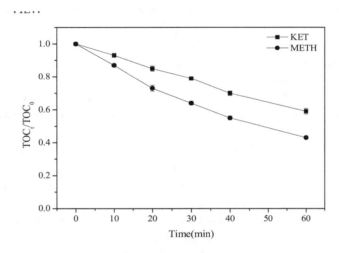

Figure 14. The mineralization of KET and METH during the UV/H$_2$O$_2$ system. Conditions: (KET)$_0$ = (METH)$_0$ = 100 µg/L, (H$_2$O$_2$)$_0$ = 500 µM, pH$_0$ = 7, T= 25 ± 1 °C.

4. Conclusions

The degradation kinetics and mechanisms of KET and METH using the UV/H$_2$O$_2$ process were investigated in this study. Their degradation in UV photolysis or H$_2$O$_2$ oxidation alone was negligible. However, 99% of KET and METH (100 µg/L) were effectively eliminated by the combination of UV and H$_2$O$_2$ within 120 and 60 min, respectively. According to the competition kinetics, the rate constants of •OH with KET and METH at pH 7 were calculated to be 4.43 × 10^9 and 7.91 × 10^9 M^{-1}·s^{-1}, respectively. The apparent rate constants of KET and METH reached the highest levels in a neutral environment. The degradation of KET and METH was significantly inhibited by HCO$_3^-$, Cl$^-$, NO$_3^-$ and HA; however, Cl$^-$ had little influence on the METH degradation. Seven reaction intermediates of METH in the UV/H$_2$O$_2$ system were identified by UPLC-MS/MS. Possible transformation mechanisms involved in the KET and METH degradation by UV/H$_2$O$_2$ oxidation system included hydrogenation, hydroxylation and electrophilic substitution. Results demonstrated that UV/H$_2$O$_2$ was an effective technique to remove KET and METH, providing a promising application for the decomposition of trace organic pollutants in natural water.

Author Contributions: Conceptualization, D.-M.G. and C.-S.G.; methodology, D.-M.G., C.-S.G., Q.-Y.F. and J.X.; software, D.-M.G., C.-S.G. and H.Z.; validation, D.-M.G., Q.-Y.F. and J.X.; formal analysis, D.-M.G. and H.Z.; resources, C.-S.G., Q.-Y.F. and J.X.; data curation, D.-M.G. and H.Z.; writing—original draft preparation, D.-M.G.; writing—review and editing, D.-M.G., C.-S.G., Q.-Y.F., H.Z. and J.X.; visualization, D.-M.G. and H.Z.; supervision, D.-M.G., Q.-Y.F. and J.X.; funding acquisition, C.-S.G. and J.X. All authors have read and agreed to the published version of the manuscript.

Acknowledgments: This study was carried out as part of the NSFC project, managed by the Jian Xu and supported by Center for Environmental Health Risk Assessment and Research, Chinese Research Academy of Environmental Sciences. We thank Wenli Qiu for her help in operating HPLC-MS. Reviewers are also thanked for the time dedicated and their comments.

References

1. Baker, D.R.; Kasprzyk-Hordern, B. Spatial and temporal occurrence of pharmaceuticals and illicit drugs in the aqueous environment and during wastewater treatment: New developments. *Sci. Total Environ.* **2013**, *454–455*, 442–456. [CrossRef]

2. Bijlsma, L.; Serrano, R.; Ferrer, C.; Tormos, I.; Hernández, F. Occurrence and behavior of illicit drugs and metabolites in sewage water from the Spanish Mediterranean coast (Valencia region). *Sci. Total Environ.* **2014**, *487*, 703–709. [CrossRef] [PubMed]

3. Wang, Z.; Xu, Z.; Li, X. Biodegradation of methamphetamine and ketamine in aquatic ecosystem and associated shift in bacterial community. *J. Hazard. Mater.* **2018**, *359*, 356–364. [CrossRef] [PubMed]

4. Du, P.; Li, K.; Li, J.; Xu, Z.; Fu, X.; Yang, J.; Zhang, H.; Li, X. Methamphetamine and ketamine use in major Chinese cities, a nationwide reconnaissance through sewage-based epidemiology. *Water Res.* **2015**, *84*, 76–84. [CrossRef] [PubMed]

5. Santos, M.E.S.; Grabicová, K.; Steinbach, C.; Schmidt-Posthaus, H.; Randák, T. Environmental concentration of methamphetamine induces pathological changes in brown trout (Salmo trutta fario). *Chemosphere* **2020**, *254*, 126882. [CrossRef] [PubMed]

6. Liao, P.H.; Hwang, C.C.; Chen, T.H.; Chen, P.J. Developmental exposures to waterborne abused drugs alter physiological function and larval locomotion in early life stages of medaka fish. *Aquat. Toxicol.* **2015**, *165*, 84–92. [CrossRef] [PubMed]

7. Awual, M.R.; Hasan, M.M. A ligand based innovative composite material for selective lead(II) capturing from wastewater. *J. Mol. Liq.* **2019**, *294*, 111679. [CrossRef]

8. Awual Rabiul, M. A novel facial composite adsorbent for enhanced copper(II) detection and removal from wastewater. *Chem. Eng. J.* **2015**, *266*, 368–375. [CrossRef]

9. Neta, P.; Huie, R.E.; Ross, A.B. Rate Constants for Reactions of Inorganic Radicals in Aqueous Solution. *J. Phys. Chem. Ref. Data* **1988**, *17*, 1027–1284. [CrossRef]

10. Russo, D.; Spasiano, D.; Vaccaro, M.; Cochran, K.H.; Richardson, S.D.; Andreozzi, R.; Puma, G.L.; Reis, N.M.; Marotta, R. Investigation on the removal of the major cocaine metabolite (benzoylecgonine) in water matrices by UV_{254}/H_2O_2 process by using a flow microcapillary film array photoreactor as an efficient experimental tool. *Water Res.* **2015**, *89*, 375–383. [CrossRef]

11. Kuo, C.; Lin, C.; Hong, P.A.K. Photocatalytic degradation of methamphetamine by UV/TiO_2—Kinetics, intermediates, and products. *Water Res.* **2015**, *74*, 1–9. [CrossRef] [PubMed]

12. Wei, C.; Yi, K.; Sun, G.; Wang, J. Synthesis of novel sonocatalyst Er3+:YAlO3/Nb2O5 and its application for sonocatalytic degradation of methamphetamine hydrochloride. *Ultrason. Sonochem.* **2018**, *42*, 57–67. [CrossRef]

13. Gu, D.; Guo, C.; Hou, S.; Lv, J.; Zhang, Y.; Feng, Q.; Zhang, Y.; Xu, J. Kinetic and mechanistic investigation on the decomposition of ketamine by UV-254 nm activated persulfate. *Chem. Eng. J.* **2019**, *370*, 19–26. [CrossRef]

14. He, X.; Pelaez, M.; Westrick, J.A.; O'Shea, K.E.; Hiskia, A.; Triantis, T.; Kaloudis, T.; Stefan, M.I.; Armah, A.; Dionysiou, D.D. Efficient removal of microcystin-LR by $UV-C/H_2O_2$ in synthetic and natural water samples. *Water Res.* **2012**, *46*, 1501–1510. [CrossRef] [PubMed]

15. Lin, C.C.; Lin, H.Y.; Hsu, L.J. Degradation of ofloxacin using UV/H_2O_2 process in a large photoreactor. *Sep. Purif. Technol.* **2016**, *168*, 57–61. [CrossRef]

16. Cheng, B.; Le, Y.; Yu, J. Preparation and enhanced photocatalytic activity of Ag@TiO2 core-shell nanocomposite nanowires. *J. Hazard. Mater.* **2010**, *177*, 971–977. [CrossRef]

17. Yu, X.; Liu, S.; Yu, J. Superparamagnetic γ-Fe2O3@SiO2@TiO2 composite microspheres with superior photocatalytic properties. *Appl. Catal. B Environ.* **2011**, *104*, 12–20. [CrossRef]

18. Ismail, L.; Ferronato, C.; Fine, L.; Jaber, F.; Chovelon, J.M. Elimination of sulfaclozine from water with SO_4^- radicals: Evaluation of different persulfate activation methods. *Appl. Catal. B Environ.* **2016**, *201*, 573–581. [CrossRef]

19. Znad, H.; Abbas, K.; Hena, S.; Awual, M.R. Synthesis a novel multilamellar mesoporous TiO 2 /ZSM-5 for photo-catalytic degradation of methyl orange dye in aqueous media. *J. Environ. Chem. Eng.* **2018**, *6*, 218–227. [CrossRef]

20. Lutterbeck, C.A.; Wilde, M.L.; Baginska, E.; Leder, C.; Machado, Ê.L.; Kümmerer, K. Degradation of cyclophosphamide and 5-fluorouracil by UV and simulated sunlight treatments: Assessment of the enhancement of the biodegradability and toxicity. *Environ. Pollut.* **2016**, *208 Pt B*, 467–476. [CrossRef]

21. Kwon, M.; Kim, S.; Yoon, Y.; Jung, Y.; Hwang, T.M.; Lee, J.; Kang, J.W. Comparative evaluation of ibuprofen removal by UV/H_2O_2 and UV/$S_2O_8^{2-}$ processes for wastewater treatment. *Chem. Eng. J.* **2015**, *269*, 379–390. [CrossRef]

22. Zuorro, A.; Fidaleo, M.; Fidaleo, M.; Lavecchia, R. Degradation and antibiotic activity reduction of chloramphenicol in aqueous solution by UV/H_2O_2 process. *J. Environ. Manag.* **2014**, *133*, 302–308. [CrossRef] [PubMed]

23. Qiu, W.; Zheng, M.; Sun, J.; Tian, Y.; Fang, M.; Zheng, Y.; Zhang, T.; Zheng, C. Photolysis of enrofloxacin, pefloxacin and sulfaquinoxaline in aqueous solution by UV/H_2O_2, UV/Fe(II), and UV/H_2O_2/Fe(II) and the toxicity of the final reaction solutions on zebrafish embryos. *Sci. Total Environ.* **2019**, *651*, 1457–1468. [CrossRef] [PubMed]

24. Sánchez-Polo, M.; Daiem, M.M.A.; Ocampo-Pérez, R.; Rivera-Utrilla, J.; Mota, A.J. Comparative study of the photodegradation of bisphenol A by HO·, $SO_4·^-$ and $CO_3·^-$/$HCO_3·$ radicals in aqueous phase. *Sci. Total Environ.* **2013**, *463–464*, 423–431.

25. Zhang, Y.; Xiao, Y.; Zhong, Y.; Lim, T. Comparison of amoxicillin photodegradation in the UV/H_2O_2 and UV/persulfate systems: Reaction kinetics, degradation pathways, and antibacterial activity. *Chem. Eng. J.* **2019**, *372*, 420–428. [CrossRef]

26. Yin, K.; Deng, L.; Luo, J.; Crittenden, J.; Liu, C.; Wei, Y.; Wang, L. Destruction of phenicol antibiotics using the UV/H_2O_2 process: Kinetics, byproducts, toxicity evaluation and trichloromethane formation potential. *Chem. Eng. J.* **2018**, *351*, 867–877. [CrossRef]

27. Moon, B.R.; Kim, T.K.; Kim, M.K.; Choi, J.; Zoh, K.D. Degradation mechanisms of Microcystin-LR during UV-B photolysis and UV/H_2O_2 processes: Byproducts and pathways. *Chemosphere* **2017**, *185*, 1039. [CrossRef]

28. Oh, B.T.; Seo, Y.S.; Sudhakar, D.; Choe, J.H.; Lee, S.M.; Park, Y.J.; Cho, M. Oxidative degradation of endotoxin by advanced oxidation process (O_3/H_2O_2 & UV/H_2O_2). *J. Hazard. Mater.* **2014**, *279*, 105–110.

29. Lutze, H.V.; Bircher, S.; Rapp, I.; Kerlin, N.; Bakkour, R.; Geisler, M.; von Sonntag, C.; Schmidt, T.C. Degradation of chlorotriazine pesticides by sulfate radicals and the influence of organic matter. *Environ. Sci. Technol.* **2015**, *49*, 1673–1680. [CrossRef]

9

Use of Ultrasound as an Advanced Oxidation Process for the Degradation of Emerging Pollutants in Water

Ana L. Camargo-Perea [1], Ainhoa Rubio-Clemente [2,3,*] and Gustavo A. Peñuela [2]

[1] Grupo de Investigación en Remediación Ambiental y Biocatálisis, Instituto de Química, Universidad de Antioquia UdeA, Calle 70, No. 52-21 Medellín, Colombia; ana.camargo@udea.edu.co

[2] Grupo GDCON, Facultad de Ingeniería, Sede de Investigaciones Universitarias (SIU), Universidad de Antioquia UdeA, Calle 70, No. 52-21 Medellín, Colombia; gustavo.penuela@udea.edu.co

[3] Facultad de Ingeniería, Tecnológico de Antioquia–Institución Universitaria TdeA, Calle 78b, No. 72A-220 Medellín, Colombia

* Correspondence: ainhoa.rubio@tdea.edu.co or ainhoarubioclem@gmail.com

Abstract: Emerging pollutants are compounds of increased environmental importance and, as such there is interest among researchers in the evaluation of their presence, continuity and elimination in different environmental matrices. The present work reviews the available scientific data on the degradation of emerging pollutants, mainly pharmaceuticals, through ultrasound, as an advanced oxidation process (AOP). This study analyzes the influence of several parameters, such as the nature of the pollutant, the ultrasonic frequency, the electrical power, the pH, the constituents of the matrix and the temperature of the solution on the efficiency of this AOP through researches previously reported in the literature. Additionally, it informs on the application of the referred process alone and/or in combination with other AOPs focusing on the treatment of domestic and industrial wastewaters containing emerging pollutants, mainly pharmaceuticals, as well as on the economic costs associated with and the future perspectives that make ultrasound a possible candidate to solve the problem of water pollution by these emerging pollutants.

Keywords: emerging pollutants; advanced oxidation process; water pollution; ultrasound

1. Introduction

Emerging contaminants (ECs) are chemical products, both natural and synthetic ones, that comprise a wide range of chemical compounds, including medical and recreational drugs, personal care products, steroids, hormones, surfactants, perfluorinated compounds, flame retardants, dyes, plasticizers and industrial additives [1–3]. The presence of ECs in the environment was not measured or controlled in the past because they did not cause concern and, in general terms, there were no studies demonstrating a health risk to humankind and living beings. Additionally, the use of ECs was not as high as it is currently; and they were not detected in water, since advances in instrumental analytical chemistry have only recently permitted their quantification at ultra-trace and trace concentrations [4,5], i.e., at concentrations from ng L^{-1} to µg L^{-1} [3,6]. Indeed, in the last years, ECs have been identified and quantified in effluents from wastewater treatment plant effluents, surface water, groundwater and even drinking water [3,5,7,8].

It is important to note that ECs can have harmful effects both on the environment where they are located and on human health. Nowadays, the toxicity ascribed to the presence of these pollutants on the environment has not been fully evaluated [8]; nevertheless, more and more eco-toxicological studies are being conducted [9]. In fact, the presence of ECs has been reported to represent a serious risk to both the environment and human health due to direct and/or indirect exposure [3,10], since they can negatively influence algae, invertebrates and fish, as well as ecosystem dynamics and community structure [11,12]. It has been found that ECs can act as endocrine disruptors and alter the reproduction cycles, water transport and osmoregulation processes of biota [13,14]. Other emerging pollutants have antimicrobial activity, leading to bacteria resistance to commonly used antibiotics [5] and, subsequently, resulting in worldwide spread of diseases. Additionally, ECs can be bioaccumulated [8], changing cellular reactions in vital organs, such as liver, kidney and gills [15]. Other studies have reported gene expression changes in organisms exposed to ECs [16].

It has been proven that some ECs are persistent pollutants that are hardly degraded by conventional processes [8,17,18]. For this reason, the implementation of new technologies to guarantee their removal is proposed [3,7,17,19].

Advanced Oxidation Processes (AOPs) have been evaluated as an option for the degradation of a variety of organic pollutants in waters [1,20]. These processes are characterized by a wide number of radical reactions, most of which involve chemical agents along with a source of ultraviolet (UV) radiation [21]. These radicals attack a large number of recalcitrant organic compounds such as ECs and, since they are not very selective, they become an excellent precursor to the conversion of a wide range of pollutants.

Several works have been carried out assisted by AOPs in order to evaluate their efficiency in degrading CEs. AOPs consist of the formation of the free hydroxyl radicals (HO•), which are capable of oxidizing toxic and/or recalcitrant organic compounds into more biodegradable and less dangerous products, such as oxidized species and short chain hydrocarbons of low molecular weight like formaldehyde and aliphatic acids [22], among other innocuous products; thus, they provide an improvement to the treatability of AOP effluents [17]. In fact, photocatalytic degradation has been conducted in the presence of UV radiation and photosensitizers including TiO_2, H_2O_2 and persulfate, among other chemical agents, obtaining very positive results [23–25]. Likewise, photo-Fenton and ozonation at basic pH have been proven to be highly efficient in the degradation of this type of pollutants [26,27]. These advanced systems, therefore, offer a solution to the problem of EC environmental accumulation and resistance to biological degradation, in contrast to other processes, such as conventional physical or chemical processes [17,25].

Nevertheless, it should be noted that, among the different AOPs used in the treatment of ECs present in water, the use of ultrasound (US) has been reported to be a highly efficient process, not only in the removal of this kind of contaminants, but also in their degradation [28,29] and the conversion of other recalcitrant pollutants [26] and microbial load [30] in water. Likewise, the use of US, as an advanced oxidation process, is environmentally "clean" since it does not require the addition of chemicals to the aqueous medium in order to achieve its EC degradation target, and does not generate waste [31] like Fenton and photo-Fenton. Consequently, the use of US waves is an alternative option for the conversion of recalcitrant ECs.

Considering the above, this article reviews and discusses the contributions of researches on the degradation of ECs, especially pharmaceuticals, due to their potential risks to human and other living beings, in aqueous media through US, as an advanced oxidation technology, considering the presence or absence of catalysts or dissolved gases, among other parameters, influencing the efficiency of the aforementioned process. Additionally, the application of this process is described focusing on domestic and industrial wastewater containing ECs, as well as the economic cost estimation associated with the future perspectives related to its implementation alone or in combination with other AOPs.

2. Ultrasound Process

The US process has been reported as a very efficient AOP for the degradation of ECs present in water [32–36]. Additionally, it can overcome the limitations ascribed to the use of other AOPs commonly used for water treatment. It is noteworthy to mention that, by using the US process, mass transfer within the reaction medium is improved, as well as the EC degradation reaction rates. Additionally, the consumption of chemicals, such as oxidizing and catalyzing agents, is reduced and no sludge is generated [22,37].

As part of this review, the state-of-the-art of the implementation of US, as an advanced oxidation process, is analyzed based on several works reported in the literature. First of all, the fundamentals of the process are described to continue with the factors mainly influencing the efficiency of the process. Afterwards, a number of examples are provided in order to have a general idea of the versatility of the advanced oxidation technology alone and/or in combination with other AOPs to efficiently degrade persistent compounds such as ECs. Finally, the capital and operation and maintenance costs are mentioned, and the future perspectives related to the application of the process are highlighted.

2.1. Operation Fundamentals

Aqueous medium sonolysis involves the production of waves through sound at a specific frequency, with compression and expansion cycles, leading to the formation of cavitation bubbles. These bubbles grow by the diffusion of vapor or gas from the liquid medium, reaching an unstable size that provokes their violent implosion, which in turn generates very high temperatures and pressures, approximately 4200 degrees K and 975 bar, producing the so-called "hot spots" that allow the decomposition of the water molecule to generate $HO\bullet$ [9], which is capable of oxidizing recalcitrant pollutants such as ECs with its high oxidation potential (2.8 V) [38], leading to the degradation of the toxic compounds and producing innocuous products, such as H_2O, carbon dioxide (CO_2) and inorganic ions.

Equations (1)–(4) show the decomposition of water and other molecules commonly dissolved in water by sonochemical waves [9,39], being the $HO\bullet$, as well as the hydroperoxyl radicals ($HO_2\bullet$), the main species that oxidizes the organic compounds present in the aqueous medium.

$$H_2O \xrightarrow{)))} H^\bullet + HO^\bullet \tag{1}$$

$$O_2 \xrightarrow{)))} 2O^\bullet \tag{2}$$

$$N_2 \xrightarrow{)))} 2N^\bullet \tag{3}$$

$$H^\bullet + O_2 \xrightarrow{)))} HO_2^\bullet \tag{4}$$

The cavitation bubbles are produced in two ways, symmetrically and asymmetrically. The difference between these is the support provided by a rigid surface (for instance, the surface of the reactor) for the bubbles to be formed. This difference has a direct influence on the way in which the bubbles implode, and thus on the release of pressure and temperature into the medium, resulting in the rupture of the water molecule and the formation of $HO\bullet$ [9]. The symmetrical bubbles release energy in all directions around their surface, while the asymmetrical ones generate an eruption of the liquid, mainly on the parts of the bubbles that are far away from the surfaces, forming long-range "micro-jets" that go to the solid surfaces [28].

There are three reaction zones in the solution during the ultrasonic treatment process: (a) inside the cavitation bubble, (b) the bubble/water interface and (c) within the bulk solution [28,32,40]. In each of these zones, different reactions occur that favor the decomposition of pollutants. Hydrophobic, non-polar and/or volatile compounds react inside the cavitation bubbles and at the bubble/water interface, while hydrophilic and/or non-volatile pollutants react within the bulk solution [28,41–43].

Inside the cavitation bubbles, the reaction of the pollutant can occur in two ways: pyrolysis of the highly volatile compounds, or chemical reaction with the free $HO\bullet$ formed. At the bubble/water interface, the reaction occurs by pyrolysis and, fundamentally, by a reaction with the $HO\bullet$ that are formed from implosion and tend to diffuse throughout the solution medium, reacting with the compounds that are present at the interface. Within the solution, decomposition occurs only by reaction with $HO\bullet$, which are released into the aqueous medium through implosion of the cavitation bubbles [9].

When free radicals reach the aqueous solution, they can recombine, as expressed in Equations (5)–(7), or react with hydroxyl ions (HO$^-$) (Equation (8)), resulting in a decrease of the system oxidation potential.

$$HO_2^\bullet + HO_2^\bullet \rightarrow H_2O_2 + O_2 + O_2\left(a^1 \, \Delta g\right) \quad k = 8.3 \times 10^5 \tag{5}$$

$$HO^\bullet + HO^\bullet \rightarrow H_2O + 1/2\left(O_2 + O_2\left(a^1 \, g\right)\right) \quad k = 5.5 \times 10^9 \tag{6}$$

$$HO^\bullet + HO_2^\bullet \rightarrow H_2O + O_2 + O_2\left(a^1 \, \Delta g\right) \quad k = 7.1 \times 10^9 \tag{7}$$

$$HO_2^\bullet + HO^- \rightarrow O_2^\bullet + H_2O \quad k = 10^{10} \tag{8}$$

However, from Equation (8), superoxide radicals (O$_2\bullet^-$) are formed, as well as from the decomposition of HO$_2\bullet$, as described by Equation (9), which also contribute to the degradation of emerging organic compounds, although in a smaller proportion than by HO$\bullet$ [38]. Additionally, in acidic medium, O$_2\bullet^-$ can react with protons (H$^+$) to form HO$_2\bullet$ (Equation (10)). Both of the free radicals can recombine, as represented in Equation (11), resulting in the production of HO$_2^-$, which in turn can be involved in HO$\bullet$ quenching (Equation (12)).

$$HO_2^\bullet \rightarrow H^+ + O_2^\bullet \quad k = 7.5 \times 10^6 \tag{9}$$

$$H^+ + O_2^\bullet \rightarrow HO_2^\bullet \quad k = 5.1 \times 10^{10} \tag{10}$$

$$HO_2^\bullet + O_2^\bullet \rightarrow HO_2^- + O_2 \quad k = 9.7 \times 10^7 \tag{11}$$

$$HO^\bullet + HO_2^- \rightarrow HO_2^\bullet + HO^- \quad k = 7.5 \times 10^9 \tag{12}$$

Hydrogen peroxide (H$_2$O$_2$) can also be formed in the US process, as described in Equation (5). In spite of the fact that H$_2$O$_2$ can scavenge HO$\bullet$ or be decomposed (Equations (13)–(15), respectively), it can be involved in the oxidation of ECs, as well as on the production of a higher amount of HO$\bullet$, when US process is combined with UV radiation.

$$HO^\bullet + H_2O_2 \rightarrow H_2O + HO_2^\bullet \quad k = 3 \times 10^7 \tag{13}$$

$$H_2O_2 \rightarrow HO_2^- + H^+ \quad k = 2 \times 10^{-2} \tag{14}$$

$$HO_2^- + H^+ \rightarrow H_2O_2 \quad k = 10^{10} \tag{15}$$

The reaction rate constants for the reactions expressed in Equations (5)–(15) were taken from Pavlovna et al. [44], demonstrating that, in general terms and according to the values of the reaction rate constants, the free radicals are easily formed through the US waves. As mentioned previously, these free radicals can react with the target pollutant; however, they can also recombine or be quenched by other compounds found in water such as the natural constituents of the matrix, making the reaction of the hydrophilic compounds within the solution less efficient and slower [45]. In this regard, in order to avoid side reactions of the US oxidation system, the optimization of the operating parameters or factors influencing the most the oxidation potential of the system must be conducted. This would subsequently allow the reduction of the economic costs associated with the studied advanced oxidation process for a more efficient degradation of the ECs of interest.

2.2. Efficiency of the Ultrasound Process

The US process must consider the control and variation of the different operating parameters, including the ultrasonic frequency, the electrical power and the pH and temperature of the solution [46,47], in order to be optimized with the subsequent reduction in the costs associated with the process performance. The nature of the contaminant of interest and the constituents of the water matrix must also be considered during the US-assisted AOP optimization procedure since they are involved in the efficiency of the process. In addition to these factors, the type and the geometry of the sonochemical reactor must be considered.

2.2.1. Reaction Zones—The Nature of the Emerging Pollutant

In the ultrasonic radiation process, as indicated above, three reaction zones are recognized for the degradation of compounds: the cavitation bubble, the bubble–water interface and the bulk solution [28,43,48]. The process by which degradation occurs differs from zone to zone. Hydrophilic substances are located within the solution, non-volatile hydrophobic compounds are mainly housed in the bubble–water interface, and volatile substances are commonly located within the cavitation bubble [22].

Inside the cavitation bubble, the degradation reaction of the contaminant occurs by pyrolysis; on the other hand, in the bubble–water interface, the main reaction mechanism is by the attack of free radicals, such as HO•, which are immediately formed by the implosion of the cavitation bubbles; finally, in the bulk solution the reaction occurs directly with the free radicals that reach this zone [28].

According to different investigations, in the US process, the degradation of volatile compounds occur in two zones: in the bubble–liquid interface, through the reaction with the HO• released from the implosion, and/or inside the bubble, directly by pyrolysis [22,41]. The rates of destruction of volatile contaminants depend on the physical and chemical conditions within the bubble, specifically the hydrophobic and volatile nature of these compounds [31,41]. On the other hand, it has been shown that the reaction rate constant of US degradation of volatile compounds decreases with increasing initial concentration, indicating that the relationship between the concentration of a volatile compound in the cavitation bubble and its concentration in the solution will influence the rate of ultrasonic reaction, considering that the collapse temperature depends on the specific heat ratio of the gas mixture [41,48–51].

Hydrophobic compounds such as carbamazepine (CBZ), which has a Henry's constant of approximately 1.08×10^{-10} atm. m^3 mol^{-1} and a moderate solubility in water [9], can be mainly housed in the bubble–water interface, but it is also found within the solution, allowing the protagonist of its degradation to be the HO•, which are immediately formed from the implosion both of the cavitation bubbles and the bubbles that travel within the solution [9].

To evaluate the zone and the way in which a compound is degraded, Nie et al. [28] have implemented the so-called "scavengers" of the HO•. In an experiment where the US process was used to degrade the pharmaceutical diclofenac (DCF), isopropyl alcohol and terephthalic acid were used to inhibit the reaction of the target compound with HO•, functioning as quenchers. The acid was considered to react with free radicals in the bulk solution, while the alcohol reacted both at the bubble–water interface and in the bulk solution. In this regard, the authors verified that when only the acid was added, the degradation of the compound was inhibited. However, when the alcohol was used exclusively as an inhibitor, degradation of the target EC was considerably reduced. It was, therefore, concluded that oxidation of DCF occurred mainly by HO• in the supercritical interface, especially when water was saturated with air and oxygen (O_2). Nonetheless, under argon (Ar)- and nitrogen (N_2)- saturated conditions, DCF degradation occurred within the cavitation bubbles and/or the bulk solution.

In a study carried out by Kidak and Dogan [52], where the degradation of alachlor through the US process was evaluated, it was concluded that due to the physical properties of the compound, such as the water solubility limit (140 mg L^{-1} at 20 °C), vapor pressure (negligible), Henry's constant (3.2×10^{-8} to 1.2×10^{-10} atm-m^3 mol^{-1}), octanol–water partition coefficient (Log K_{ow} = 2.63–3.53) and its positive ionization, the compound was housed in the bubble–liquid interface, indicating that the degradation was due to the HO• recently formed from the implosion of the cavitation bubbles. The degradation obtained of the target compound was near 100% with a frequency of 575 kHz and an electrical power of 90 W.

Adityosulindro et al. [53] evaluated the degradation of ibuprofen (IBU) in order to ascertain the reaction zone in which the degradation of IBU was established, and whether it was due exclusively to HO•. For this purpose, they tested the sequestration of these radicals through two compounds, n-butanol, which is a short chain alcohol with partial solubility in water that is expected to react with the radicals housed in the bubble–liquid interface; and acetic acid, which should react with the free radicals in the bulk solution due to it is a completely miscible compound. The results obtained indicated that, indisputably, IBU reacted with the HO• recently formed during the implosion of the cavitation bubbles, which means that it is a compound housed in the interfacial zone [53]. The same conclusion was reached by Méndez-Arriaga et al. [42], who attributed the degradation of IBU to the HO• recently produced, since IBU is considered to be housed at the bubble–water interface due to its Henry's constant (1.5×10^{-7} atm m^3 mol^{-1}), low solubility in water (21 mg L^{-1}) and octanol–water partition coefficient (3.9).

In the case of acetaminophen (ACP), a polar compound with high solubility (12.5 mg mL^{-1}), Villaroel et al. [54] reported that this contaminant was degraded in a greater proportion within the bulk, estimating that its behavior would be that of a hydrophilic substrate. Nonetheless, in this investigation, it was concluded that ACP can be housed both in the bulk solution and in the bubble–water interface, attributing its degradation to the HO• formed during the implosion of the cavitation bubbles. Based on the aforementioned authors' estimations, the hydrophilic or hydrophobic behavior of the target compound was more related to the initial pH value of the solution at which the study was carried out.

2.2.2. Ultrasonic Frequency

The frequency with which ultrasonic waves are produced can range from 20 to 10,000 kHz, and the US process is divided into three regions: low, high and very high frequency [22]. In Table 1, the frequency ranges used in the ultrasonic oxidation process are listed.

Table 1. Frequency ranges used in the ultrasonic process. Taken from [22].

Name	Ultrasound Range (kHz)
Very high	5000–10,000
High	200–1000
Low	20–100

Ultrasonic frequency is a fundamental parameter in the performance of US process, since the size and duration of the cavitation bubble, the violence of the implosion and, therefore, the production of HO• depend considerably on it [9,55].

The number of cavitation bubbles and bubble collapses increases with rising frequency. However, it is important to note that the bubbles generated at high frequencies are small, and release less energy than low frequency bubbles generated by a single pulse [6,56,57]. In addition, the escape of more HO• is inferred, before recombining, when faster collapses occur [9,58]. In this sense, the optimal frequency is determined by the integral efficiency of the energy discharge, which depends on the quantity, size and lifetime of the bubbles. It is noteworthy to mention that the optimal frequency varies according to the different compound to be treated [52,59].

Rao et al. [9] chose two frequency values (200 and 400 kHz) to determine the optimal one for the degradation of CBZ. The first of these values was more effective for the degradation of the target compound. This result was ascribed to the differences in calorimetric powers obtained for both frequencies under the same electrical power (100 W), resulting in a higher calorimetric power for the 200 kHz frequency. This can be attributed to what was previously explained, i.e., each EC requires an optimal frequency at which its degradation will be favored, which depends on its physicochemical properties. This optimal frequency will also be influenced by the geometry of the reactor since, as mentioned above, it will depend on the formation of symmetrical or asymmetrical cavitation bubbles.

On the other hand, in the research carried out by Güyer and Ince [23], different levels of ultrasonic frequency were evaluated in the US process of the DCF. The results obtained allowed the conclusion that the maximal rates of DCF degradation were reached at a frequency of 861 kHz and the minimal ones at 1145 kHz (carrying out tests with values of 577, 861 and 1145 kHz). The improvement between the 577 and 861 kHz was due to the fact that the latter reduced the size of the bubbles, leading to a greater number of bubbles and active oscillations, which contributed to the generation of HO• improvement. However, the highest frequency evaluated this efficiency was reduced due to the fact that the "optimal" frequency related to the reactor configuration was surpassed [53,56].

2.2.3. Electrical Power

The electrical power supplied to the ultrasonic transducer is a critical parameter that can largely determine the performance of the US process [9].

For Jiang et al. [41], the increase in ultrasonic power in the degradation of volatile compounds such as chlorobenzene, 1, 4-dichlorobenzene and 1-chloronaphthalene caused an increase in the cavitation energy, decreasing the cavitation limit and increasing the amount of bubbles produced. This resulted in a rise in the rate of degradation of this type of compounds, considering that the bubbles formed had enough energy to pyrolyze the tested pollutants. This is justified by the fact that volatile compounds are pyrolyzed within the cavitation bubbles, so the more bubbles formed, the more spaces for these compounds to react.

In a study carried out by Tran et al. [18], sonochemical efficiency was evaluated by means of calorimetric tests to determine the optimal power and to propose an experimental design in order to degrade the drug CBZ. It was determined that powers between 20 and 40 W favored sonochemical efficiencies, unlike what happened with powers of 10 W. This finding was attributed to the fact that as the power increased, so did the ultrasonic energy of the reactor, which caused the pulsation and collapse of the bubbles to be generated at a faster rate, resulting in a greater number of cavitation bubbles.

It is important to note that the effect of ultrasonic power and oxidizing species can be influenced by bubble dynamics [18]. The results reported by Gogate et al. [60] indicated that the size, number, lifetime and pressure of the bubbles were a complex function of the power dissipation rate. The research conducted by these authors explains the results obtained by Tran et al. [18], since they found that by increasing the power, the number of cavitation bubbles rose and, consequently, the production of HO• increased. In this way, the degradation of the target compound, CBZ in this case, was directly increased. Similar results were observed in the work carried out by Madhavan et al. [61] for DCF, who studied the degradation of this compound under a frequency of 213 kHz, a temperature of 25 °C, a variation of power density between 16–55 mW mL^{-1} and a concentration of the pharmaceutical compound of interest of 0.07 mM. The same conclusion was also reached by Rao et al. [9], who studied the degradation of CBZ at pH 6, a frequency of 200 kHz and a power variation between 20 and 100 W, obtaining a higher degradation of CBZ at 100 W.

In the work carried out by Naddeo et al. [6], the degradation of DCF during the US process was evaluated. It was determined that, by increasing the power density from 100 to 400 W L^{-1}, the concentration of the contaminant decreased, making US the most efficient process. This result supports the theory developed in the work described above, i.e., the greater the potency, the greater the degradation percentage of the compound under study.

On the other hand, the combination of different levels of the parameters that influence the degradation of a compound in the US process must be considered. In this context, when the response surface methodology was used to determine the optimal operating levels of CBZ ultrasonic treatment, [18] it was observed that the treatment time had a more statistically significant impact on the efficiency of CBZ removal in comparison with the electrical power, as efficient degradation of the contaminant at lesser powers (10–40 W) required more treatment time. This fact is ultimately reflected in the use of electricity and, therefore, in higher operating costs associated with the application of the oxidation process.

Meanwhile, Kidak and Dogan [52] stated that increasing electrical power also increased the number of the bubbles formed, and that better results were expected in the degradation of the pollutants of interest. This assertion was supported by the results obtained in the experimentation with alachlor, where near 100% degradation was obtained through the US process (initial concentration of alachlor 100 μg L^{-1}, frequency of 575 kHz and powers of 45, 60 and 90 W). In addition, they observed an increase in the reaction rate constants as the ultrasonic power increased.

However, in the work carried out by Ince [62], it was evidenced that the degradation of paracetamol (PCT), also known as ACP, DCF and IBU was reduced when operating with a high frequency (861 kHz). The author attributed that fact to the formation of clouds of bubbles when exceeding the threshold power (optimal), which increased the sound waves and, as a result, decreased the cavitation activity. The same author pointed out that below the power threshold, when the power was increased, the efficiency of the process rose.

Adityosulindro et al. [48] evaluated the degradation of IBU by the US process and the influence of the power density in the conversion of the target pharmaceutical. It was determined that increasing the power in a range between 25–100 W L^{-1}, over 180 min of treatment, contributed to a greater formation of HO•. However, the authors stated that above a critical or optimal power density value, a cloud of bubbles would be formed, dispersing the formation of sound waves, which would in turn decrease the efficiency of the process [53].

2.2.4. Solution pH

The pH of the solution is a fundamental parameter in oxidation-reduction reactions. In the US process, the pH indicates the hydrophobic or hydrophilic nature of the target compound behavior, depending on whether the structure in which the pollutant is found is ionic or molecular. This property will allow the position to be determined in which the contaminant is housed in the US process, i.e., in the bulk solution (hydrophilic, non-volatile compounds), in the bubble–water interface (semi-volatile hydrophobic compounds), or within the cavitation bubble (hydrophobic, volatile compounds) [9]. This position, in turn, will determine whether the degradation pathway of the contaminant is by pyrolysis or by reaction with the HO• formed by implosion of the cavitation bubbles.

In the research carried out by Tran et al. [18], CBZ was degraded through the US process, considering the effects of the electric power, initial drug concentration, treatment time and pH of the solution (7–10). In this work, no significant influence on the part of pH was evidenced through an experimental factorial design, as a response surface methodology. This can be attributed to the fact that no tests were performed with acid pH values, which influence the structural form in which the compound is found in the aqueous medium and, therefore, the reaction zone in which it is found. Specifically, the pH values will favor or disfavor the hydrophobicity of the compound, with more hydrophobic compounds that are closer to the bubble–water interface reacting with the HO• that have just been formed from the implosion of the cavitation bubbles, whereas those compounds further away from the bubble–liquid (hydrophilic) interface possibly reacting with the HO• that reach the solution.

On the other hand, in the work carried out by Rao et al. [9], the influence of pH on the degradation of CBZ was evaluated, using levels between 2.0 and 11.0. The results showed that, at pH values between 4.5 and 11.0, the degradation remained constant and decreased in equal proportion, but with pH values close to 2.0 there was a small decrease in the degradation efficiency. This was ascribed to the fact that CBZ reacts at the bubble–water interface whenever hydrophobicity is favored—a result that was achieved with pH values between 4.5 and 11.0—whereas with pH values close to 2.0 the ionic structure of the compound, and thus its hydrophilicity, was favored. The compounds that can lodge very close to the cavitation bubbles can react with a greater amount of HO• than those ones that are in the bulk solution, which must wait for these oxidizing agents to reach them, being able to react with another compound along the way, such as the natural constituents of the aqueous matrix tested.

Meanwhile, Huang et al. [46] evaluated the degradation of DCF through US process in the presence of Zn^0, performing an analysis of the influence of pH on this type of process. It was found that at pH higher than 2 the degradation of DCF was very small, while at equal or lower values, the degradation of the tested compound reached percentages higher than 80%. The authors attributed this behavior to the fact that the pKa of DCF is 4.15, considering that aqueous media with a pH lower than this value will manage to maintain the molecular structure of this compound, and concluded that this form favored the absorption reaction of DCF by Zn^0.

In the degradation of IBU through the US process, the influence of pH was evaluated, experimenting with values higher and lower than the pKa of the compound (4.9). It was found that, at lower values (2.6 and 4.3), the compound remained unproned and its degradation slightly increased, while the opposite occurred with an alkaline pH value (8.0), where IBU degradation was affected. However, the authors argued that under its ionic form, IBU should accumulate less at the bubble–water interface, which is where the HO• attack mainly occurs [53].

Al-Hamadani et al. [31] evaluated the degradation of sulfamethoxazole (SFX) and IBU under three pH conditions: acid (3.5), below the pKa values of the target compounds; basic (7), above the pKa values; alkaline (9.5), well above these values. The results showed degradations near 100% of the compounds in 1 h of treatment for a pH below pKa, while degradation was significantly affected above these values. This is attributed to the molecular form of the compounds, i.e., when the pH of the solution was below pKa, the hydrophobicity of the drugs and, therefore, their position in the bubble–water interface is improved, favoring a rapid reaction with the HO• recently formed during the implosion of the cavitation bubbles.

2.2.5. Constituents of the Water Matrix

Various investigations related to the degradation of ECs in water through AOPs have been carried out in aqueous matrices with different constituents. On one hand, some researches have been developed with synthetic waters which, in general, involve the use of distilled water doped with the chemical components offering the specific characteristics with which the researcher wishes to work. On the other hand, there are works operating with real wastewater or in which the efficiency of the process for natural surface and drinking water is evaluated.

The research carried out by Tran et al. [27] identified the levels of the operating parameters at which IBU could be degraded by 65% through a sono-electrolytic process under controlled conditions in synthetic water, using a statistical optimization procedure. These same conditions were evaluated with sewage from a municipal treatment plant, with organic and inorganic compounds, as well as microbial load, which was doped with a specific concentration of IBU. The result obtained was a greater degradation of the compound of interest (90%) than that statistically estimated with synthetic water. This result was ascribed to the apparent presence of the chloride ion (Cl^-), which favored electrolysis, and might also favor the formation of hypochlorous acid (HClO), which can improve IBU oxidation. This demonstrates the importance of studying the organic and inorganic content of the water to be treated, as this may favor or limit the degradation of the target compounds.

It has been reported that Cl^- have different effects on the elimination of ECs present in water treated by means of AOPs [63,64]. Rao et al. [9] evaluated the degradation of CBZ (0.025 mM) using 200 kHz 100 W US. These authors investigated the presence of different inorganic anions to determine their influence on the process. The anions evaluated were Cl^-, SO_4^{2-} and NO_3^-, and it was found that Cl^- slightly restricted the degradation of the investigated drug, while the others did not have a significant impact on the degradation of the compound of interest. This slight inhibition in CBZ degradation due to the presence of Cl^- can be attributed to the reaction of this ion with the HO• dispersed in the solution, resulting in the formation of $ClOH•^-$.

In the work reported by Adityosulindro et al. [53] on the Fenton, US oxidation system and US-Fenton process, the efficiency of the degradation of IBU in distilled water and in wastewater from a municipal treatment plant was compared. The results showed a negligible difference between the degradation capabilities of all the evaluated processes in both distilled water and wastewater. In this context, the authors stated that the organic and inorganic content of the sewage effluent did not compete with IBU for HO• and that the latter was capable of reacting first with the oxidizing agent. It is important to highlight that the experimentation was carried out at acid pH, which could favor the location of IBU in the interface zone, making it more competitive when reacting with the HO• formed from the implosion of the cavitation bubbles.

Rao et al. [9] compared the degradation of CBZ contained in synthetic water with that of an effluent from a municipal wastewater treatment plant, evaluating the efficiency of two processes: ultrasonic irradiation alone, and in combination with photolysis using UV radiation emitting at a wavelength of 254 nm. The results showed that, for the US process, the constituents of the real wastewater matrix had no influence on the degradation of CBZ when compared with the results for distilled water. On the other hand, in the combined process, the wastewater matrix increased the efficiency of the studied drug degradation. This can be attributed to the photolysis of certain compounds contained in the wastewater that provide the oxidizing agent and favor the degradation of CBZ. As a matter of fact, the referred authors gave the example of nitrate ions (NO_3^-).

In the research conducted by Villaroel et al. [54], the influence of ionic constituents of water on the degradation of ACP (82.69 μmol L^{-1} and 1.65 μmol L^{-1}), at a power of 60 W and ultrasonic frequency of 600 kHz, was evaluated. The results obtained in distilled water and in synthetic water containing calcium ions (Ca^{2+}), magnesium ions (Mg^{2+}), sulphates ions (SO_4^{2-}), bicarbonates ions (HCO_3^-), Cl^-, potassium ions (K^+) and fluorides ions (F^-) were compared. The results indicated that, for the lowest concentration of ACP, a more pronounced acceleration of degradation was observed when this occurred in water with similar ion content than in distilled water. The authors attributed this to the high content of HCO_3^-, which was likely to be the protagonist in the formation of the carbonate radical (HCO_3•) when reacting with HO• radicals, being HCO_3•, a contributor to the degradation of the target EC.

With regard to the use of dissolved gases and their influence on the degradation of organic ECs, in the work conducted by Nie et al. [28], whose objective was to degrade DCF through the US process, it was observed that under saturated air, O_2 and Ar, a complete mineralization of nitrogen and a partial mineralization of carbon was achieved. When oxygenation was added to the reaction solution, HO_2• was formed. Although these radicals do not have an oxidation potential as high as HO•, as mentioned previously, HO_2• can contribute to the degradation of the compounds of interest [6].

On the other hand, it must be highlighted that when chlorine atoms are part of the target EC structure, they are transformed to Cl^- through the reaction of the pollutant of interest with HO• or by pyrolysis in the US process [41,49]. Therefore, the release of Cl^- occurs during the sonochemical degradation of chlorinated compounds, which was attributed to the rapid excision of the carbon–chlorine bonds by high temperature combustion occurring within the cavitation bubbles or at the bubble–liquid interface. Cl^-, as indicated above, can reduce the oxidation potential of the process.

Under this scenario, studies aiming at examining the efficiency of the US process are required to be conducted by using real matrices due to the natural constituents of the water matrices can positively or negatively influence the degradation percentages and reaction rates of the ECs of interest.

2.2.6. Temperature of the Solution

According to some authors, temperature variation in the US process directly influences cavitation intensity due to the changes in the physicochemical properties of the compound and the type of cavities formed, which can affect the kinetic velocity constant of the degradation reaction [65].

Al-Hamadani et al. [31] indicated that certain parameters were affected by increasing the temperature in the US process. First, it was found that cavitation energy decreased, as well as the threshold limit of the energy required to produce cavitation. In addition, it was found that the amount of dissolved gas was reduced, leading to the transfer of organic molecules from the bulk solution to the bubble–water interfacial region. Finally, the vapor pressure increased, causing the cavitation bubbles to contain more water vapor. Furthermore, the aforementioned authors, who evaluated the degradation of SFX and IBU through US, evidenced the temperature influence on the oxidation process. Temperatures between 15 and 55 °C were tested and it was concluded that, when this parameter was increased, the degradation of the studied compounds rose, as a rise in the temperature of the bulk caused the cavitation threshold to lower, which contributed to the formation of a greater number of cavitation bubbles and, therefore, to a greater amount of HO•. However, these authors pointed out that other works have shown an adverse effect of temperature on the degradation of the contaminant. These findings can be attributed to the fact that the surface tension and viscosity of the solution increase, generating cavitation bubbles with less intensity due to a rise in the vapor pressure of the liquid.

2.3. Application of Ultrasound Process to Water Treatment

Table 2 compiles several relevant research works related to the treatment of ECs through ultrasound as an AOP alone or in combination with other physical-chemical and advanced oxidation technologies.

Table 2. Summary of works related to the removal of emerging pollutants through the ultrasound process and its combination with other physical-chemical and advanced oxidation processes.

Process	Ref.	Pollutant/Type of Water	Operating Conditions	Found Results
US	[28]	DCF/Synthetic water	Co DCF: 0.05 mM. Frequency: 585 kHz. Power intensity 160 W L^{-1} pH: 7 Situations: air saturation, argon, oxygen and nitrogen. Temperature: 4 °C Glass cylindrical reactor of 750 mL connected to transducer Working volume: 500 mL. Treatment time: 60 min. HO• scavenger agents: Isopropyl alcohol and terephthalic acid. Co H$_2$O$_2$: 0.5 and 5 mM.	The elimination of DCF (without scavenger) and the formation of chloride ions were established as first-order reactions. Dichlorination rates, under all gas saturation conditions, were 1 to 2 times higher than DCF degradation rates. Dichlorination was a major reaction pathway during ultrasonic degradation of DCF; it developed within the solution by HO• attacks. There was only a partial mineralization in the 4 gas saturation conditions. The lowest peroxide concentration allowed a higher rate of degradation of the DCF.
US	[52]	Alachlor/ Synthetic water	Co Alachlor: 100 μg L^{-1} Frequency: 575, 861 y 1141 kHz. Electric power: 45, 60 and 90 W. Reactor: Glass cylindrical reactor of 500 mL Temperature: 25 °C. Treatment time: 90 min. pH: 7	Alachlor degradation was a pseudo-first order kinetics. A 100% degradation of alachlor and a mineralization of 25% was achieved, in 60 minutes of treatment, with a frequency of 575 kHz and a power of 90 W. The intermediate products from degradation of each tested power were analyzed, identifying their abundance in the samples.
US	[66]	Rosaniline (PRA) and ethyl violet (EV)	[PRA] and [EV]: 10 ppm Frequency: 350 kHz Electrical Power: 60 W. Treatment time: 30 min. Presence of ions: Cl$^-$, NO$_3^-$, SO$_4^{2-}$, CO$_3^{2-}$.	A complete degradation of EV and PRA was observed with a first order pseudo velocity constant. A good COD removal of 97% and 92%, respectively, was observed for EV and PRA after 3 h. The rate constants were higher with the addition of chloride ions in the case of EV and were not altered in the case of PRA. The improved degradation of EV in the presence of chloride is probably due to the salting effect and the reaction of the secondary radicals. EV degradation decreased from 100% to 80% with an increase in carbonate ion concentration from 0 to 100 ppm. In the case of PRA, a significant improvement in degradation was observed with the addition of CO$_3^{2-}$.

Table 2. *Cont.*

Process	Ref.	Pollutant/Type of Water	Operating Conditions	Found Results
US	[33]	Benzophenone-3 (BP-3)/ Synthetic water	Treatment time: 10 min Frequency: 574, 856 and 1134 kHz. Electrical Power: 100–200 W L^{-1}. [BP-3]: 1 ppm. Temperature: 25 ± 2 °C. Relationship of pulse time and silence time: PT/ST.	574 kHz or a lower frequency value is optimal for degradation of BP-3. The optimum power density level was 200 W L^{-1}. A maximum degradation level of 79.2% was obtained for EP = 200 W L^{-1}, a PT/ST ratio of 10 and frequency 574 kHz. The degradation was almost the same for all PT/ST ratios from 3 to 12.
US	[34]	Triclosan (TCS)/ Synthetic water	Treatment time: 60 min. Frequency: 215, 373, 574, 856 and 1134 kHz. Electrical Power: 40, 76, 140 and 200 W L^{-1} [TCS]: 1 mg L^{-1}. Temperature: 25 ± 2 °C. Treatment volume: 300 mL.	The 574 kHz frequency had the highest degradation rates. With 574 kHz, at 40 W L^{-1}, 88% of TCS degraded in 60 min, while at 140 W L^{-1}, TCS degraded completely in less than 25 min. The highest TCS degradation rate was obtained at the highest power density level of the equipment, 200 W L^{-1}. It was shown that the only variable that had statistical significance and an effect on degradation after 10 min was the power density.
US	[35]	Bisphenol-A/ Synthetic water	Frequency: 300 kHz. Electrical Power: 80 W. Treatment volume: 300 mL. [BPA]: 0.12 and 300 μM. pH: 8.3 [HCO_3^-]: 12–500 mg L^{-1} Temperature: 21 °C. Addition: Cl^-, SO_4^{2-} and HPO_4^{2-} [6 mM].	The addition of HCO_3^-, in the range of 12–500 mg L^{-1} did not have a significant effect on the BPA degradation rate. The bicarbonate concentration had a significant effect for the 0.12 BPA concentration: a higher bicarbonate concentration produced higher initial decomposition rates. Solutions containing ions other than bicarbonate showed significantly lower degradation rates. The bicarbonate/carbonate solution produced a significantly improved degradation rate of BPA.
US	[54]	Acetaminophen (ACP)/ Synthetic water and mineral water	Frequency: 600 kHz. Electrical Power: 20–60 W. Treatment volume: 300 mL. [ACP]: 82.69 μM. pH: 3–12. Temperature: 20 ± 1 °C. Addition: glucose, oxalic acid, propan-2-ol and hexan-1-ol.	The ultrasonic degradation in acidic medium (pH 3.0–5.6) is greater than that obtained in basic aqueous solutions (pH 9.5–12.0). The degradation of ACP would increase if its hydrophobicity is favored. The degradation rate increases with increasing acoustic power. The substrate degradation rate increases with increasing initial substrate concentration to a plateau. The presence of organic compounds negatively affects the sonochemical degradation efficiency of ACP, except glucose. A positive effect of mineral water was observed when the ACP concentration decreased 50 times (1.65 μM).
US	[36]	1-H- Benzotriazole (1HB)	[1HB]: 41.97–167.88 μM. Presence of oxygen, nitrogen, ozone and radical scavengers	With the increase in concentration, the degradation rate of 1HB also increased by 40%. A high applied ultrasonic power improved the degree of elimination of 1HB. The initial degradation rate accelerated in the presence of ozone and oxygen, but was inhibited by nitrogen. The most favorable pH for degradation was an acid medium. The removal of more than 90% of the contaminant was achieved

Table 2. *Cont.*

Process	Ref.	Pollutant/Type of Water	Operating Conditions	Found Results
US/Electro-oxidation (EO)	[27]	IBU/ Synthetic water and sewage	Co IBU Synthetic: 10 mg L^{-1} Increase in conductivity Na$_2$SO$_4$ 0.01 mol L^{-1}. Co IBU Municipal: 20, 100 μg L^{-1} and 10 mg L^{-1}. pH residual municipal: 6.6. Frequency: 520 kHz. Electric power: 10–40 W. Current densities: 3.6–35.7 mA cm^{-2}. Cylindrical reactor with a cathode and an anode immersed in the solution. Temperature: 5–40 °C. Working volume: 3 L. Treatment time: 30–180 min.	The best constant for speed and efficiency of degradation was obtained with the US/EO, process, followed by EO alone and then US alone. 84.74% elimination of the IBU was achieved with US/EO. In the EO process, HO• can be generated on the surface of the electrode, then the US increases the mass transfer between these and the contaminants. Between 10–40 °C there were no significant differences in the degradation of IBU. Intensity of the current and treatment time are the most influential factors. Optimum conditions are: 110 min treatment, 4.09 A and 20 W. In municipal sewage, 90% of IBU was removed.
US O$_3$ O$_3$/US US/UV O$_3$/UV US/O$_3$/UV	[62]	Azo dyes (AD), Endocrine Disrupting Compounds (EDC) and pharmaceuticals (PHAC)/ Synthetic water	Reactor 1: horn-type sonicator. Capacity of 100 mL. Frequency 20 kHz. Power: 0.46 W mL^{-1}. Reactor 2: plate-type sonicator. Frequency: 577, 866, 1100 kHz. Power intensity: 0.23 w mL^{-1}. Use US + O$_3$. Reactor 3: Ultrasonic bath. Frequency: 200 kHz. Power: 0.07 W mL^{-1}. Reactor 4: tailor-made hexagonal glass reactor coupled with 3 UV lamps (254 nm). Frequency: 520 kHz. Power: 0.19 W mL^{-1}.	AD degradation is faster by O$_3$/US. The UV/US process was very effective in degrading AD. With the addition of H$_2$O$_2$ a better discoloration was obtained. The rate of AD decomposition is faster in the presence of solid particles. EDCs had better degradation at alkaline pH and low frequency. At acidic pH, degradation was improved by adding Fenton or O$_3$ processes. For PHAC, ultrasonic processes were more efficient at high frequencies and acid pH.
US/Zn^0	[46]	DCF/Synthetic water	Co DCF: 10 mg L^{-1}. Reactor: Beakers, ultrasound probe. Working volume: 100 mL. pH: 2–7. Frequency: 20 kHz Power: 30–300 W. Treatment time: 30 min. Addition of Zn^0	At acid pH, the US process accompanied with Zn^0 was more efficient, while adding Zn^0 alone and experimenting with the US alone did not result in further degradation of DCF. At pH higher than 2 the DCF was not eliminated. At pH 2, degradation of 80.92% was achieved in 15 min. Process of US/Zn^0. There were no significant differences in degradation at different Zn^0 concentrations and different power densities. Dichlorination was the degradation pathway. The main aspect of this reaction, together with the Zn^0 reduction, was the O$_2$•$^-$.
US Fenton/US	[53]	IBU/Synthetic water and municipal sewage	Co IBU: 20 mg L^{-1}. pH: 2–8. Power density: 25–100 W L^{-1}. Frequency: 12–862 kHz. Addition of H$_2$O$_2$. Addition of Iron (Fe). HO scavenger agents: n-butanol and acetic acid. Reactor: 1 L glass. Ultrasound probe, cup horn type. Temperature: 25 °C.	At alkaline pH the degradation rate decreased significantly. The addition of H$_2$O$_2$ did not contribute to thedegradation of IBU by the US process. The sono-Fenton process was more efficient in eliminating the IBU than both processes separately. In the sono-Fenton process no significant influence on the degradation of the IBU was achieved by varying the power density in the studied range. In the municipal sewage the degradation was more effective with the combined processes, with results similar to those obtained with synthetic water. However, the efficiency of the individual US process decreased.

Table 2. *Cont.*

Process	Ref.	Pollutant/Type of Water	Operating Conditions	Found Results
US US/UV	[9]	CBZ/Synthetic water	Co CBZ: 0.00625–0.1 mM. Sonolytic Reactor: 500 mL Cylindrical glass beaker Frequency: 200 and 400 kHz. Power: 20–100 W. Temperature: 20 °C. pH: 2–11. Photolytic reactor: Camera with two low-pressure Hg lamps, 253.7 nm. Combined reactor: Assembly of the sonolytic reactor inside the photolytic reactor.	CBZ degradation follows a pseudo-first order kinetics. Faster degradation rate and greater removal with a frequency of 200 kHz. When methanol was applied as HO• sequestering agent, there was no significant drug removal. The HO• was the protagonist of the degradation. As electrical power increased, CBZ degradation increased. SO_4^{2-} and NO_3^- hindered the transfer of electrons during oxidation. The degradation of CBZ with UV radiation alone was negligible. The UV/US process achieved the highest CBZ removal. Twenty-one reaction intermediates were detected.
US/Single-walled carbon nanotubes	[31]	SFX and IBU/ Synthetic water	Co SFX and IBU: 10 μM. Single-walled carbon nanotubes (SCN). Stainless steel reactor. Frequency: 1000 kHz Power: 180 W pH: 3.5–7–9.5. Temperature: 15 to 55 °C. Reaction time: 60 min. Working volume: 1 L.	As the temperature increased, the cavitation threshold decreased, bubble formation increased together with the amount of HO•. At pH values below the pKa of the compounds, complete degradation was obtained within 50–60 minutes. At higher pH values, complete degradation was not achieved. In the presence of the SCN the degradation and the speed constant of the same was favored. The adsorption capacity of the SCN favored the removal of the compounds.
US/EO	[29]	CBZ /Synthetic water	Working volume: Reactor 1: 1 L and Reactor 2: 100 L. Cathode and anode in the form of expanded metal plates. Anode: Ti/PbO_2 Cathode: Ti Electric current: 1–15 A. Type of water: Potable (from the tap). Co CBZ: 10 mg L^{-1}. Na_2SO_4: 0.01 mol L^{-1}. Temperature: 20 °C. Ceramic transducer: diameter 4 cm. Frequency: 520 kHz. Power: between 10 and 40 W. Reaction time: between 90 and 180 min.	The combined US/EO process offered the best kinetic velocity constant. The degree of synergy, in the combination of the processes, rose with the increase in US power. As the current intensity increased, the depurative capacity rose. CBZ degradation was greater when the two processes (US and EO) were implemented simultaneously than separately. There was a 99.5% degradation of CBZ with the combined process.
US/O_2 /Fe	[67]	Metazachlor (MTZ)/Synthetic water	Generator US: 20 kHz. Titanium alloy probe. Co: 10 μM MTZ. pH: 3.0. Temperature: 22 °C. Presence or absence of dissolved oxygen. Presence or absence of nitrogen. Treatment time: 120 min. Addition of powdered ferric oxyhydroxide 50 mg L^{-1}.	MTZ degradation followed a pseudo-first order kinetics. The saturation of water with oxygen favored the degradation of MTZ. Excess oxygen can capture H• and avoid recombination with HO•. With the addition of ferric oxide and the recombination of HO• to produce H_2O_2, the Fenton process is generated in the middle of sonolysis. The application of US made the iron leaching process three times faster than conventional mechanical agitation, allowing better contact between the liquid and solid phases. 97% of MTZ was degraded with the addition of ferric oxide. The velocity constant was twice than that of US process alone.

Table 2. *Cont.*

Process	Ref.	Pollutant/Type of Water	Operating Conditions	Found Results
US/Additives	[68]	Oxacillin (OXA)/ Synthetic water	Working volume: 250 mL Electrical power: 60 W. Frequency: 275 kHz. Temperature: 20 °C. Mannitol and calcium carbonate were used as additives	In the presence of additives, OXA was efficiently removed. The sonochemical process was able to completely degrade the antibiotic, generating solutions without Antimicrobial Activity. The contaminant did not mineralize even after 360 min.
US/O$_3$	[32]	Benzophenone-3 (Bp3)/ Synthetic water	Frequency: 20 kHz. Electrical power: 55.9 W. Temperature: 25 °C. Working volume: 200 mL [Bp3]: 3.9 mg L^{-1}. pH: 2, 6.5 and 10. O$_3$: 0.5 mL min^{-1}. N$_2$ y O$_2$: 800 mL min^{-1}. Presence of nitrate, chloride and bicarbonate ions [5 mmol L^{-1}].	Increasing the electrical power also increases the degradation of Bp3. At a lower pH (2) a more effective degradation of Bp3 was observed. PKa Bp3: 8.06. The presence of O$_2$, O$_3$ and the combined process of US/O$_3$ improved the degradation of Bp3. Being faster US/O$_3$. Bicarbonate ions accelerated the degradation of Bp3.

Due to the demonstrated efficiency ascribed to the use of US-assisted AOPs in the degradation of ECs in water, it has been widely applied for tackling the problem of water pollution with these pollutants of growing concern [32,36,64]. As stated previously, it is highlighted that the water matrix is a topic of utmost importance when it comes to the evaluation of the pollutant removal capability through AOPs. In fact, in the literature, different works have been reported based on the elimination through US waves of various ECs commonly present in water matrices of different nature, from drinking water effluents to natural surface water, with domestic and industrial wastewaters being highly studied [40,69,70] due to the vast variety of compounds that can be found in these kinds of aqueous matrices.

For instance, Cetinkaya et al. [69] investigated the decolorization of textile waters using the sono-Fenton process, obtaining better results at pH 3, achieving 96% of color removal. The influence of ferrous ions (Fe^{2+}) concentration was analyzed, testing its variation between 0.05 g L^{-1} and 0.2 g L^{-1}. A color removal of 90% and 99% was observed with the lowest and the highest Fe^{2+} concentration, respectively. These results indicated that the sono-Fenton process required small amounts of Fe^{2+} to achieve high removals of the dyes. Additionally, H$_2$O$_2$ consumption was reduced by about 30% with the sono-Fenton process compared to the classic Fenton process. Furthermore, authors optimized operating parameters involved in the investigated AOP, achieving the highest removal of color at a frequency of 35 kHz, pH 3, 0.05 g L^{-1} of Fe^{2+}, 1.65 g L^{-1} of H$_2$O$_2$ and a treatment time of 60 min.

The removal of tetracycline (TC) has also been evaluated by Nasseri et al. [40] in a wastewater effluent by applying the US process. Some of the natural characteristics of the studied wastewater were: pH 7.9, chemical organic demand (COD) of 25 mg L^{-1}, HCO$_3^-$ content of 164 mg L^{-1}, Cl$^-$ of 92 mg L^{-1}, NO$_3^-$ of 24 mg L^{-1} and Na$^+$ of 50 mg L^{-1}. A lower removal rate of TC, but in the same order of magnitude, in wastewater (1.25×10^{-2} min^{-1}) compared to that one obtained in ultrapure water (1.75×10^{-2} min^{-1}) was observed. These results may be ascribed to the negative influence of the water constituents, as explained previously; in this case, due to the high levels of organic matter, in terms of COD, which can prevent the formation of OH• and, subsequently, reduce the rate of TC degradation.

In turn, Serna-Galvis et al. [71] experimented with wastewater from El Salitre Treatment Plant, located in Bogotá (Colombia), with the objective of applying the sono-photo-Fenton/Oxalic Acid AOP for the removal of the following pharmaceuticals: DCF, CBZ, venlafaxine, ciprofloxacin, norfloxacin, valsartan, losartan, irbesartan, SFX, clarithromycin, azithromycin, erythromycin, metronidazole, trimethoprimine and clinimetropimine, as well as cocaine and its main metabolite benzoylecgonine. The operating conditions were: 300 mL of working volume, 88 W L^{-1} of power density, 375 kHz of frequency, 20 °C of temperature, a UVA lamp of 4 W, a Fe^{2+} content of 5 mg L^{-1} and an oxalic acid concentration of 2 mg L^{-1}. It was observed that the application of the sonochemical process alone led to the release of contaminants from suspended solids. The addition of Fe^{2+}, UVA light and oxalic acid to the US process significantly increased the elimination of the studied ECs in the effluent, thanks to the production of additional HO• through reactions between iron and the sonogenerated H_2O_2. It is important to note that the presence of oxalic acid makes iron more available for the formation of additional free radicals within the solution, causing the improvement of EC degradation.

With the aim of comparing the findings of degradation reported by US in wastewater, the work conducted by Vilardi et al. [70], where the efficiency of conventional and heterogeneous Fenton for the degradation of contaminants present in the wastewater of a tannery in terms of COD, total phenolic compounds (TP) and Cr(VI), is presented. The authors carried out the experimentation at large laboratory scale using a reactor with a volume of 7.4 L. It was concluded that the heterogeneous Fenton process was significantly more efficient with respect to the conventional one for the elimination of COD and TP, once the optimal values of the operating parameters were found. The percentages of COD and TP removal for the heterogeneous Fenton were 75.5 ± 2.1% and 85.1 ± 0.7%, respectively. Likewise, it was observed that a smaller amount of iron sludge was produced due to the heterogeneous Fenton process (17.5%) compared to that one achieved through the conventional Fenton process (21.6%), which is a key aspect for the feasible implementation of the process at industrial scale.

Although the heterogeneous Fenton process implemented above was demonstrated to produce relatively low amounts of sludge, a more environmentally safe process must be required to overcome the pollution of aqueous resources with recalcitrant contaminants. In this regard, the use of US as an AOP alone or in combination with other advanced oxidation technologies seem to be an attractive treatment option.

3. Future Perspectives

Although the application of US alone as an advanced oxidation technology to overcome the critical situation ascribed to ECs in aqueous environments has been demonstrated to be efficient, the coupling of US with other AOPs could improve the mineralization of emerging organic compounds [72,73] within a further reduced time of treatment. For this reason, the use of US hybrid techniques has been recently studied to improve EC mineralization results [6,23,45,73]. A clear example of this is the combination of sonolysis with the Fenton process. This combination, which is so-called sono-Fenton, could stimulate a faster conversion and/or mineralization of ECs. This is achieved through: firstly, higher generation of HO• [74]; secondly, an improved mixture and contact between HO• and the pollutants of interest [72,73], and thirdly, improved generation of Fe^{2+} [75].

Different strategies in addition to the combination of the Fenton process with sonolysis have been tested in the last years. An example of this is the work developed by Tran et al. [20], where the electro-oxidation (EO) process was combined with US. This combination was based on the fact that, initially, the formation of HO• is achieved on the wall of an electrode made up of a non-active material through the EO process, and the chemical exchange of these HO• with contaminants could then be improved due to the formation of the US waves and cavitation bubbles resulting from the US process. In this study, a higher kinetic velocity constant and a greater efficiency in the removal of IBU was obtained with the combined process of EO/US in comparison with the results obtained in each process independently. As a result, 90% of the IBU contained in samples of municipal sewage was removed using optimal parameter levels, such as the treatment time, the current intensity and the US power, which were determined through the response surface methodology. The beneficial results of the exposure of electrochemical cells to the effects of US power are related to the improved mass transport, increased current efficiencies, and continuous electrode surface activation [20,76,77]. These effects can be attributed to the rapid generation and collapse of the micro-bubbles within the electrolyte medium or near the electrode surface [20,77,78].

In turn, Ince's 2018 study [62] evaluated the degradation of toxic ECs through US in combination with other AOPs. In this study PCT, DCF and IBU were analyzed, finding that the degradation of the selected ECs was more efficient at high frequencies and acid pH. Degradation was further improved with the presence of solid catalysts, which provided surfaces that enhanced the formation of cavitation bubbles and, therefore, the performance of the oxidation processes. In the referred research, the use of iron nano- and micro-particles resulted in a higher rate of DCF elimination by using nano-particles [62], which was attributed to the synergy of US with these particles through the enrichment of massive surfaces with excessive sorption sites and cavitation nuclei. In addition, reactions at the bubble–liquid interface were intensified by the distortion of asymmetric shapes, the degree of which increases as particle size decreases [62]. On the other hand, the coupling of an ozonation system with UV radiation and sonication, with the optional addition of $FeSO_4$, completely degraded DCF [62]. Finally, this work compared the efficiency of the following AOPs: US, O_3/US, UV/US and O_3/US/UV. High removals of the drugs of interest were found in all the tested processes, reaching about 100% elimination accompanied by a mineralization between 40 and 60% of all the ECs with the combination of US, O_3 and UV radiation.

In the work developed by Rao et al. [9], sonolytic and photolytic AOPs were combined for the degradation of CBZ. The result was a significant improvement in the drug degradation compared to the results obtained when the processes were individually implemented. The reason for this fact was related to the formation of H_2O_2 resulting from the recombination of HO• from sonolysis. This oxidizing agent can be photolized by UV light and more HO• can be produced, which are the main contributors to CBZ degradation.

CBZ removal was also studied by Mohapatra et al. [79], through the US process, Fenton and ferro-sonication (a combination of $FeSO_4$ with the US process). It was found that the most efficient AOP was the Fenton process, with elimination percentages between 84–100%; this was followed by ferro-sonication, with values between 62–93%, while sonolysis only achieved CBZ elimination percentages between 22%–51%. The authors concluded that the higher the radiation intensity (5.8, 12.4 and 16 W cm^{-2}), the greater the elimination of the target drug. Moreover, according to their research, the resulting ranges of efficiency between one process and another were because $FeSO_4$ contributed to the formation of a greater amount of HO•.

Although Fenton process has been proven to be an efficient technology for the degradation of some ECs [70,79], residual sludge is produced, especially when the homogeneous Fenton process is applied [70]. In this regard, further studies are needed to give an alternative use to such as sludge, contributing to the so-called principles of the circular economy. In this regard, Vilardi et al. [80] treated a tannery wastewater with mixed-iron coated olive stone bio-sorbent particles in combination with H_2O_2. They found a COD removal efficiency of 58.4% and a TP removal of 59.2%, at H_2O_2/COD (w/w) equal to 0.875. The coated olive stones were regenerated with sodium hydroxide (NaOH) and oxalic acid ($C_2H_2O_4$) solutions after five cycles in order to enable their reuse.

In addition, considerate the circular economy principles, economic costs analysis must be carried out in order to discern whether an AOP tested at laboratory or pilot plant can be scale up for industrial application in real water effluents.

4. Cost Consideration

As reviewed, the efficiency of ultrasound has been demonstrated to degrade any kind of recalcitrant pollutants. However, there are limitations related to the economic costs associated with the use of this advance oxidation technology for the treatment of water containing toxic pollutants [39]. One such limitation is the cost, which can be divided into two groups: the capital or inversion costs, which consists of those costs associated with the manufacture of the sonochemical reactors and can be amortized over a span of years at a considered amortization rate [39], and the operation and maintenance costs. The economic cost estimation linked to the operation and maintenance labor include the part replacements, which mainly consists of the transducer element replacement and the tip or electronic circuit replacements. In fact, according to Mahamuni and Adewuyi [39], the part replacement costs are assumed to be 0.5% of the capital costs. Labor and analytical costs must also be considered when operation and maintenance costs are estimated. Labor costs include inspection, repair and replacement based on hours of service life of control panels, leakages and pressure gauge, among others. In turn, analytical costs consist of the costs related to the analysis of samples and, subsequently, the costs associated with the reactants and chemicals used for the sample analysis. Additionally, electrical costs, which can be based on the power consumption of the referred AOP, are of utmost importance since they are usually very high, especially in those countries where the cost of each kWatt is high. Hence, the use of renewable resources for generating electrical energy is an attractive option that is emerging for the advanced oxidation system to be implemented. As a matter of fact, Rubio-Clemente et al. [81] assessed the efficiency of the UV/H_2O_2 system powered by a photovoltaic (PV) system in a photochemical reactor at laboratory scale. According to the results reported by the authors, similar efficiency was observed between the oxidation system powered with energy from the electrical grid and that one generated using the PV cells implemented.

Another alternative for reducing the economic costs related to the use of ultrasound for treating polluted water is utilizing hybrid oxidation techniques by combining US with other AOP, including the use of oxidizing or catalyzing agents, such as ozone (O_3), H_2O_2, iron, titanium dioxide (TiO_2), wolfram trioxide (WO_3), zinc oxide (ZnO), etc., and electrochemistry to name just a few. In this regard, Expósito et al. [25] evaluated the efficiencies of mineralization in terms of total organic carbon (TOC) and CBZ removal by using the US/UV/H_2O_2/Fe oxidation process at laboratory scale in a thin film UV reactor coupled to a 24 kHz 200 W direct immersion horn-type sonicator, obtaining efficiencies around 90%, which are higher than the efficiencies reached by the processes alone. In fact, a synergistic effect higher than 55% was found between the US process and UV irradiation.

However, although application of US hybrid techniques in some occasions can be more attractive for water treatment, Mahamuni and Adewuyi [39] reported that the costs associated with these treatment techniques are one to two orders of magnitude higher than when US is implemented alone. This can be ascribed to the costs linked to the additional chemicals used for the hybrid process to occur, i.e., the use of oxidizing agents such as O_3 and H_2O_2, or the catalyzing agents as iron salts, TiO_2, ZnO or WO_3, among others, as well as the adjustment of the pH of the solution if needed. Moreover, when US is used along with UV radiation, the costs associated with the replacements of the lamps and the electrical consumption of the lamps must be considered, as well as those ones related to the O_3 generator repair when O_3 is combined with US.

With this in mind, it can be concluded that the cost estimation studies based on pilot plants would be of high importance for to discern both the capital and the operation and maintenance costs related to the implementation of the US process. Furthermore, although high efficiencies can be obtained in a short period of time by using hybrid techniques with US, the economic costs associated with it are higher; therefore, further studies are needed to discern if the combination of US with another AOP is worth to be implemented under any circumstances. On the other hand, the type of pollutant plays a crucial role on the cost estimation procedures, since treating water containing hydrophobic pollutants has lower costs ascribed in comparison with those ones for treating compounds of hydrophilic nature [39].

5. Conclusions

After a critical review of the results found in the literature concerning the US process for the elimination of ECs, it is important to highlight the following conclusions:

- The US process is environmentally clean, as it does not produce chemical residues or sludge in comparison with other AOPs, such as Fenton and photo-Fenton processes, and other advanced oxidation technologies using catalysts, including TiO_2, ZnO and WO_3, among others.

- The nature of the pollutant is an issue of utmost concern when evaluating the efficiency of the ultrasound process, since hydrophobic, non-polar and/or volatile compounds react inside the cavitation bubbles and at the bubble/water interface, while hydrophilic and/or non-volatile pollutants react within the bulk solution.

- On the other hand, the operating parameters, such as the pH and the temperature of the solution, ultrasonic frequency, electrical power, dissolved gases and the nature and concentration of the pollutant, must be evaluated under a wide range, since the efficiency of the process depends on them. In this regard, the considered operating factors should be optimized in order to maximize the degradation of the pollutant of interest and minimize the operation and maintenance costs.

- The degradation efficiency of aqueous pollutants also depends, to a large extent, on the type of sonoreactor and the geometry of the system. Therefore, the optimization of the sonoreactor, in terms of geometry and type, is recommended to be carried out especially when scaling the US-assisted AOP up.

- Further researches are needed for evaluating the efficiency of the referred process in real water matrices since, as reviewed, aqueous matrix background can highly influence the efficiency of the oxidation system and, subsequently, the degradation of the pollutant to be studied.

- The combination of ultrasound with other advanced oxidation or conventional processes used for water treatment can offer a high percentage of removal and mineralization of the compound under study. However, the associated economic costs are commonly higher than when US is applied alone. Therefore, further studies based on the efficiency about the cost estimation of the US oxidation process alone and in combination with other AOPs are required, especially in pilot plants, to obtain a closer point of view for the advanced oxidation technology scale-up.

Author Contributions: Conceptualization, A.L.C.-P., A.R.-C., G.A.P.; investigation, A.L.C.-P., A.R.-C., G.A.P.; writing—original draft preparation, A.L.C.-P.; writing—review and editing, A.R.-C., G.A.P.; supervision, A.R.-C., G.A.P. All authors have read and agreed to the published version of the manuscript.

Acknowledgments: To the Universidad de Antioquia for its commitment to education in the country and its stimulus "Student Instructor" that contributed to the development of this article. To the Research Group "Diagnóstico y Control de la Contaminación-GDCON" for its support in the development of this review.

References

1. Gil, M.J.; Soto, A.M.; Usma, J.I.; Gutiérrez, O.D. Contaminantes emergentes en aguas, efectos y posibles tratamientos. *Producción + Limpia* **2012**, *7*. Available online: https://www.hit2lead.com/ (accessed on 9 April 2020).

2. Zhang, Y.; Geißen, S.U.; Gal, C. Carbamazepine and diclofenac: Removal in wastewater treatment plants and occurrence in water bodies. *Chemosphere* **2008**, *73*, 1151–1161. [CrossRef] [PubMed]

3. Gogoi, A.; Mazumder, P.; Tyagi, V.K.; Tushara Chaminda, G.G.; An, A.K.; Kumar, M. Occurrence and fate of emerging contaminants in water environment: A review. *Groundw. Sustain. Dev.* **2018**, *6*, 169–180. [CrossRef]

4. Aristizabal-Ciro, C.; Botero-Coy, B.; López, F.; Peñuela, G.A. Monitoring pharmaceuticals and personal care products in reservoir water used for drinking water supply. *Environ. Sci. Pollut. Res.* **2017**, *24*, 7335–7347. [CrossRef] [PubMed]

5. Rozman, D.; Hrkal, Z.; Váňa, M.; Vymazal, J.; Boukalová, Z. Occurrence of pharmaceuticals in wastewater and their interaction with shallow aquifers: A case study of Horní Beřkovice, Czech Republic. *Water* **2017**, *9*, 218. [CrossRef]

6. Naddeo, V.; Belgiorno, V.; Ricco, D.; Kassinos, D. Degradation of diclofenac during sonolysis, ozonation and their simultaneous application. *Ultrason. Sonochem.* **2009**, *16*, 790–794. [CrossRef] [PubMed]

7. Hai, F.I.; Yang, S.; Asif, M.B.; Sencadas, V.; Shawkat, S.; Sanderson-Smith, M.; Gorman, J.; Xu, Z.Q.; Yamamoto, K. Carbamazepine as a possible anthropogenic marker in water: Occurrences, toxicological effects, regulations and removal by wastewater treatment technologies. *Water* **2018**, *10*, 107. [CrossRef]

8. Emmanouil, C.; Bekyrou, M.; Psomopoulos, C.; Kungolos, A. An Insight into Ingredients of Toxicological Interest in Personal Care Products and A Small–Scale Sampling Survey of the Greek Market: Delineating a Potential Contamination Source for Water Resources. *Water* **2019**, *11*, 2501. [CrossRef]

9. Rao, Y.; Yang, H.; Xue, D.; Guo, Y.; Qi, F.; Ma, J. Sonolytic and sonophotolytic degradation of Carbamazepine: Kinetic and mechanisms. *Ultrason. Sonochem.* **2016**, *32*, 371–379. [CrossRef]

10. Cleuvers, M. Mixture toxicity of the anti-inflammatory drugs diclofenac, ibuprofen, naproxen, and acetylsalicylic acid. *Ecotoxicol. Environ. Saf.* **2004**, *59*, 309–315. [CrossRef]

11. Jarvis, A.L.; Bernot, M.J.; Bernot, R.J. The effects of the psychiatric drug carbamazepine on freshwater invertebrate communities and ecosystem dynamics. *Sci. Total Environ.* **2014**, *496*, 461–470. [CrossRef] [PubMed]

12. Almeida, Â.; Calisto, V.; Esteves, V.I.; Schneider, R.J.; Soares, A.M.V.M.; Figueira, E.; Freitas, R. Presence of the pharmaceutical drug carbamazepine in coastal systems: Effects on bivalves. *Aquat. Toxicol.* **2014**, *156*, 74–87. [CrossRef] [PubMed]

13. Han, S.; Choi, K.; Kim, J.; Ji, K.; Kim, S.; Ahn, B.; Choi, K.; Khim, J.S.; Zhang, X.; Giesy, J.P. Endocrine disruption and consequences of chronic exposure to ibuprofen in Japanese medaka (Oryzias latipes) and freshwater cladocerans Daphnia magna and Moina macrocopa. *Aquat. Toxicol.* **2010**, *98*, 256–264. [CrossRef] [PubMed]

14. Gonzalez-Rey, M.; Bebianno, M.J. Does non-steroidal anti-inflammatory (NSAID) ibuprofen induce antioxidant stress and endocrine disruption in mussel Mytilus galloprovincialis? *Environ. Toxicol. Pharmacol.* **2012**, *33*, 361–371. [CrossRef]

15. Schmidt, W.; O'Rourke, K.; Hernan, R.; Quinn, B. Effects of the pharmaceuticals gemfibrozil and diclofenac on the marine mussel (Mytilus Spp.) and their comparison with standardized toxicity tests. *Mar. Pollut. Bull.* **2011**, *62*, 1389–1395. [CrossRef] [PubMed]

16. Guiloski, I.C.; Ribas, J.L.C.; da Silva Pereira, L.; Neves, A.P.P.; Silva de Assis, H.C. Effects of trophic exposure to dexamethasone and diclofenac in freshwater fish. *Ecotoxicol. Environ. Saf.* **2015**, *114*, 204–211. [CrossRef]

17. Rubio-Clemente, A.; Torres-Palma, R.A.; Peñuela, G.A. Removal of polycyclic aromatic hydrocarbons in aqueous environment by chemical treatments: A review. *Sci. Total Environ.* **2014**, *478*, 201–225. [CrossRef]

18. Tran, N.; Drogui, P.; Zaviska, F.; Brar, S.K. Sonochemical degradation of the persistent pharmaceutical carbamazepine. *J. Environ. Manag.* **2013**, *131*, 25–32. [CrossRef]

19. González, K.; Quesada, I.; Julcour, C.; Delmas, H.; Cruz, G.; Jáuregui, U.J. El empleo del ultrasonido en el tratamiento de aguas residuales. *Rev. CENIC Cienc. Químicas* **2010**, *41*, 1–11.

20. Tran, N.; Drogui, P.; Brar, S.K. Sonoelectrochemical oxidation of carbamazepine in waters: Optimization using response surface methodology. *J. Chem. Technol. Biotechnol.* **2015**, *90*, 921–929. [CrossRef]

21. Ikehata, K.; Naghashkar, N.J.; El-Din, M.G. Degradation of Aqueous Pharmaceuticals by Ozonation and Advanced Oxidation Processes: A Review. *Ozone Sci. Eng.* **2006**, *28*, 353–414. [CrossRef]

22. Torres-Palma, R.A.; Serna-Galvis, E.A. Chapter 7 Sonolysis. In *Advanced Oxidation Processes for Waste Water Treatment*; Ameta, S.C., Ameta, R., Eds.; Academic Press: Cambridge, MA, USA, 2018; pp. 177–213. [CrossRef]

23. Güyer, G.T.; Ince, N.H. Degradation of diclofenac in water by homogeneous and heterogeneous sonolysis. *Ultrason. Sonochem.* **2011**, *18*, 114–119. [CrossRef] [PubMed]

24. Lin, L.; Wang, H.; Xu, P. Immobilized TiO$_2$-reduced graphene oxide nanocomposites on optical fibers as high performance photocatalysts for degradation of pharmaceuticals. *Chem. Eng. J.* **2017**, *310*, 389–398. [CrossRef]

25. Expósito, A.J.; Patterson, D.A.; Monteagudo, J.M.; Durán, A. Sono-photo-degradation of carbamazepine in a thin falling film reactor: Operation costs in pilot plant. *Ultrason. Sonochem.* **2017**, *34*, 496–503. [CrossRef]

26. Kakavandi, B.; Ahmadi, M. Efficient treatment of saline recalcitrant petrochemical wastewater using heterogeneous UV-assisted sono-Fenton process. *Ultrason. Sonochem.* **2019**, *56*, 25–36. [CrossRef]

27. Tran, N.; Drogui, P.; Nguyen, L.; Brar, S.K. Optimization of sono-electrochemical oxidation of ibuprofen in wastewater. *J. Environ. Chem. Eng.* **2015**, *3*, 2637–2646. [CrossRef]

28. Nie, E.; Yang, M.; Wang, D.; Yang, X.; Luo, X.; Zheng, Z. Degradation of diclofenac by ultrasonic irradiation: Kinetic studies and degradation pathways. *Chemosphere* **2014**, *113*, 165–170. [CrossRef]

29. Tran, N.; Drogui, P.; Brar, S.K.; De Coninck, A. Synergistic effects of ultrasounds in the sonoelectrochemical oxidation of pharmaceutical carbamazepine pollutant. *Ultrason. Sonochem.* **2017**, *34*, 380–388. [CrossRef]

30. Rubio-Clemente, A.; Chica, E.; Peñuela, G. Total coliform inactivation in natural water by UV/H$_2$O$_2$, UV/US, and UV/US/H$_2$O$_2$ systems. *Environ. Sci. Pollut. Res.* **2019**, *26*, 4462–4473. [CrossRef]

31. Al-Hamadani, Y.A.J.; Chu, K.H.; Flora, J.R.V.; Kim, D.H.; Jang, M.; Sohn, J.; Yoon, Y. Sonocatalytical degradation enhancement for ibuprofen and sulfamethoxazole in the presence of glass beads and single-walled carbon nanotubes. *Ultrason. Sonochem.* **2016**, *32*, 440–448. [CrossRef]

32. Zúñiga-Benítez, H.; Soltan, J.; Peñuela, G.A. Application of ultrasound for degradation of benzophenone-3 in aqueous solutions. *Int. J. Environ. Sci. Technol.* **2016**, *13*, 77–86. [CrossRef]

33. Vega, L.P.; Gomez-Miranda, I.N.; Peñuela, G.A. Benzophenone-3 ultrasound degradation in a multifrequency reactor: Response surface methodology approach. *Ultrason. Sonochem.* **2018**, *43*, 201–207. [CrossRef]

34. Vega, L.P.; Soltan, J.; Peñuela, G.A. Sonochemical degradation of triclosan in water in a multifrequency reactor. *Environ. Sci. Pollut. Res. Int.* **2019**, *26*, 4450–4461. [CrossRef] [PubMed]

35. Pétrier, C.; Torres-Palma, R.; Combet, E.; Sarantakos, G.; Baup, S.; Pulgarin, C. Enhanced sonochemical degradation of bisphenol-A by bicarbonate ions. *Ultrason. Sonochem.* **2010**, *17*, 111–115. [CrossRef] [PubMed]

36. Zuñiga, H.; Soltan, J.; Peñuela, G.A. Ultrasonic degradation of 1-H-Benzotriazole in water. *Water Sci. Technol.* **2014**, *70*, 152–159. [CrossRef]

37. Ince, N.H.; Tezcanli, G.; Belen, R.K.; Apikyan, İ.G. Ultrasound as a catalyzer of aqueous reaction systems: The state of the art and environmental applications. *Appl. Catal. B Environ.* **2001**, *29*, 167–176. [CrossRef]

38. Litter, M.; Quici, N. Photochemical Advanced Oxidation Processes for Water and Wastewater Treatment. *Recent Pat. Eng.* **2010**, *4*, 217–241. [CrossRef]

39. Mahamuni, N.N.; Adewuyi, Y.G. Advanced oxidation processes (AOPs) involving ultrasound for waste water treatment: A review with emphasis on cost estimation. *Ultrason. Sonochem.* **2010**, *17*, 990–1003. [CrossRef]

40. Nasseri, S.; Mahvi, A.H.; Seyedsalehi, M.; Yaghmaeian, K.; Nabizadeh, R.; Alimohammadi, M.; Safari, G.H. Degradation kinetics of tetracycline in aqueous solutions using peroxydisulfate activated by ultrasound irradiation: Effect of radical scavenger and water matrix. *J. Mol. Liq.* **2017**, *241*, 704–714. [CrossRef]

41. Jiang, Y.; Pétrier, C.; David Waite, T. Kinetics and mechanisms of ultrasonic degradation of volatile chlorinated aromatics in aqueous solutions. *Ultrason. Sonochem.* **2002**, *9*, 317–323. [CrossRef]

42. Méndez-Arriaga, F.; Torres-Palma, R.A.; Pétrier, C.; Esplugas, S.; Gimenez, J.; Pulgarin, C. Ultrasonic treatment of water contaminated with ibuprofen. *Water Res.* **2008**, *42*, 4243–4248. [CrossRef] [PubMed]

43. Chiha, M.; Merouani, S.; Hamdaoui, O.; Baup, S.; Gondrexon, N.; Pétrier, C. Modeling of ultrasonic degradation of non-volatile organic compounds by Langmuir-type kinetics. *Ultrason. Sonochem.* **2010**, *17*, 773–782. [CrossRef] [PubMed]

44. Pavlovna, I.; Vladimirovna, S.; Mihailovich, I.; Alekseevna, N.; Evgenevna, O.; Olegovna, O. Mechanism of chemiluminescence in Fenton reaction. *J. Biophys. Chem.* **2012**, *3*, 88–100. [CrossRef]

45. Naddeo, V.; Belgiorno, V.; Kassinos, D.; Mantzavinos, D.; Meric, S. Ultrasonic degradation, mineralization and detoxification of diclofenac in water: Optimization of operating parameters. *Ultrason. Sonochem.* **2010**, *17*, 179–185. [CrossRef]

46. Huang, T.; Zhang, G.; Chong, S.; Liu, Y.; Zhang, N.; Fang, S.; Zhu, J. Effects and mechanism of diclofenac degradation in aqueous solution by US/Zn$_0$. *Ultrason. Sonochem.* **2017**, *37*, 676–685. [CrossRef]

47. Hartmann, J.; Bartels, P.; Mau, U.; Witter, M.; Tümpling, W.V.; Hofmann, J.; Nietzschmann, E. Degradation of the drug diclofenac in water by sonolysis in presence of catalysts. *Chemosphere* **2008**, *70*, 453–461. [CrossRef]

48. Song, W.; Teshiba, T.; Rein, K.; O'Shea, K.E. Ultrasonically Induced Degradation and Detoxification of Microcystin-LR (Cyanobacterial Toxin). *Environ. Sci. Technol.* **2005**, *39*, 6300–6305. [CrossRef]

49. Drijvers, D.; Van Langenhove, H.; Vervaet, K. Sonolysis of chlorobenzene in aqueous solution: Organic intermediates. *Ultrason. Sonochem.* **1998**, *5*, 13–19. [CrossRef]

50. Zhang, G.; Hua, I. Cavitation chemistry of polychlorinated biphenyls: Decomposition mechanisms and rates. *Environ. Sci. Technol.* **2000**, *34*, 1529–1534. [CrossRef]

51. Hoffmann, M.H. Die in Zentraleuropa verwilderten und kultivierten nordamerikanischen Astern. *Feddes Repert.* **1996**, *107*, 163–188. [CrossRef]

52. Kidak, R.; Dogan, S. Degradation of trace concentrations of alachlor by medium frequency ultrasound. *Chem. Eng. Process. Process Intensif.* **2015**, *89*, 19–27. [CrossRef]

53. Adityosulindro, S.; Barthe, L.; González-Labrada, K.; Jáuregui, U.J.; Delmas, H.; Julcour, C. Sonolysis and sono-Fenton oxidation for removal of ibuprofen in (waste) water. *Ultrason. Sonochem.* **2017**, *39*, 889–896. [CrossRef] [PubMed]

54. Villaroel, E.; Silva-Agredo, J.; Petrier, C.; Taborda, G.; Torres-Palma, R.A. Ultrasonic degradation of acetaminophen in water: Effect of sonochemical parameters and water matrix. *Ultrason. Sonochem.* **2014**, *21*, 1763–1769. [CrossRef] [PubMed]

55. Petrier, C.; Jeunet, A.; Luche, J.L.; Reverdy, G. Unexpected frequency effects on the rate of oxidative processes induced by ultrasound. *J. Am. Chem. Soc.* **1992**, *114*, 3148–3150. [CrossRef]

56.   Pétrier, C.; Francony, A. Ultrasonic waste-water treatment: Incidence of ultrasonic frequency on the rate of phenol and carbon tetrachloride degradation. *Ultrason. Sonochem.* **1997**, *4*, 295–300. [CrossRef]

57.   Beckett, M.A.; Hua, I. Impact of Ultrasonic Frequency on Aqueous Sonoluminescence and Sonochemistry. *J. Phys. Chem. A* **2001**, *105*, 3796–3802. [CrossRef]

58.   Petrier, C.; David, B.; Laguian, S. Ultrasonic degradation at 20 kHz and 500 kHz of atrazine and pentachlorophenol in aqueous solution: Preliminary results. *Chemosphere* **1996**, *32*, 1709–1718. [CrossRef]

59.   Ziylan, A.; Koltypin, Y.; Gedanken, A.; Ince, N.H. More on sonolytic and sonocatalytic decomposition of Diclofenac using zero-valent iron. *Ultrason. Sonochem.* **2013**, *20*, 580–586. [CrossRef]

60.   Gogate, P.R.; Sutkar, V.S.; Pandit, A.B. Sonochemical reactors: Important design and scale up considerations with a special emphasis on heterogeneous systems. *Chem. Eng. J.* **2011**, *166*, 1066–1082. [CrossRef]

61.   Madhavan, J.; Kumar, P.S.S.; Anandan, S.; Zhou, M.; Grieser, F.; Ashokkumar, M. Ultrasound assisted photocatalytic degradation of diclofenac in an aqueous environment. *Chemosphere* **2010**, *80*, 747–752. [CrossRef]

62.   Ince, N.H. Ultrasound-assisted advanced oxidation processes for water decontaminaration. *Ultrason. Sonochem.* **2018**, *40*, 97–103. [CrossRef] [PubMed]

63.   Grebel, J.E.; Pignatello, J.J.; Mitch, W.A. Effect of Halide Ions and Carbonates on Organic Contaminant Degradation by Hydroxyl Radical-Based Advanced Oxidation Processes in Saline Waters. *Environ. Sci. Technol.* **2010**, *44*, 6822–6828. [CrossRef] [PubMed]

64.   Yang, Y.; Pignatello, J.J.; Ma, J.; Mitch, W.A. Comparison of Halide Impacts on the Efficiency of Contaminant Degradation by Sulfate and Hydroxyl Radical-Based Advanced Oxidation Processes (AOPs). *Environ. Sci. Technol.* **2014**, *48*, 2344–2351. [CrossRef] [PubMed]

65.   Golash, N.; Gogate, P.R. Degradation of dichlorvos containing wastewaters using sonochemical reactors. *Ultrason. Sonochem.* **2012**, *19*, 1051–1060. [CrossRef] [PubMed]

66.   Rayaroth, M.P.; Aravind, U.K.; Aravindakumar, C.T. Effect of inorganic ions on the ultrasound initiated degradation and product formation of triphenylmethane dyes. *Ultrason. Sonochem.* **2018**, *48*, 482–491. [CrossRef]

67.   Kask, M.; Krichevskaya, M.; Bolobajev, J. Sonolytic degradation of pesticide metazachlor in water: The role of dissolved oxygen and ferric sludge in the process intensification. *J. Environ. Chem. Eng.* **2019**, *7*, 103095. [CrossRef]

68.   Serna-Galvis, E.A.; Silva-Agredo, J.; Giraldo-Aguirre, A.L.; Flórez-Acosta, O.A.; Torres-Palma, R.A. High frequency ultrasound as a selective advanced oxidation process to remove penicillinic antibiotics and eliminate its antimicrobial activity from water. *Ultrason. Sonochem.* **2016**, *31*, 276–283. [CrossRef]

69.   Cetinkaya, S.G.; Morcali, M.H.; Akarsu, S.; Ziba, C.A.; Dolaz, M. Comparison of classic Fenton with ultrasound Fenton processes on industrial textile wastewater. *Sustain. Environ. Res.* **2018**, *28*, 165–170. [CrossRef]

70.   Vilardi, G.; Rodríguez-Rodríguez, J.; Ochando-Pulido, J.M.; Verdone, N.; Martinez-Ferez, A.; Di Palma, L. Large Laboratory-Plant application for the treatment of a Tannery wastewater by Fenton oxidation: Fe(II) and nZVI catalysts comparison and kinetic modelling. *Process Saf. Environ. Prot.* **2018**, *117*, 629–638. [CrossRef]

71.   Serna-Galvis, E.A.; Botero-Coy, A.M.; Martínez-Pachón, D.; Moncayo-Lasso, A.; Ibáñez, M.; Hernández, F.; Torres-Palma, R.A. Degradation of seventeen contaminants of emerging concern in municipal wastewater effluents by sonochemical advanced oxidation processes. *Water Res.* **2019**, *154*, 349–360. [CrossRef]

72.   Bagal, M.V.; Gogate, P.R. Degradation of diclofenac sodium using combined processes based on hydrodynamic cavitation and heterogeneous photocatalysis. *Ultrason. Sonochem.* **2014**, *21*, 1035–1043. [CrossRef] [PubMed]

73.   Liang, J.; Komarov, S.; Hayashi, N.; Kasai, E. Improvement in sonochemical degradation of 4-chlorophenol by combined use of Fenton-like reagents. *Ultrason. Sonochem.* **2007**, *14*, 201–207. [CrossRef] [PubMed]

74.   Neppolian, B.; Jung, H.; Choi, H.; Lee, J.H.; Kang, J.W. Sonolytic degradation of methyl tert-butyl ether: The role of coupled fenton process and persulphate ion. *Water Res.* **2002**, *36*, 4699–4708. [CrossRef]

75.   Lin, J.G.; Ma, Y.S. Oxidation of 2-Chlorophenol in Water by Ultrasound/Fenton Method. *J. Environ. Eng.* **2000**, *126*, 130–137. [CrossRef]

76.   Birkin, P.R.; Silva-Martinez, S. A study on the effect of ultrasound on electrochemical phenomena. *Ultrason. Sonochem.* **1997**, *4*, 121–122. [CrossRef]

77.   Klima, J.; Bernard, C.; Degrand, C. Sonoelectrochemistry: Transient cavitation in acetonitrile in the neighbourhood of a polarized electrode. *J. Electroanal. Chem.* **1995**, *399*, 147–155. [CrossRef]

78. Macounova, K.; Klima, J.; Bernard, C.; Degrand, C. Ultrasound-assisted anodic oxidation of diuron. *J. Electroanal. Chem.* **1998**, *457*, 141–147. [CrossRef]
79. Mohapatra, D.P.; Brar, S.K.; Tyagi, R.D.; Picard, P.; Surampalli, R.Y. A comparative study of ultrasonication, Fenton's oxidation and ferro-sonication treatment for degradation of carbamazepine from wastewater and toxicity test by Yeast Estrogen Screen (YES) assay. *Sci. Total Environ.* **2013**, *447*, 280–285. [CrossRef]
80. Vilardi, G.; Ochando-Pulido, J.M.; Stoller, M.; Verdone, N.; Di Palma, L. Fenton oxidation and chromium recovery from tannery wastewater by means of iron-based coated biomass as heterogeneous catalyst in fixed-bed columns. *Chem. Eng. J.* **2018**, *351*, 1–11. [CrossRef]
81. Rubio-Clemente, A.; Chica, E.; Peñuela, G.A. Photovoltaic array for powering advanced oxidation processes: Sizing, application and investment costs for the degradation of a mixture of anthracene and benzo[a]pyrene in natural water by the UV/H$_2$O$_2$ system. *J. Environ. Chem. Eng.* **2018**, *6*, 2751–2761. [CrossRef]

10

Removal Characteristics of Effluent Organic Matter (EfOM) in Pharmaceutical Tailwater by a Combined Coagulation and UV/O$_3$ Process

Jian Wang [1,2], Yonghui Song [2,*], Feng Qian [2,3,*], Cong Du [2,3], Huibin Yu [2,3]
and Liancheng Xiang [2,3]

[1] State Key Joint Laboratory of Environment Simulation and Pollution Control, School of Environment,
Tsinghua University, Beijing 100084, China; jian-wan14@mails.tsinghua.edu.cn
[2] State Key Laboratory of Environmental Criteria and Risk Assessment, Chinese Research Academy of
Environmental Sciences, Beijing 100012, China; ducongducong@126.com (C.D.); yhbybx@163.com (H.Y.);
xianglc@craes.org.cn (L.X.)
[3] Department of Urban Water Environmental Research, Chinese Research Academy of Environmental Sciences,
Beijing 100012, China
* Correspondence: songyh@craes.org.cn (Y.S.); qianfeng@craes.org.cn (F.Q.)

Abstract: A novel coagulation combined with UV/O$_3$ process was employed to remove the effluent organic matter (EfOM) from a biotreated pharmaceutical wastewater for harmlessness. The removal behavior of EfOM by UV/O$_3$ process was characterized by synchronous fluorescence spectroscopy (SFS) integrating two-dimensional correlation (2D-COS) and principal component analysis (PCA) technology. The highest dissolved organic carbon (DOC) and ratio of UV$_{254}$ and DOC (SUVA) removal efficiency reached 55.8% and 68.7% by coagulation-UV/O$_3$ process after 60 min oxidation, respectively. Five main components of pharmaceutical tail wastewater (PTW) were identified by SFS. Spectral analysis revealed that UV/O$_3$ was selective for the removal of different fluorescent components, especially fulvic acid-like fluorescent (FLF) component and humus-like fluorescent (HLF) component. Synchronous fluorescence/UV-visible two-dimensional correlation spectra analysis showed that the degradation of organic matter occurred sequentially in the order of HLF, FLF, microbial humus-like fluorescence component (MHLF), tryptophan-like fluorescent component (TRLF), tyrosine-like fluorescent component (TYLF). The UV/O$_3$ process removed 95.6% of HLF, 80.0% of FLF, 56.0% of TRLF, 50.8% of MHLF and 44.4% of TYLF. Therefore, the coagulation-UV/O$_3$ process was proven to be an attractive way to reduce the environmental risks of PTW.

Keywords: pharmaceutical wastewater; UV/O$_3$; coagulation; fluorescence spectrum; removal characteristics

1. Introduction

Pharmaceutical wastewater is one of the important sources for emerging pollutants such as hormones, antibiotics and non-biodegradable organic intermediates [1–3]. The pharmaceutical residue usually entered the aquatic environment via sewage, and even low concentration can impact the drinking water and human health [4]. Despite undergoing biological treatment, the pharmaceutical residue cannot be completely metabolized [2,5,6]. Some nonbiodegradable and toxic substances still exist in biologically treated effluent (pharmaceutical tail wastewater, PTW). Therefore, intensive treatment must be carried out to realize harmlessness and reduce environmental risks [7].

After biological treatment, the biodegradability of PTW was low and not suitable for continued biological treatment [3,7,8]. The BOD$_5$/COD (B/C) of PTW was only close to 0.1, meaning that the organic matter from PTW was difficult to biodegrade. Therefore, the effect of biological treatment would be very poor in the further processing, physical chemistry and some advanced oxidation techniques should be considered. Among these, coagulation and UV/O$_3$ were the most two common technologies to remove specific organic pollutants from wastewater, such as contaminated groundwater, drinking water, industrial wastewater and landfill leachate [9–11]. Combined application of coagulation and UV/O$_3$ are suitable to treat PTW, since it has a certain removal effect on most organic matter ranging from low to high molecular weight. Coagulation treatment can effectively remove suspended particles, colloidal particles and dissolved organic matter (DOM) in sewage [12]. UV radiation can stimulate ozone oxidation to produce highly reactive hydroxyl radicals, which have the ability to oxidize and remove almost any organic contaminants [13]. However, studies on using UV/O$_3$, coagulation or combined coagulation-UV/O$_3$ to remove organic components of PTW have rarely been reported. Understanding the removal characteristics of effluent organic matter (EfOM) of PTW was an important basis for examining the effect of coagulation-UV/O$_3$ treatment. Therefore, it was necessary to find alternative characterization methods to evaluate the removal of EfOM during the oxidation process.

DOM, as the most important part of EfOM, can determine the coagulant, disinfection by-products, membrane fouling, type of microbial activity, and removal effect of contaminants [14,15]. The fluorescent component of DOM can be analyzed by fluorescence spectroscopy technology. The outstanding advantage of fluorescence spectroscopy is that it can quickly obtain measured DOM characteristic information with high sensitivity and has the characteristics of non-destructiveness and low cost [16,17]. Many fluorescent components in DOM can be analyzed and determined by combining fluorescence excitation-emitter matrix with parallel factors, self-organizing mapping algorithms or area integral methods [18]. Among them, synchronous fluorescence spectroscopy (SFS) was constituted by the measured fluorescence intensity signal and the corresponding excitation wavelength (or emission wavelength), based on simultaneous wavelength scanning of excitation and emission monochromators [19]. It has been widely used in the simultaneous analysis of heterogeneous mixtures, due to its advantages of simplified spectrum, reduced light scattering and improved selectivity. However, partial overlap of SFS wavelength can reduce its selectivity in DOM analysis. This problem can be solved through the combination of SFS and two-dimensional correlation (2D-COS) by extending peaks in the second dimension. Additionally, 2D-COS can be used as strong evidence for detecting correlations between features of dynamic spectra at two different wavelengths. It showed that combined SFS with 2D-COS can effectively describe the removal characteristics of DOM in PTW. Dynamic spectral changes are triggered by external disturbances, including various physical, chemical, and biological phenomena. The use of 2D-COS can explain the fundamental mechanisms of complex and heterogeneous material changes by enhancing the resolution of the spectrum and identifying the sequence of any subtle spectral changes in response to external disturbances [20,21].

Although the above methods have been widely used, there are few studies on the application of SFS technology and 2D-COS in the field of DOM structure in biotreated pharmaceutical wastewater by using coagulation-UV/O$_3$ combination method. The purpose of this paper is to (1) study the dynamic spectral changes of DOM in PTW after UV/O$_3$ oxidation using SFS technology, and (2) characterize the dynamic spectra and their relationship with different wavelengths by using 2D-COS combined with SFS technology, so as to reveal the degradation characteristics of different organic components in PTW.

2. Material and Methods

2.1. Pharmaceutical Tail Wastewater

The PTW was collected from the secondary settling tank of a pharmaceutical company wastewater treatment plant in Northeast China. The pharmaceutical company is a comprehensive enterprise, focusing on the production of chemosynthetic drugs and bio-fermentation drugs. The average daily production of pharmaceutical wastewater is 30,000 m^3. After being pretreated in a regulation tank, the wastewater was hydrolyzed and acidified in a two-stage hydrolysis acidification tank, and then entered into a one-unit activated sludge reactor (UNITANK) for biological treatment. The water quality index of pharmaceutical wastewater before and after biotreatment was provided in Table 1. All samples were filtered through a 0.45 μm glass fiber membrane prior to measurement, except for turbidity.

Table 1. Characteristics of the pharmaceutical wastewater before and after biotreatment.

Value	Value	
	Before Biotreatment	After Biotreatment
pH	5.3–6.1	6.8–7.2
DOC (mg/L)	739–892	77–126
COD (mg/L)	3331–5183	201–343
SCOD	2861–4658	180–277
NH_4^+-N (mg/L)	61–73	13–15
B/C	0.25–0.32	0.09–0.13
UV_{254} (/cm)	-	0.906–1.31
Turbidity (NTU)	178–582	46–52
Conductivity (mS/cm)	-	14.95–15.57

2.2. Coagulation and UV/O₃ Treatment

The coagulation test was carried out in a six-unit agitator (ZR4-6, Zhongrun technology development Co., Ltd. Shenzhen, China). The 1000 mL pharmaceutical tail wastewater was placed in a cylindrical beaker with a volume of 1.5 L. The initial pH of the wastewater was adjusted by 0.1 mol/L of NaOH and HCl. In order to fully dissolve and react the added coagulant in the wastewater, the procedure of the agitator was set as follows: quick stirring for 2 min with a speed of 250 rpm, slow stirring for 10 min at a speed of 50 rpm. At last, the supernatant of treated wastewater was collected after 1 h of static treatment for the relevant water quality analysis and three-dimensional fluorescence analysis. Polyferric sulfate $((Fe_2(OH)_n(SO4)_{3-n/2})_m$, PFS) was chosen as the coagulant in this study. The coagulant dosage is 0.4 g/L (according to the results of preliminary laboratory research), and the initial solution pH is 7.0.

Individual UV irradiation, O_3 oxidation and UV/O_3 oxidation experiments were processed in a closed double-layer cylindrical glass reactor with diameter of 8 cm, height of 13.5 cm and effective volume of 750 mL. The low-pressure UV lamp with the power of 15 W was used in this study to emit monochromatic light at emission spectra $\lambda = 254$ nm. The UV lamp was placed in a 3.3 cm diameter quartz tube, which was placed in the center of the reactor. Both ozone generator (CFS20) and ozone concentration detector (UV_{300}) used in this study were purchased from Beijing Shanmei Shuimei Environmental Protection Technology Co., Ltd. (Beijing, China), and the ozone gas flow rate was controlled at 48–50 L/h to maintain the ozone concentration at around 10 mg/L (according to the results of preliminary laboratory research) during the O_3 oxidation and UV/O_3 treatment test. For comparison, the pH value of the biochemical treated pharmaceutical wastewater was adjusted to 7.0 by 0.1 mol/L H_2SO_4 and NaOH solution prior to individual UV irradiation, O_3 oxidation alone and UV/O_3 treatment, consistent with the pH value of the coagulation treatment. All the tests were carried out at room temperature (25 ± 2 °C).

2.3. Water Quality and Spectral Analysis

The turbidity was measured using a WGZ-1 turbidity meter from Xinrui Instrument Co., Ltd. (Shanghai, China). The dissolved organic carbon (DOC) was determined by Shimadzu TOC-VCPH analyzer (Shimadzu, Kyoto, Japan). Chemical oxygen demand (COD) was measured by spectrophotometry (DRB200, HACH, Loveland, CO, USA). OxiTop® system was used to determine BOD_5. The 254 nm UV absorbance (UV_{254}) was measured by UV-Vis spectrophotometer (UV-6100, METASH, Shanghai, China) with UV absorption spectrum in the range of 200–600 nm. The NH_4^+-N was determined by Nessler spectrophotometry. The analysis methods of the above conventional indicators are all standard analytical methods published by the Ministry of Environmental Protection

(HJ 535-2009). The SUVA value is defined as the ratio of the absorbance at 254 nm to the solution DOC concentration.

Synchronous fluorescence spectroscopy (SFS) was performed using a Hitachi fluorescence spectrophotometer (F-7000) from Hitachi, Tokyo, Japan. The response time of the spectrometer was set to 0.5 s and the photomultiplier tubes (PMT) voltage was 400 V. The wavelength range was from 260 nm to 550 nm by simultaneously scanning the excitation wavelength (ex) and emission wavelength (em). The passband was 0.2 nm and at the same time we kept the wavelength difference constant, $\Delta\lambda = \lambda_{em} - \lambda_{ex} = 18$ nm. The scanning speed was set to 240 nm/min. All spectral values were deducted from their respective program blanks. Fluorescence spectroscopy was usually performed within one day of sampling.

Two-dimensional correlation fluorescence spectroscopy (2D-COS) was analyzed by using the "2D-Shige" software developed by Kwansei-Gakuin University. Samples were taken at different time points from the UV/O$_3$ oxidation test, the sequence of time can be regarded as the specified external disturbance. Therefore, a matrix depending on sampling units at different points in time can be generated.

The synchronous correlation spectra can be calculated by the following formula [22]:

$$\phi(x_1, x_2) = \frac{1}{m-1}\Sigma_{j=1}^{m}I_j(x_1, t)I_j(x_2, t) \tag{1}$$

The asynchronous correlation spectrum is determined by the following formula [22]:

$$\varphi(x_1, x_2) = \frac{1}{m-1}\Sigma_{j=1}^{m}I_j(x_1, t)\Sigma_{k=1}^{m}N_{jk}I_j(x_2, t) \tag{2}$$

where x represented an index variable (number of sampling points) of the synchronous fluorescence spectrum caused by disturbance variable t. $I_{(x, t)}$ was the analytical spectrum of m evenly distributed on t (between T_{min} and T_{max}). N_{jk} including the j column and original element k was called the discrete Hilbert-Noda transformation matrix and was defined as follows [22]:

$$N_{jk} = \{^{0}_{\frac{1}{\pi(k-j)}} \quad if = k; \; otherwise \tag{3}$$

The fluorescence intensity $\phi(x_1, x_2)$ of the synchronous fluorescence spectrum can represent a simultaneous or consistent change in the two independent spectral intensities determined by the perturbation variable t between T_{min} and T_{max} at x_1 and x_2.

The spectral intensity of the asynchronous two-dimensional $\varphi(x_1, x_2)$ represented a continuous or continuous but inconsistent change in the spectral intensity measured at x_1 and x_2, respectively.

3. Results and Discussion

3.1. Changes in DOC and SUVA by Coagulation and UV/O$_3$ Process

DOC and SUVA are two important indicators for characterizing the organic matter in water treatment. The lower the SUVA value, the lower the risk of producing disinfection byproducts. As showed in Figure 1, after the coagulation pretreatment stage, the removal rate of DOC and SUVA reached 37.5% and 24.4%, respectively. After the oxidation stage, the DOC and SUVA was further reduced, the highest DOC and SUVA removal efficiency reached 55.8% and 68.7% by coagulation-UV/O$_3$ process after 60 min oxidation.

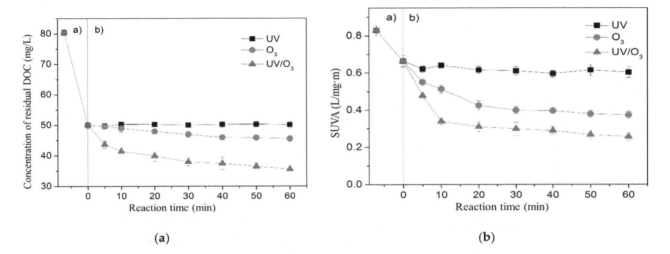

Figure 1. Effect of coagulation-UV, coagulation-O_3 and coagulation-UV/O_3 process on dissolved organic carbon (DOC) in pharmaceutical tail wastewater (PTW) (**a**: coagulation stage; **b**: oxidation stage).

The DOC concentration slightly changed during individual O_3 oxidation, indicating that the O_3 treatment mainly changed the structure of the organic matter by direct oxidation to form intermediate products instead of mineralizing organics into CO_2 and H_2O. Therefore, the 30.5% of SUVA was reduced by O_3 oxidation but only 5% of DOC was removed at the same time. In comparison, the ·OH formed by UV excitation in O_3 oxidation enhanced the degradation and mineralization of the active compound, thereby UV/O_3 achieved better DOC removal efficiency than O_3 oxidation.

Since the removal of organic matter was not achieved by UV irradiation alone, the effectiveness of UV/O_3 on PTW treatment demonstrated a synergistic effect of the combination of UV and O_3. In the UV/O_3 treatment process, the reduction in DOC content included direct and indirect effects. Direct effect was manifested in the photochemical mineralization of photosensitive substances (such as aromatic ring), and indirect effects represented the utilization of reactive oxygen species (OH), whose major production pathways require O_2 [23]. It can be seen in the first 20 min of the oxidation reaction that the SUVA value of the wastewater was rapidly decreased. Then, as time goes on, the decline rate of SUVA slowed down. At the end of the reaction (60 min), the removal rate of SUVA by O_3 or UV/O_3 reached 55% and 68.7%, respectively. These results indicate that organic components with UV absorption were easily oxidized by the O_3 or UV/O_3 process. The removal of SUVA (about 68.7%) was even higher than the removal rate of DOC. This indicates the preferential removal of aromatic luminescent chromophore or partially aromatic structure that can be converted to a non-UV absorbing compound by photochemical reaction. Many photochemical products were reported to have molecular weights as low as those of organic acids, alcohols, aldehydes, and inorganic carbons [24,25].

3.2. SFS Characteristics and Component Identification

3.2.1. SFS Characteristics and Fluorescent Component Removal

In order to reveal the removal mechanism of organics in PTW in the process of UV/O_3 treatment, the SFS characteristics of PTW taken at different time points during the oxidation treatment were analyzed. Three main peaks and two broad shoulders were shown in SFS (Figure 2a), including tyrosine-like fluorescent component (TYLF, λ = 265–300 nm), tryptophan-like fluorescent component (TRLF, λ = 300–360 nm) [26,27], microbial humus-like fluorescence component (MHLF, λ = 360–420 nm), fulvic acid-like fluorescent component (FLF, λ = 420–460 nm) and humus-like fluorescent component (HLF, λ = 460–520 nm) [28]. It can be seen that the fluorescence intensity in the whole wavelength range decreased as the oxidation reaction proceeded. Similar to the fluorescence spectrum, the absolute fluorescence loss of the solution during the reaction can be taken as a function of wavelength and compared over different irradiation oxidation times (Figure 2b).

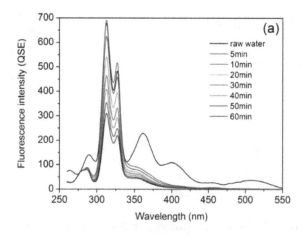 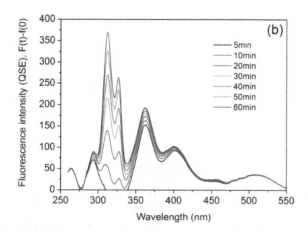

Figure 2. Changes in spectral responses of PTW with different irradiation times in UV/O$_3$ system: (**a**) synchronous fluorescence spectroscopy (SFS) ($\Delta\lambda$ = 18 nm) and (**b**) absolute fluorescence losses of excitation wavelength function.

As expected, the initial loss did not occur in the TRLF region, but in the TYLF, MHLF and the FLF component. The TYLF, MHLF and FLF components decreased rapidly and TRLF increased slightly in the first 5 min of UV/O$_3$ irradiation. This was likely to be due to the conversion of MHLF and FLF into TRLF during the irradiation oxidation.

3.2.2. SFS Component Identification and Principle Component Analysis

Principle component analysis (PCA) was carried out in this study to identify the component based on SFS at dierent times. The independent fluorescent components in PTW can be distinguished by cluster scores at dierent spectroscopic wavelength through PCA analysis. The Kaiser-Meyer-Olkin (KMO) test and Bartlett's test of sphericity (P) were two indexes to test the correlation between variables in the relevant array. In this study, the KMO value was 0.843, $p <$ 0.001, indicating that SFS was well suited for PCA application [29].

After performing PCA analysis on the SFS of eight time points, two principal components (PC) were generated with the cumulative variance contribution rate of 99.69%, which can reflect most of the characteristics of the SFS (Figure 3). The scores curve of dierent wavelengths can show the characteristics of the spectral waveform of each principal component. Therefore, fluorescent groups that contributed variance to SFS signal can be identified [30]. PC1 with a variance of 98.8% showed three main peaks and four shoulders (Figure 3a). Same as the SFS characteristics, five fluorescent components were identified by wavelength at PC1, including TYLF (at 289 nm), TRLF (at 312 nm, 329 nm, 359 nm), MHLF (at 380 nm), FLF (at 450 nm) and HLF (at 470 nm). PC2 (variance of 0.89%) showed four main peaks and two shoulders (Figure 3b).

The main peak of TYLF at 292 nm exhibited a red-shift of 3 nm and its wavelength migration was longer than that in PC1, indicating that the polarity of the organic component was enhanced and the hydrophobicity was reduced. The MHLF peak was transferred to a shorter wavelength (360 nm) compared to PC1. The peak of FLF component had a red-shift of 5 nm at the wavelength of 455 nm. The TRLF component was identified at the wavelengths of 312 nm and 329 nm, no red-shift and blue-shift occurred compared PC1.

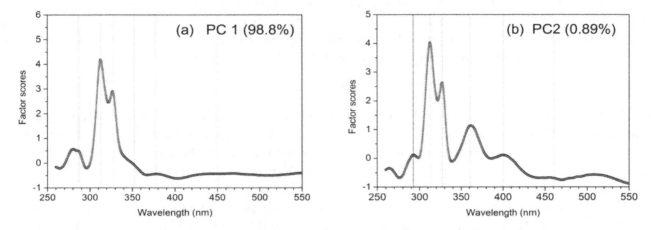

Figure 3. Scores plots of principle component 1 (PC1) **(a)** and principal component 2 (PC2) **(b)** for spectral wavelengths in UV/O_3 system.

3.2.3. Organic Component Spectral Area Integral

The SFS can be divided into five synchronous fluorescence regions based on excitation wavelength, which were corresponding to TYLF, TRLF, MHLF, FLF and HLF. The area integral of wavelength and fluorescence intensity can represent the relative abundance of homologous component. According to the area integral, the removal efficiency of fluorescence components can be studied. The distributions of the abundance of the DOM components were shown in Figure 4. It turned out that the DOC concentration showed a significant positive correlation to the variation of TYLF ($r = 0.940$, $p = 0.002$), TRLF ($r = 0.988$, $p = 0.0003$) and HLF ($r = 0.929$, $p = 0.003$), but a weak positive correlation to the MHLF ($r = 0.430$, $p = 0.335$) and FLF ($r = 0.499$, $p = 0.254$), which indicated that TYLF, TRLF and HLF could more accurately indicate DOC changes than MHLF and FLF. This was caused by the selective removal of fluorescent component in UV/O_3 irradiation oxidation process.

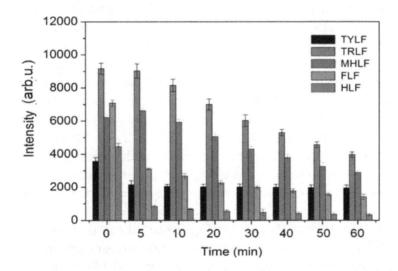

Figure 4. Distributions of the abundance of the dissolved organic matter (DOM) components in UV/O_3 system.

It can be seen in Figure 4 that the FLF, HLF and TYLF components decreased rapidly due to the UV/O_3 oxidation. Consequently, 82.2% of HLF, 40.4% of FLF and 40.0% of TYLF were removed in the first 5 min, indicating that the FLF, HLF and TYLF can be effectively removed by UV/O_3. This well explained the quick degradation of DOC and SUVA in the first reaction time, which was discussed before. However, in the subsequent reactions, the removal rate of FLF, HLF and TYLF gradually slowed down, and the TRLF and MHLF components began to decline. After 60 min of UV/O_3 oxidation, the maximum removal efficiency was obtained by HLF component of 95.6%, followed by FLF of 80.0%, TRLF of 56.0%, MHLF of 50.8% and TYLF of 44.4%. This indicated that the UV/O_3 oxidation system could effectively remove the different fluorescent component in PTW, especially the HLF and FLF, while it took more time to completely remove the TYLF, TRLF and MHLF components.

3.3. Synchronous Fluorescence/UV-Visible Two-Dimensional Correlation Spectra Analysis

To reduce spectral overlap and reveal the DOM transformation process, 2D-COS was used to analyze two-dimensional fluorescence spectra [26]. Two-dimensional synchronous fluorescence spectrum of PTW was a symmetric spectrum about diagonals, in which there were four main self-peaks at wavelengths of 313 nm, 326 nm, 360 nm, and 400 nm (Figure 5a). The results are consistent with the PCA analysis. The self-peak represented the overall sensitivity of the corresponding spectral region after the change in spectral intensity as an external disturbance. During the reaction, the intensity of the self-peak at 313 nm and 326 nm was higher than that at 360 nm and 400 nm, indicating that the fluorescence intensity of the TYLF and TRLF component was clearly higher than that of the MHLF. The solid cross peaks indicated a positive correlation of fluorescence components at 313 nm, 326 nm, 360 nm, and 400 nm. Indirectly this confirmed the fluorescence components of DOM in PTW were synchronously removed in the process of the coagulation-UV/O_3 process. In the asynchronous fluorescence spectrum (Figure 5b), two negative cross peaks appeared at 289/313 nm and 289/326 nm, while five positive cross peaks were found at: 313/360, 313/400, 313/501, 326/360 and 326/400 nm. The positive or negative asynchronous crossover peak could provide information on the disturbance order of spectral band along the external variables. According to Nado rules, the change order of the spectral bands was: 501, 400, 360, 326, 313, 289 nm. This indirectly confirmed the removal characteristics of each component in the tail water treatment process of the synthetic pharmaceutical park, which had been discussed before.

The 2D-COS was used to study, in more detail, the characteristics of different absorption wavelengths of organic matter in the PTW during the oxidation process, the synchronous and asynchronous 2D-COS analysis using UV-visible absorption spectra with different oxidation time was shown in Figure 5c. Strong self-peaks at 225 nm and 255 nm and weak self-peak at 350 nm were observed in the synchronous 2D-UV-visible correlation spectrum and the intensity of the first two absorption bands at 225 nm and 255 nm decreased significantly with the irradiation time. This result was consistent with our previous observations of the overall trend of higher absorption losses at shorter wavelengths. In the asynchronous 2D-UV-visible correlation spectrum, three absorption bands of the oxidative chemical reaction were observed, and the wavelength ranges were: 200–240, 240–270, and 270–320 nm, respectively (Figure 5d). The sequence relationship of spectral variation characteristics with oxidation time can be derived: 270–320, 200–240, 240–270 nm. This indicated that the oxidation reaction first removed the macromolecular organic matter with a conjugated bond in the wavelength range of 270 nm–320 nm, and then removed the organic matter at 240–270 nm.

In order to further reveal the characteristics of organic matter removal and degradation, synchronous fluorescence/UV-visible two-dimensional correlation spectra were obtained. From the synchronous two-dimensional correlation map (Figure 5e), it can be found that the aromatic group at the wavelength of 255 nm in UV-visible spectrum was corresponding to the variation of TRLF, MHLF, FLF and HLF components at 313 nm, 326 nm, 360 nm and 400 nm in synchronous fluorescence. This indicated that the TRLF, MHLF, FLF and HLF components containing aromatic groups were degraded during the oxidation process. At the same time, the wavelength in the range of 400–500 nm in the UV-visible spectrum was opposite to the variation trend of the TRLF and MHLF components, which represented a humus-like substance with a high degree of humification and low lignin content that was generated after the oxidation reaction. Through analysis of asynchronous two-dimensional correlation map (Figure 5f), UV-visible spectrum at 285 nm (organic matter containing unsaturated conjugated double-bond structure) has a negative peak with v1/v2 at 360 nm (FLF) and 400 nm (HLF) in synchronous fluorescence, respectively. It was shown that the organic compounds with unsaturated conjugated double-bond structure in HLF and FLF was first degraded by the oxidation process. At the same time, the UV-visible spectrum at 255 nm showed a positive peak with *v1/v2* at 313 nm (TRLF) and 326 nm (MHLF) in the synchronous fluorescence, respectively, indicating the MHLF- and TRLF-containing aromatic groups were removed sequentially in subsequent oxidation. The spectral characteristics can explain that the degradation of organic matter occurred sequentially in the order of HLF→FLF→MHLF→TRLF.

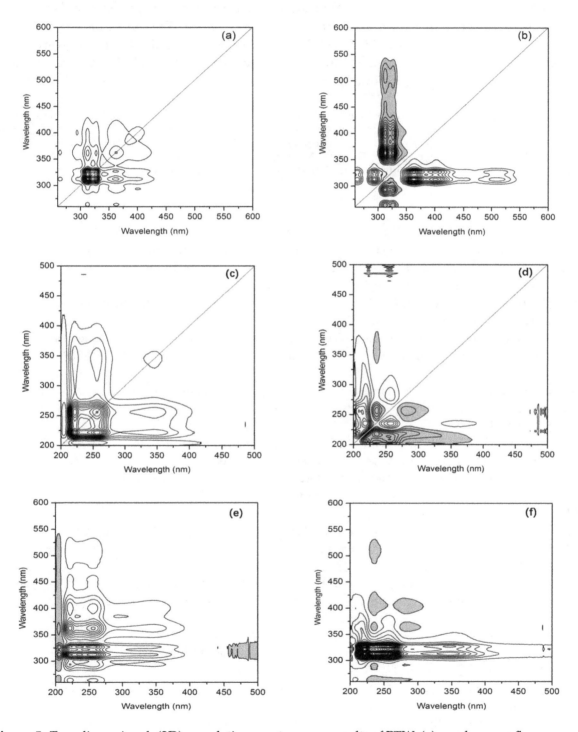

Figure 5. Two-dimensional- (2D)-correlation spectroscopy results of PTW: (**a**) synchronous fluorescence map, (**b**) asynchronous fluorescence map, (**c**) synchronous UV-visible map, (**d**) asynchronous UV-visible map, (**e**) synchronous fluorescence/UV-visible two-dimensional correlation map, (**f**) asynchronous fluorescence/UV-visible two-dimensional correlation map. The solid and the gray represent the positive and negative signs, respectively.

The organic matter in PTW can be effectively removed by coagulation-UV/O_3 pretreatment. After 60 min reaction, the removal rate of SUVA and DOC reached 68.7% and 55.8%, respectively. The UV can significantly enhance the mineralization of organic matters by O_3, therefore the treatment effect was significantly better than the O_3 oxidation alone. TRLF, TYLF, MHLF, FLF and HLF components were identified by the SFS combined with 2D-COS and PCA analysis. All the results consistently show that HLF and FLF were the main components of DOM in the tail water of synthetic pharmaceutical parks. UV/O_3 was selective for the removal of different fluorescent components, which can quickly remove FLF and HLF from the PTW and convert them into TRLF. The order of degradation of the different fluorescent components was HLF→FLF→MHLF→TRLF→TYLF. SFS combined with 2D-COS and PCA can quickly and effectively reveal the spectral dynamics of DOM in UV/O_3 treatment system and the removal and degradation characteristics of different organic components.

Author Contributions: Conceptualization, J.W. and F.Q.; methodology, J.W., H.Y. and C.D.; software, J.W., F.Q., C.D. and H.Y.; validation, Y.S. and L.X.; formal analysis, J.W., F.Q.; investigation, J.W., C.D., H.Y. and F.Q.; resources, Y.S.; data curation, Y.S.; writing—original draft preparation, J.W.; writing—review and editing, J.W. and F.Q.; visualization, J.W. and C.D.; supervision, Y.S. and L.X.; project administration, Y.S.; funding acquisition, Y.S. and F.Q. All authors have read and agreed to the published version of the manuscript.

References

1.   Qian, F.; He, M.; Song, Y.; Tysklind, M.; Wu, J. A bibliometric analysis of global research progress on pharmaceutical wastewater treatment during 1994–2013. *Environ. Earth Sci.* **2015**, *73*, 4995–5005. [CrossRef]

2.   Balcıoğlu, I.A.; Ötker, M. Treatment of pharmaceutical wastewater containing antibiotics by O_3 and O_3/H_2O_2 processes. *Chemosphere* **2003**, *50*, 85–95. [CrossRef]

3.   López-Fernández, R.; Martínez, L.; Villaverde, S. Membrane bioreactor for the treatment of pharmaceutical wastewater containing corticosteroids. *Desalination* **2012**, *300*, 19–23. [CrossRef]

4.   Rivera-Utrilla, J.; Sánchez-Polo, M.; Ferro-García, M.; Prados-Joya, G.; Ocampo-Perez, R. Pharmaceuticals as emerging contaminants and their removal from water. A review. *Chemosphere* **2013**, *93*, 1268–1287. [CrossRef] [PubMed]

5.   Sirés, I.; Brillas, E.; Sadornil, I.S. Remediation of water pollution caused by pharmaceutical residues based on electrochemical separation and degradation technologies: A review. *Environ. Int.* **2012**, *40*, 212–229. [CrossRef] [PubMed]

6.   Ikehata, K.; Naghashkar, N.J.; El-Din, M.G. Degradation of Aqueous Pharmaceuticals by Ozonation and Advanced Oxidation Processes: A Review. *Ozone Sci. Eng.* **2006**, *28*, 353–414. [CrossRef]

7.   Gadipelly, C.; Pérez-González, A.; Yadav, G.D.; Ortiz, I.; Ibáñez, R.; Rathod, V.K.; Marathe, K.V. Pharmaceutical Industry Wastewater: Review of the Technologies for Water Treatment and Reuse. *Ind. Eng. Chem. Res.* **2014**, *53*, 11571–11592. [CrossRef]

8.   Tang, C.-J.; Zheng, P.; Chen, T.-T.; Zhang, J.-Q.; Mahmood, Q.; Ding, S.; Chen, X.-G.; Chen, J.-W.; Wu, D.-T. Enhanced nitrogen removal from pharmaceutical wastewater using SBA-ANAMMOX process. *Water Res.* **2011**, *45*, 201–210. [CrossRef]

9.   Biń, A.K.; Sobera-Madej, S. Comparison of the Advanced Oxidation Processes (UV, UV/H_2O_2 and O_3) for the Removal of Antibiotic Substances during Wastewater Treatment. *Ozone Sci. Eng.* **2012**, *34*, 136–139. [CrossRef]

10.   Huber, M.M.; Göbel, A.; Joss, A.; Hermann, N.; Löffler, D.; McArdell, C.S.; Ried, A.; Siegrist, H.; Ternes, T.; Von Gunten, U. Oxidation of Pharmaceuticals during Ozonation of Municipal Wastewater Effluents: A Pilot Study. *Environ. Sci. Technol.* **2005**, *39*, 4290–4299. [CrossRef]

11.   Gong, J.; Liu, Y.; Sun, X. O_3 and UV/O_3 oxidation of organic constituents of biotreated municipal wastewater. *Water Res.* **2008**, *42*, 1238–1244. [CrossRef] [PubMed]

12.   Qian, F.; Sun, X.; Lei, J. Removal characteristics of organics in bio-treated textile wastewater reclamation by a stepwise coagulation and intermediate GAC/O_3 oxidation process. *Chem. Eng. J.* **2013**, *214*, 112–118. [CrossRef]

13.   Sirtori, C.; Zapata, A.; Oller, I.; Gernjak, W.; Agüera, A.; Malato, S. Decontamination industrial pharmaceutical wastewater by combining solar photo-Fenton and biological treatment. *Water Res.* **2009**, *43*, 661–668. [CrossRef] [PubMed]

14. Zhu, G.; Yin, J.; Zhang, P.; Wang, X.; Fan, G.; Hua, B.; Ren, B.; Zheng, H.; Deng, B. DOM removal by flocculation process: Fluorescence excitation–emission matrix spectroscopy (EEMs) characterization. *Desalination* **2014**, *346*, 38–45. [CrossRef]

15. Meng, F.; Huang, G.; Li, Z.; Li, S. Microbial Transformation of Structural and Functional Makeup of Human-Impacted Riverine Dissolved Organic Matter. *Ind. Eng. Chem. Res.* **2012**, *51*, 6212–6218. [CrossRef]

16. Carstea, E.; Bridgeman, J.; Baker, A.; Reynolds, D. Fluorescence spectroscopy for wastewater monitoring: A review. *Water Res.* **2016**, *95*, 205–219. [CrossRef]

17. Miano, T.; Senesi, N. Synchronous excitation fluorescence spectroscopy applied to soil humic substances chemistry. *Sci. Total Environ.* **1992**, *117*, 41–51. [CrossRef]

18. Henderson, R.; Baker, A.; Murphy, K.R.; Hambly, A.; Stuetz, R.M.; Khan, S.J. Fluorescence as a potential monitoring tool for recycled water systems: A review. *Water Res.* **2009**, *43*, 863–881. [CrossRef]

19. Baker, A.; Cumberland, S.; Bradley, C.; Buckley, C.; Bridgeman, J. To what extent can portable fluorescence spectroscopy be used in the real-time assessment of microbial water quality? *Sci. Total Environ.* **2015**, *532*, 14–19. [CrossRef]

20. Jung, Y.M.; Noda, I. New Approaches to Generalized Two-Dimensional Correlation Spectroscopy and Its Applications. *Appl. Spectrosc. Rev.* **2006**, *41*, 515–547. [CrossRef]

21. Ozaki, Y.; Czarnik-Matusewicz, B.; Sasic, S. Two-Dimensional Correlation Spectroscopy in Analytical Chemistry. *Anal. Sci.* **2002**, *17*, i663–i666.

22. Noda, I.; Liu, Y.; Ozaki, Y. Two-Dimensional Correlation Spectroscopy Study of Temperature-Dependent Spectral Variations of N-Methylacetamide in the Pure Liquid State. 2. Two-Dimensional Raman and Infrared—Raman Heterospectral Analysis. *J. Phys. Chem.* **1996**, *100*, 8674–8680. [CrossRef]

23. Lou, T.; Xie, H. Photochemical alteration of the molecular weight of dissolved organic matter. *Chemosphere* **2006**, *65*, 2333–2342. [CrossRef]

24. Pullin, M.J.; Bertilsson, S.; Goldstone, J.V.; Voelker, B.M. Effects of sunlight and hydroxyl radical on dissolved organic matter: Bacterial growth efficiency and production of carboxylic acids and other substrates. *Limnol. Oceanogr.* **2004**, *49*, 2011–2022. [CrossRef]

25. Vidali, R.; Remoundaki, E.; Tsezos, M. Humic Acids Copper Binding Following Their Photochemical Alteration by Simulated Solar Light. *Aquat. Geochem.* **2009**, *16*, 207–218. [CrossRef]

26. Hur, J. Microbial Changes in Selected Operational Descriptors of Dissolved Organic Matters From Various Sources in a Watershed. *Water Air Soil Pollut.* **2010**, *215*, 465–476. [CrossRef]

27. Yu, H.; Song, Y.; Tu, X.; Du, E.; Liu, R.; Peng, J. Assessing removal efficiency of dissolved organic matter in wastewater treatment using fluorescence excitation emission matrices with parallel factor analysis and second derivative synchronous fluorescence. *Bioresour. Technol.* **2013**, *144*, 595–601. [CrossRef]

28. Pan, H.; Yu, H.; Wang, Y.; Liu, R.; Lei, H. Investigating variations of fluorescent dissolved organic matter in wastewater treatment using synchronous fluorescence spectroscopy combined with principal component analysis and two-dimensional correlation. *Environ. Technol.* **2017**, *39*, 1–8. [CrossRef]

29. Yu, H.; Song, Y.; Liu, R.; Pan, H.; Xiang, L.; Qian, F. Identifying changes in dissolved organic matter content and characteristics by fluorescence spectroscopy coupled with self-organizing map and classification and regression tree analysis during wastewater treatment. *Chemosphere* **2014**, *113*, 79–86. [CrossRef]

30. Guo, X.-J.; Yuan, D.-H.; Jiang, J.-Y.; Zhang, H.; Deng, Y. Detection of dissolved organic matter in saline–alkali soils using synchronous fluorescence spectroscopy and principal component analysis. *Spectrochim. Acta Part A Mol. Biomol. Spectrosc.* **2013**, *104*, 280–286. [CrossRef]

11

A Pilot Study Combining Ultrafiltration with Ozonation for the Treatment of Secondary Urban Wastewater: Organic Micropollutants, Microbial Load and Biological Effects

Cátia A. L. Graça [1], Sara Ribeirinho-Soares [2], Joana Abreu-Silva [3], Inês I. Ramos [4],
Ana R. Ribeiro [1], Sérgio M. Castro-Silva [5], Marcela A. Segundo [4], Célia M. Manaia [3,*],
Olga C. Nunes [2,*] and Adrián M. T. Silva [1,*]

[1] Laboratory of Separation and Reaction Engineering-Laboratory of Catalysis and Materials (LSRE-LCM),
    Faculdade de Engenharia, Universidade do Porto, Rua Dr. Roberto Frias, 4200-465 Porto, Portugal;
    catiaalgraca@fe.up.pt (C.A.L.G.); ritalado@fe.up.pt (A.R.R.)
[2] LEPABE—Laboratory for Process Engineering, Environment, Biotechnology and Energy,
    Faculdade de Engenharia, Universidade do Porto, Rua Dr. Roberto Frias, 4200-465 Porto, Portugal;
    saramariasoares@hotmail.com
[3] Universidade Católica Portuguesa, CBQF—Centro de Biotecnologia e Química Fina—Laboratório Associado,
    Escola Superior de Biotecnologia, Rua Diogo Botelho 1327, 4169-005 Porto, Portugal; jsilva@porto.ucp.pt
[4] LAQV, REQUIMTE, Departamento de Ciências Químicas, Faculdade de Farmácia, Universidade do Porto,
    Rua de Jorge Viterbo Ferreira 228, 4050-313 Porto, Portugal; iibmramos@gmail.com (I.I.R.);
    msegundo@ff.up.pt (M.A.S.)
[5] Adventech-Advanced Environmental Technologies, Centro Empresarial e Tecnológico, Rua de Fundões 151,
    3700-121 São João da Madeira, Portugal; sergio.silva@adventech.pt
*   Correspondence: cmanaia@porto.ucp.pt (C.M.M.); opnunes@fe.up.pt (O.C.N.); adrian@fe.up.pt (A.M.T.S.)

Abstract: Ozonation followed by ultrafiltration (O_3 + UF) was employed at pilot scale for the treatment of secondary urban wastewater, envisaging its safe reuse for crop irrigation. Chemical contaminants of emerging concern (CECs) and priority substances (PSs), microbial load, estrogenic activity, cell viability and cellular metabolic activity were measured before and immediately after O_3 + UF treatment. The microbial load was also evaluated after one-week storage of the treated water to assess potential bacteria regrowth. Among the organic micropollutants detected, only citalopram and isoproturon were not removed below the limit of quantification. The treatment was also effective in the reduction in the bacterial loads considering current legislation in water quality for irrigation (i.e., in terms of enterobacteria and nematode eggs). However, after seven days of storage, total heterotrophs regrew to levels close to the initial, with the concomitant increase in the genes 16S rRNA and *intI*1. The assessment of biological effects revealed similar water quality before and after treatment, meaning that O_3 + UF did not produce detectable toxic by-products. Thus, the findings of this study indicate that the wastewater treated with this technology comply with the water quality standards for irrigation, even when stored up to one week, although improvements must be made to minimise microbial overgrowth.

Keywords: advanced oxidation; membrane technology; micropollutants; biological contaminants; cytotoxicity; wastewater reuse

1. Introduction

Urban wastewater reuse is considered an important strategy when addressing water scarcity issues [1]. This is a common practice in some countries, where the treated wastewater is mostly directed for agricultural irrigation [2]; however, urban wastewater often contains a variety of contaminants, such as salts, metals, metalloids, pathogens, and organic micropollutants, such as residual drugs, endocrine-disrupting chemicals, and residues from personal care products, among others [3,4]. Moreover, there is growing evidence that conventional urban wastewater treatment plants (UWWTPs) are not completely effective in eliminating bacteria and chemical micropollutants [5,6], rendering the effluent unsuitable for crops irrigation. Failure to properly treat and manage wastewater can generate adverse health effects, accumulation of heavy metals in crops, and the production of low-quality agricultural goods [3]. A new regulation on minimum preconditions for water reuse for agricultural irrigation has entered into force in the EU, which encompasses coordinated water-quality monitoring requisites for the safe reuse of treated urban wastewater [7]. These new rules will be put into practice in 2023 and are expected to promote water reuse. This regulation also demands an established water reuse risk management plan that should consider the environmental quality standards for priority substances and certain other pollutants, as well as additional requirements, such as heavy metals, pesticides, disinfection by-products, pharmaceuticals, and other substances of emerging concern, including micropollutants and microplastics. It also addressed the identification of some preventive measures that can be taken to limit risks, namely additional disinfection or pollutant removal measures.

Advanced oxidation processes (AOPs) and technologies (AOTs), such as ozonation, have emerged as effective tertiary treatments for the removal of both chemical and biological contaminants in UWWTPs [8,9]. Ozonation is among the few AOTs that have been applied to large-scale water treatment, due to its strong oxidation ability and broad-spectrum disinfection [10]. Ozone can react either by direct oxidation of organic pollutants (mostly at acidic conditions), or via hydroxyl radical formation (mainly produced under alkaline conditions) [10]. Studies employing ozone-based AOTs in UWWTP effluents have yielded remarkable results regarding the simultaneous removal of CECs and the reduction in the microbial load at different ozone doses and contact times [11–16]; however, bacterial regrowth in stored treated wastewater has been observed [14–16], which might be the result of the bacteria's ability to repair injuries, promoting fast regrowth, when stress levels are lowered. This may jeopardize water quality in the long term, thus prompting its immediate reuse rather than storing this water. Additionally, the use of chlorine as the traditional disinfection agent in stored water may not ensure its safety, because injured bacteria can also survive and regrow at low chlorine doses [17]. A suitable approach would be a physical separation step, using membrane-like technology. Although ozone may damage cell components, such as lipids, proteins and DNA, membrane filtration acts via size exclusion and adsorption, retaining microorganisms [18]. Among the available options in the market for full-scale applications, ultrafiltration (UF) membranes are favourable alternatives for bacteria removal due to their small pore size (0.01 to 0.1 μm). Moreover, studies have shown that UF is preferred to other filtration alternatives to avoid the regrowth of antibiotic-resistant bacteria (ARB) [19,20]. For example, Hembach et al., 2019 [18] reported the efficiency of UF in the disinfection of a secondary effluent of a UWWTP, and the results were compared with those obtained with single ozonation. The authors reported that UF (using a membrane pore size of 20 nm) was not able to remove the entire bacterial community, whereas ozonation presented limited effectiveness on the reduction in the same contaminants when using an ozone concentration optimised for micropollutant removal. Thus, these authors suggested further investigations coupling both technologies to achieve both micropollutant removal and bacteria mitigation, which was the target of the present study.

Thus, the present study investigated the potential of using UF in combination with ozonation, operating in continuous mode at a pilot scale, for the treatment of the secondary effluent of a UWWTP. Parameters commonly legislated in different countries were considered when assessing the suitability of treated wastewater for reuse in irrigation (Portuguese laws, US EPA, FAO guidelines and WHO). Moreover, envisaging higher quality criteria, the following parameters were also included in this work: (i) priority substances and CECs identified in Directive 2013/39/EU and Decision 495/2015/EU [21,22], respectively; (ii) load of selected microbial groups; and (iii) potential estrogenic activity, cytotoxicity, and cell viability (biological effects). All these parameters were analysed in both freshly collected and

O_3 + UF treated wastewater to assess treatment efficiency. Biological effects are particularly important to evaluate, due to the possibility of formation of toxic by-products after ozonation. Moreover, microbiological indicators were re-examined after a 7-day storage period to assess potential bacteria regrowth. Regarding other studies coupling O_3 to UF, only a few evaluate the feasibility of this system for urban wastewater reclamation [23–26] and, as far as it is known, none of those comprise the simultaneous evaluation of physico-chemical parameters, removal of priority substances and CECs, microbial inactivation and regrowth, and investigation of biological effects, which are important parameters for safe wastewater reuse, this work bringing a valuable contribution to the knowledge on this field.

2. Materials and Methods

2.1. Chemicals and Materials

All reference and isotopically labelled internal standards for liquid chromatography (>98% purity) were acquired from Sigma-Aldrich (Steinhein, Germany). Ethanol 99.5% (HPLC grade) was obtained from Fisher Scientific U.K. Ltd. (Loughborough, UK). Acetonitrile (MS grade) was purchased from VWR International (Fontenay-sous-Bois, France), whereas formic and sulphuric acid were obtained from Merck (Darmstadt, Germany). Multichannel tubular ceramic membranes with a selective layer of α-Al_2O_3 (nominal pore size of 10 nm) were provided by Rauschert Distribution GmbH, Inopor® (Schesslitz, Germany). Membrane dimensions were 305 mm in length with 15 mm glazed ends. The external diameter was 25 mm, and it contained 19 internal channels of 3.5 mm diameter each.

For microbial culture analyses, water samples were filtered through cellulose nitrate membranes (0.22 μm pore size, 47 mm diameter), provided by Sartorius (Gottingen, Germany). For DNA-based analyses, water samples were filtered through track-etched polycarbonate membranes (0.22 μm pore size, 47 mm diameter) from Whatman® Nuclepore™, provided by VWR (Alfragide, Portugal).

For cell culture experiments, dimethyl sulfoxide (DMSO; ≥99.9%), Triton™ X-100, and thiazolyl blue tetrazolium (MTT) were purchased from Sigma-Aldrich (Steinhein, Germany). Dulbecco's modified Eagle medium (DMEM; ref: 31966-021), heat-inactivated foetal bovine serum (FBS), penicillin-streptomycin (PenStrep), and trypsin-EDTA (1X) were purchased from Gibco® through Life Technologies™ (Warrington, UK). Murine fibroblasts L929 were obtained from the American Type Culture Collection (ATCC, Wesel, Germany). Caco-2 cell line was also purchased from ATCC and used between passage number 35 and 42. LDH Cytotoxicity Detection Kit was acquired from Takara Bio Inc. (Shiga, Japan). The XenoScreen YES/YAS assay kit for estrogenic activity assessment was acquired from Xenometrix® (Allschwil, Switzerland).

The ultrapure water used in the experiments and analytical methods was supplied by a Milli-Q water system (18.2 MΩ cm).

2.2. Secondary Effluent and Treated Samples

The secondary effluent used in the advanced treatment assays was collected at three different dates (between September and October 2019) from a full-scale UWWTP located in northern Portugal. In this UWWTP, the water line treatment includes a preliminary step (trash racking and dredging) followed by decantation, biological treatment with activated sludge, and a final decantation stage before discharging the effluent to the river. In this study, freshly collected samples of this UWWTP secondary effluent were divided into two aliquots, one of which was immediately analysed (WW) and another was directed to the O_3 + UF treatment unit. Details of the analytical methods employed to characterise the UWWTP secondary effluent (WW) are given in Section 2.4, and its chemical and biological characterisation can be found in Tables 1 and 2. Samples collected after O_3 + UF treatment (TWW_0) were also immediately processed for microbiological analyses and DNA extraction. In addition, aliquots of TWW_0 were stored for seven days in sterile glass bottles under dark conditions and at room temperature (herein named as TWW_7) to assess possible bacterial regrowth in a hypothetical storage scenario for wastewater reuse.

Table 1. Characterisation of the urban wastewater treatment plant (UWWTP) secondary effluent, before (WW) and immediately after treatment (TWW$_0$), and standards of water for irrigation (Decree-Law 236/98) and wastewater reuse in irrigation without restriction, for urban wastewaters which treatment includes a disinfection step (Decree-Law 119/2019) and for wastewater reuse in the Eastern Mediterranean Region—WHO, 2016.

Parameters	UWWTP Secondary Effluent (WW)	After O$_3$ + UF Treatment (TWW$_0$)	Decree-Law 236/98 [27] MVR	Decree-Law 119/2019 [28] PV	WHO 2016 [29] MVR
Al (mg/L)	9.55×10^{-5}	6.10×10^{-5}	5.0	5	5.0
As (mg/L)	1.12×10^{-5}	$<5 \times 10^{-6}$	0.1	n.a	0.1
Ba (mg/L)	4.25×10^{-5}	1.52×10^{-5}	1.0	n.a	n.a
Be (mg/L)	$<5 \times 10^{-6}$	$<5 \times 10^{-6}$	0.5	0.1	0.1
B (mg/L)	1.29×10^{-4}	1.06×10^{-4}	0.3	variable	n.a
Cd (mg/L)	$<5 \times 10^{-6}$	$<5 \times 10^{-6}$	0.01	n.a	0.1
Pb (mg/L)	6.73×10^{-6}	6.55×10^{-6}	5.0	n.a	5.0
Cl$^-$ (mg/L)	80.8	79.5	70	n.a	142 [b]
Co (mg/L)	$<5 \times 10^{-6}$	$<5 \times 10^{-6}$	0.05	0.05	0.05
Cu (mg/L)	1.26×10^{-5}	6.28×10^{-5}	0.2	n.a	0.2
Total Cr (mg/L)	$<5 \times 10^{-6}$	$<5 \times 10^{-6}$	0.1	n.a	0.1
Sn (mg/L)	$<5 \times 10^{-6}$	$<5 \times 10^{-6}$	2.0	n.a	n.a
Fe (mg/L)	1.06×10^{-4}	2.36×10^{-5}	5.0	2.0	5.0
F$^-$ (mg/L)	<DL	<DL	1.0	2.0	1.0
Li (mg/L)	1.98×10^{-5}	1.96×10^{-5}	2.5	2.5	2.5
Mn (mg/L)	4.56×10^{-5}	3.77×10^{-5}	0.2	0.2	0.2
Mo (mg/L)	2.45×10^{-5}	8.60×10^{-5}	0.005	0.01	0.01
Ni (mg/L)	$<5 \times 10^{-6}$	$<5 \times 10^{-6}$	0.5	n.a	0.2
NO$_3^-$ (mg/L)	0.9 ± 0.4	7.70	50	n.a	9.5 [b]
Salinity (µS/cm)	848	782	1000	variable	700 [b]
TDS (mg/L)	335	191	640	n.a	450 [b]
SAR (meq/L)	2.49	1.50	8	variable	3.0
Se (mg/L)	$<5 \times 10^{-6}$	$<5 \times 10^{-6}$	0.02	0.02	0.02
TSS (mg/L)	24.50	0.00	60	≤10	20 [c]
SO$_4^{2-}$ (mg/L)	45.2	50.0	575	n.a	n.a
V (mg/L)	$<5 \times 10^{-6}$	$<5 \times 10^{-6}$	0.1	n.a	0.1
Zn (mg/L)	4.70×10^{-5}	2.61×10^{-5}	2	n.a	2.0
pH	7.0 ± 1.0	8.0 ± 0.2	6.5–8.4	n.a	6.5–8.4
E. coli (log CFU/100 mL)	6.67	<DL	2.0	≤10	2.3 [c]
Intestinal parasite eggs [a]	0.00	0.00	n.a	≤1	n.a

DL stands for detection limit; MVR stands for maximum value recommended; n.a stands for not applicable/available; SAR stands for sodium adsorption ratio; PV stands for parametric value; TDS stands for total dissolved solids; TSS stands for total suspended solids. [a] Analysed by an external laboratory—the maximum value allowed (MVA) for this parameter in the Decree-Law 236/98 is 1. [b] Value up to which there is no restriction to use in irrigation. [c] Permitted limit for greywater reuse in irrigation of vegetables likely to be eaten uncooked.

Table 2. Additional analyses made to the UWWTP secondary effluent, before (WW) and immediately after treatment (TWW$_0$).

Additional Analyses	UWWTP Secondary Effluent (WW)	After O$_3$ + UF Treatment (TWW$_0$)	Becerra et al., 2015 [30]		Decree-Law 119/2019 [28]
			MVA	MVR	PV
Dissolved organic carbon (DOC, mg/L)	11.0 ± 0.8	9.6 ± 0.8	n.a.	n.a.	n.a.
Biological oxygen demand (BOD$_5$, mg/L)	15.1 ± 1.1	0	10 [b]	n.a	≤10 [c]
Chemical oxygen demand (COD, mg/L)	22.7 ± 0.7	5.4 ± 0.8	60–200	n.a	n.a.
Turbidity (NTU)	3.25 ± 0.15	0.28 ± 0.02	2	n.a	≤5
NH$_4^+$	<DL	0.59	n.a	n.a	10
PO$_4^{3-}$	<DL	<DL	n.a	n.a	n.a

DL stands for detection limit; MVA stands for maximum value allowed; MVR stands for maximum recommended value; PV stands for parametric value; n.a stands for not applicable/available.

2.3. Experimental Setup and Procedure

A scheme of the experimental apparatus is depicted in Figure 1. Ozonation was performed in a packed-bed column (2.2 I.D × 70 cm height) with a useful volume of approximately 0.35 L and containing glass Raschig rings (6 mm I.D × 6 mm height), because the water–ozone mass transfer achieved in the column packed with these Raschig rings was up to 3 times higher than that in a bubble column [31]. Firstly, the reactor was filled with ultrapure water (through a peristaltic pump) to regulate the desired concentration of ozone in the liquid phase. Ozone was produced from pure oxygen in a BMT 802X ozone generator and bubbled at the bottom of the column. The ozone concentration in the gas inlet was regulated by adjusting the oxygen gas flow rate with a mass flow controller and the electric intensity of the ozone generator (BMT 802X). The concentration of ozone in the liquid phase (dissolved ozone) was measured with an ATI model Q45H dissolved ozone analyser placed at the exit of the column. High ozone doses and contact time increase the capital and operating costs, therefore a low ozone dose (0.9 ± 0.1 gO$_3$/gDOC) and a short hydraulic retention time (HRT: 8 min obtained with a liquid flow rate of 46 mL min^{-1}) were investigated. These experimental conditions were selected in preliminary tests and fixed for all the subsequent experiments.

After a period, ultrapure water in the inlet liquid stream was replaced by the UWWTP effluent to start the ozonation experiments. Samples of ozonised wastewater were only collected after a period of two residence times (~16 min), in order to ensure that the steady state was achieved (i.e., when the outlet wastewater achieved a constant concentration of pollutants in two subsequent measurements). Then, the ozonised effluent was directed to the feed tank of the UF pilot reactor, aiming for the physical removal of microbial cells. Fifteen litres of ozonised effluent was pumped to the UF pilot through a peristaltic pump (Varmec®) and filtered through the 10 nm α-Al$_2$O$_3$ membrane operating in cross-flow mode (1 bar of transmembrane pressure). The UF pilot was designed in a way that the liquid flow of ozonised wastewater was automatically regulated to maintain the pressure constant inside the membrane housing compartment. The concentrate was recirculated to the feed tank [32], while a composite sample of the permeate was collected and split for microbiological and chemical analysis (TWW$_0$ immediately after O$_3$ + UF treatment and TWW$_7$ after being stored for seven days). UF was performed after O$_3$ and not the other way around, because by doing so, the membrane fouling is minimised [33,34]. At the end of the treatment, the membrane was left with H$_2$O$_2$ (30% w/v) overnight, followed by abundant washing with boiling water and autoclaved before starting another experiment. This cleaning procedure was defined to restore the membrane permeance and sterility.

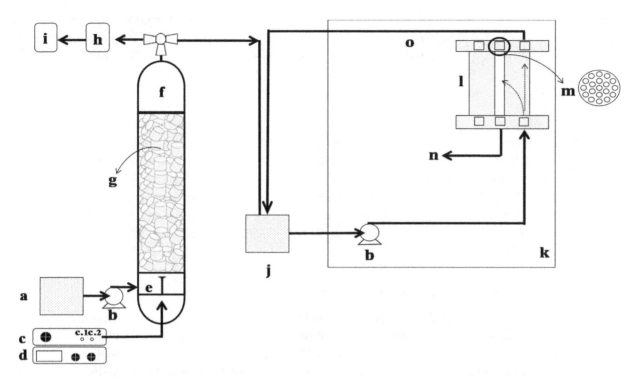

Figure 1. Scheme of the experimental apparatus. (**a**) feed tank containing deionised water or UWWTP effluent; (**b**) peristaltic pump; (**c**) ozone generator (c.1—O_2 entrance; c.2—O_3 exit); (**d**) mass flow controller; (**e**) ozone diffuser; (**f**) packed-bed column; (**g**) Raschig rings; (**h**) ozone analyser; (**i**) ozone destroyer; (**j**) feed to the ultrafiltration (UF) pilot; (**k**) UF pilot system; (**l**) membrane housing; (**m**) 19 channel ceramic membrane (top view); (**n**) permeate stream; (**o**) concentrate stream.

2.4. Chemical Analyses

The anionic and cationic contents (Cl^-, NO_3^-, SO_4^{2-}, Na^+, K^+) in water samples were determined by ion chromatography, as, using a Metrohm 881 Compact IC Pro apparatus equipped with a Metrosep C4 Cationic Exchange Column (250 mm × 4.0 mm) for the quantification of cations and a Metrosep A Supp 7 Anionic Exchange Column (250 mm × 4.0 mm) for quantification of anions. The content of metals was determined by using an inductively coupled plasma-optical emission spectrometer (ICP-OES, thermo scientific, model iCAP 7000 Series). The pH and conductivity of water were measured with pHenomenal® pH 1100L apparatus (VWR, Germany) and a conductivity meter (Crison GLP 31), respectively. Other relevant parameters (referred to as "additional analyses" in Table 2) were considered to assess the quality of water for irrigation: dissolved organic carbon (DOC) determined in a TOC-L analyser (Shimadzu TOC-5000A); turbidity measured with a turbidimeter (Hanna instruments, model HI88703); chemical oxygen demand (COD) determined by the closed reflux method (EPA standard method 5220D); and biochemical oxygen demand measured according to the EPA standard method 5210B (respirometric method) for a 5 day period (BOD_5). These analyses were performed as recommended in the standard methods for the examination of water and wastewater [35].

Moreover, the concentration of target organic micropollutants was determined using ultra-high performance liquid chromatography with tandem mass spectrometry (UHPLC-MS/MS) with Shimadzu Corporation apparatus (Tokyo, Japan) consisting of a triple quadrupole mass spectrometer detector (Ultra-Fast Mass Spectrometry series LCMS-8040) with an ESI (Electrospray Ionisation) source operating in both positive and negative ionisation modes. The mobile phase and operating conditions of the UHPLC-MS/MS system for the detection and quantification of the target pollutants are described elsewhere [16,36]. Prior to UHPLC-MS/MS analysis, WW and TWW_0 samples were pre-concentrated and cleaned up by solid-phase extraction (SPE) using Oasis® HLB (Hydrophilic-Lipophilic-Balanced sorbent, 150 mg, 6 mL) cartridges (Waters, Milford, Massachusetts, USA), according to the methodology described elsewhere [37]. For internal calibration, isotopically labelled internal standards were added to the samples before SPE. The preconcentration procedure was performed in duplicate for all the samples. This methodology allows to determine a total of 14 organic micropollutants.

2.5. Microbial Culture Analyses

Volumes ranging from 100 mL to 1 mL of WW, TWW_0 or TWW_7 samples or of serial 10-fold dilutions thereof were filtered in triplicate and placed onto the appropriate culture media of the target microbial group: Plate Count Agar (PCA, VWR International (Pennsylvania, USA)) (30 °C, 48 h) for culturable heterothrops, m-Faecal Coliform Agar (mFC, Thermo Fisher Scientific, Massachusetts, USA) (37 °C, 24 h) for enterobacteria, Slanetz Bartley Agar (Thermo Fisher Scientific, Massachusetts, USA) (37 °C, 48 h) for enterococci, and Rose Bengal Chloramphenicol Agar (Thermo Fisher Scientific, Massachusetts, U.S.A.) (25 °C, 5 days) for fungi. Results were expressed as colony forming units per 100 mL of sample (CFU/ 100 mL).

2.6. DNA Extraction, 16S rRNA and Intl1 Genes Quantification

Volumes of 100 mL of WW, 2 L of TWW_0, and 800 mL to 1 L of TWW_7 were vacuum-filtrated and processed in three independent samplings as biological replicates. DNA extraction was performed using the DNeasy® PowerWater® Kit (QIAGEN, Hilden, Germany) according to Rocha et al., 2020 [38] and with two additional steps suggested in the manufacturer's troubleshooting guide: after adding the lysis solution, a heating step at 65 °C for 10 min was included in the protocol; and to ensure the removal of residual ethanol before DNA elution, the centrifugation step was conducted in a clean collection tube for an additional minute. DNA samples were stored at −20 °C until quantitative PCR (qPCR) analysis.

The 16S rRNA gene (a marker for total bacteria) and the *intI*1 gene encoding a class 1 integron-integrase (a marker of anthropogenic impact) were quantified based on qPCR to assess the removal efficiency of bacteria after treatment [39,40]. Gene-specific primer sequences are listed in previous studies [41,42] and provided as supplementary information in Table S1. Gene quantification was based on SYBR Green qPCR assays in a StepOnePlus™ Real-Time PCR System (Life Technologies, USA) and interpolation to the standard curve run in each assay, as described elsewhere [39,43].

The data that met the quality criteria described in Rocha et al., 2018 [44] were expressed as the ratio of gene copy number per 100 mL of water sample (WW, TWW_0, and TWW_7). The secondary wastewater effluent (WW) was used as reference to assess the removal efficiency of both 16S rRNA and *intI*1 genes in treated samples, immediately after treatment (TWW_0) and after storage for 7 days (TWW_7). The duration of 7 days was selected to allow enough time for eventual injured cells surviving the treatment to fully recover, as we have verified in previous works with other treatment solutions [14,15].

2.7. Biological Effect Assays

2.7.1. Cell Culture and Incubation with Water Samples

Murine fibroblasts L929 and Caco-2 cells were cultured in Dulbecco's modified Eagle medium (DMEM) with D-glucose (4.5 g L^{-1}), sodium pyruvate (0.11 g L^{-1}), L-alanyl-L-glutamine (0.86 g L^{-1}) and further supplemented with 10% (*v/v*) heat inactivated foetal bovine serum (FBS), and 5% (*v/v*) of PenStrep (37 °C, 5% CO_2 and 95% of humidity). For cell viability and cytotoxicity assessment, the cells were detached from the culture flask as described elsewhere [45]. After cell counting in Neubauer chamber (Boeco, Germany), the suspension was centrifuged at 300 g for 5 min, and the cell pellet was suspended in culture medium to a final concentration of 5×10^4 cells per well. Cells were then seeded in a 96-well microplate (100 µL per well) and cultured for 24 h at 37 °C (5% CO_2 and 95% humidity).

2.7.2. Thiazolyl Blue Tetrazolium Reduction (MTT) and Lactate Dehydrogenase (LDH) Assays

Cellular metabolic activity was evaluated as indicator of cytotoxicity by the thiazolyl blue tetrazolium reduction (MTT) assay, whereas cell membrane integrity was evaluated through the lactate dehydrogenase (LDH) assay, providing information about cell viability. Briefly, test water samples were filtered using Corning® syringe filters (Sigma-Aldrich®, St. Louis, MO, USA) with 0.20 μm pore diameter and diluted 1:10 and 1:5 in DMEM. After discarding culture supernatant, 100 μL of diluted samples were added to cell layers and incubated at 37 °C (5% CO_2 and 95% humidity). After 24 h, the supernatant was removed for LDH assay, while the remaining content of the wells was used for MTT assay. For MTT assay, absence of cytotoxicity (100%) was estimated by replacing water test sample by culture medium. For LDH assay, the absence of cell viability (100%) was estimated by replacing water test sample by 1% (v/v) Triton X-100 solution prepared in culture medium.

2.7.3. Yeast Estrogen Screen (YES) Assay for Estrogenic Activity Assessment

WW and TWW_0 samples were filtered through 0.21 μm hydrophilic membranes and analysed directly, without any preconcentration. The YES assay and data analysis were performed according to the kit manufacturer's instructions. Calibration was established using standard solutions of the natural estrogen 17-β-estradiol (E2), at concentrations between 10^{-6}–10^{-9} mol L^{-1}. E2 also worked as positive control while ultrapure water was used as negative control. E2 standard solutions were prepared in DMSO (<1% in the assay medium), therefore a solvent blank was also assayed. Samples, standards, and control solutions were transferred to a 96-well microplate, mixed with assay medium, and inoculated with the transformed yeast cells. The mixture was then incubated for 48 h at 31 °C under orbital shaking. Spectrophotometric measurements at 570 nm (β-galactosidase expression) and 690 nm (yeast growth) were carried out using a Cytation3® microplate reader (Bio-Tek Instruments, Winooski, USA). The potential estrogen agonistic activity was estimated through the calculation of the parameters growth factor (G) and induction ratio (IR). The G parameter was calculated as the ratio of absorbance values measured at 690 nm for the sample and for the solvent $(A_{690})_{sample}/ (A_{690})_{solvent}$. The IR parameter was calculated as $(1/G) \times ((A_{570} - A_{690})_{sample}/(A_{570} - A_{690})_{solvent})$.

3. Results and Discussion

3.1. Micropollutant Removal, Mineralisation, and Other Physico-Chemical Parameters

Under the regulation on minimum requirements for water reuse in agricultural irrigation, the environmental quality standards for priority substances and certain other pollutants should be targeted [7,21]. Moreover, the same regulation refers to additional requirements for risk assessment, including micropollutants. From the chemical organic micropollutants analysed in fresh (WW) and O_3+UF treated water samples (TWW_0), only 9 out of 14 were detected. The antiplatelet clopidogrel, the herbicide isoproturon, the anti-inflammatory diclofenac, the industrial compound PFOS (perfluorooctanesulfonic acid), and the lipid regulator bezafibrate were detected with a frequency of 100% in WW samples during the sampling campaign (Figure 2). Alachlor was also detected in all WW samples but below the limit of quantification ($LOQ_{alachlor}$ < 25 ng L^{-1}), whereas warfarin, citalopram, and clofibric acid were detected only in some samples. According to the Directive 2013/39/EU and Decision 495/2015/EU [21,22], alachlor, isoproturon and PFOS are considered PSs, whereas the others are considered CECs. After treatment, most micropollutants presented values below LOD—Limit Of Detection. Only alachlor, clopidogrel, citalopram and isoproturon were detected: the first two were below the LOQ—Limit Of Quantification (25 and 5 ng L^{-1}, respectively), whereas the latter two were found at concentrations up to 529 and 10.6 ng L^{-1}, respectively. In fact, isoproturon was the micropollutant with the lowest removal percentage (i.e., 80% of maximum removal). All priority substances (alachlor, isoproturon and PFOS) were below their environmental quality standards defined in the EU Directive 2013/39 [21], complying with the requirements of the EU Regulation 2020/741 [7].

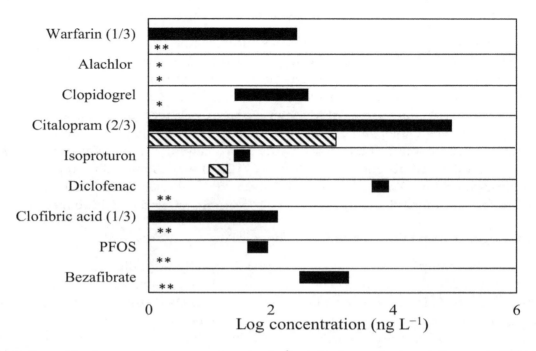

Figure 2. Logarithmic range of concentrations (ng L^{-1}) of the detected micropollutants in WW (black bar) and TWW$_0$ (striped bar) for samples with concentrations above LOQ. The frequency of occurrence was 100% (3/3) for all compounds, except when indicated in brackets after the compound name. * <LOQ and ** <LOD (compounds with concentrations < LOD before treatment are not shown in this figure for the sake of simplicity).

DOC and pH values did not remarkably vary after treatment (Tables 1 and 2, respectively). Values for DOC are not regulated and both pHs (before and after treatment) comply with the maximum value allowed (MVA). Thus, considering that regulations of water quality for irrigation often do not inform about adequate levels of organic matter, it can be assumed that the achieved values of DOC and micropollutants in treated water do not invalidate its use for irrigation. Moreover, the available literature mentioning the monitoring of DOC in water for irrigation recommends the evaluation of DOC when COD and BOD$_5$ are at the so-called alarming levels (>60 mgO$_2$ L^{-1} and >10 mgO$_2$ L^{-1}, respectively) [28,30,46], which is not the case of TWW$_0$ (Table 2).

In the combined process, ozonation was expected to be mainly responsible for the removal of micropollutants and dissolved organic matter rather than UF [18]. These results are coherent with other studies performing solely ozonation, in which the authors attributed the low yield of mineralisation to the formation of recalcitrant organic intermediates deriving from the organic micropollutants or, more likely, from the oxidation of dissolved organic matter naturally present in the wastewater [14,16]. For instance, using a similar experimental apparatus for the continuous ozonation of a secondary-UWWTP effluent (without UF), Moreira et al., 2016 [14] reported a DOC removal of ~30% (retention time of 26 min), whereas Iakovides et al., 2019 [16] obtained a DOC removal of ~10% (with similar ozone dosage and retention time).

Regarding other physico-chemical parameters, TWW$_0$ presents values below the maximum recommended in the Portuguese Laws of (i) water for irrigation [31] and (ii) treated wastewater for reuse [28,30]. The only exception is for the concentration of chloride in Table 1 (ca. 80 mg L^{-1} before and after treatment) which is slightly higher than the maximum value recommended (MVR) of 70 mg L^{-1} in the oldest law [27], which is not included in the newest one [28]. It is worth mentioning that this maximum value recommended for chloride was stipulated considering the sensitivity of tobacco crops; therefore, TWW$_0$ might not be appropriate for irrigation of this specific crop, but not necessarily inappropriate in the case of crops tolerant to these concentrations of chloride. For instance, some crops of fruits and vegetables are highly tolerant to chloride, such as Rangpur lime and cauliflower, for which the water for irrigation can contain up to 600 and 710 mg L^{-1} of chloride, respectively [47]. In fact, TWW_0 can be applied for irrigation according to the WHO (World Health Organization) and FAO (Food and Agriculture Organization of the United Nations) guidelines of water quality for surface irrigation, where the allowed chloride concentration is up to 142 mg L^{-1} (Table 1) [29], i.e., well above the value determined for the wastewater in this study (ca. 80 mg L^{-1}). The value of salinity (782 μS/cm) is slightly higher than that recommended by FAO and WHO [28] for the use of water for irrigation with no restriction (<700 μS/cm), but this value is not defined in Portuguese guidelines. Another interesting observation is the increase in the nitrate concentration after treatment, although still below the maximum value recommended [27], which can be attributed to the oxidation of nitrogen-containing substances that are likely to be present in the secondary effluent of UWWTPs [48]. Sulphate and copper contents also suffer a slight increase after treatment, which can be due to their release from sediments/soil particles after ozonation [49,50].

Future work must consider the energy demand of these processes [51] and life cycle assessment (LCA) [52–54] for the elimination of micropollutants from urban wastewater—these studies being particularly scarce with data at full scale. For instance, it has been concluded that ozonation has a lower energy demand compared to the use of membranes or UV/H_2O_2 [9]. Conversely, the electrical energy demand of ozonation is higher than those determined for powdered activated carbon (PAC) addition or granular activated carbon (GAC) filtration, but always being a plant-specific issue [51]. Performing LCA, it was suggested that ozonation has a better overall environmental performance than the photo-Fenton process [53], whereas reverse osmosis causes higher environmental burdens than ozonation due to the high energy and material consumption [52]. In these processes, generated impacts result mainly from the production of energy needed (and the respective energy mix) and from the use of some specific reagents [54].

3.2. Microbial Inactivation and Regrowth

As expected, a reduction in the load of the microbiological groups analysed was observed immediately after treatment (Figure 3). Reductions of nearly 3.5 log-units of 16S rRNA gene (indicative of the abundance of total bacteria) and 3.7 log-units of culturable heterotrophs occurred. The abundance of *intI*1 followed a similar trend, with a reduction of ~4.6 log-units immediately after treatment, whereas enterobacteria, enterococci, and fungi, with reductions higher than five log-units, reached values below the detection limit (0.33 CFU per 100 mL). Microbial inactivation can be transient [14,55,56], therefore further assays testing the regrowth capacity after seven days of storage of the treated wastewater were performed (TWW_7 samples). It is known that bacterial reactivation is influenced by factors such as storage conditions, in particular temperature, availability of nutrients, ultraviolet light, and assimilable organic carbon content, among others [57,58]. Therefore, the conditions to perform this assessment were selected to mimic the most common real storage conditions, i.e., room temperature (25 ± 2 °C) and absence of light to minimise DNA repair mechanisms [59]. The abundance of the 16S rRNA and *intI*1 genes, as well as the heterotrophic counts, recovered to values close to those observed in WW samples. The same pattern was observed for fungi, although with a lower regrowth extent (~1.6 log-units). The transient effect of single ozone-based processes for the treatment of UWWTP effluents was reported before [14–16]. In fact, even when operating with close ozone doses (0.75 gO₃/gDOC) and higher HRT (10–60 min) to those used here (0.9 gO₃/gDOC, HRT 8 min), reactivation of all the microbial groups analysed in the current study has been described in the literature [14,15]. In contrast, in the present study, regrowth of faecal indicators (enterobacteria and enterococci) was not observed in TWW_7 samples. Notwithstanding, from a microbiological quality point of view, both TWW_0 and TWW_7 comply with the biological parameters included in the quality standards of water for crops irrigation, both in Portugal [27] and United States [60], or the Portuguese/European Union quality standards of wastewater reuse in irrigation without restriction [28,61]. In fact, faecal coliforms or *Escherichia coli* (enterobacteria) and nematode eggs are the only biological parameters included in these quality standards, for which values were found below the stipulated thresholds (Table 1).

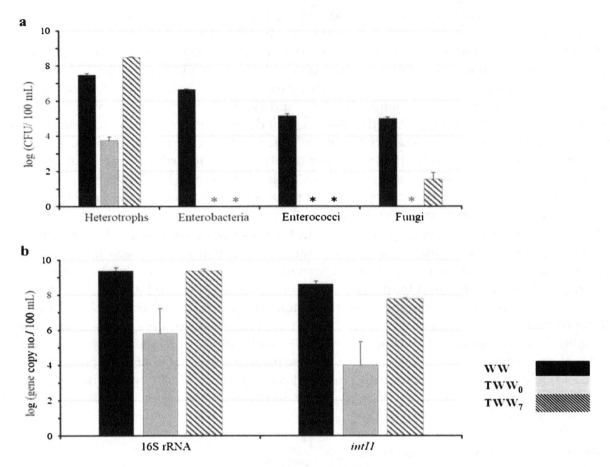

Figure 3. Microbiological water quality. (**a**) Culturable heterotrophs, enterobacteria, enterococci, and fungi, expressed as log (CFU/100 mL of sample); and (**b**) qPCR-based quantification of 16S rRNA and intI1 genes, expressed as log (gene copy number/ 100 mL of sample). * below the detection limit (0.33 CFU/100 mL).

Based on the abundance of enterobacteria in wastewater immediately after ozonation (10^2–10^3 CFU 100 mL $^{-1}$) or after 3 day storage (10^3–10^4 CFU 100 mL^{-1}) reported by Moreira et al. (2016) [14] and Iakovides et al., 2019 [16], the utilisation of ozonation alone would not produce wastewater compatible with its further use in irrigation. In contrast, the combination of UF with O_3 utilised here improved the efficiency of the treatment. The membrane fouling observed during the filtration process, which was evidenced by the permeate flow decrease from ~60 mL min^{-1} to ~16 mL min^{-1}, was most likely derived from bacteria that survived ozonation, cell debris and undissolved (in)organic matter. Nevertheless, the total suspended solid (TSS) value after O_3 was unquantifiable. In spite of the considerable improvements demonstrated in this study, the post-storage increase in total heterotrophs and genes shows that there is still room for additional tuning of the process to prevent the possible contamination of the permeate tank with spores of heterotrophic bacteria or fungi.

3.3. Evaluation of Biological Effects

Cytotoxic and cell viability effects of wastewater collected before (WW) and after treatment with O_3 + UF (TWW$_0$) were evaluated for skin (L929) and digestive epithelium (Caco-2) cell models by performing complementary MTT and LDH assays (Table 3). Considering that cell viability upon exposure to water samples depends on the final composition of the growth medium [62], test samples were diluted 5 and 10 times in culture medium before incubation with cell layers. Similar cytotoxicity (MTT) and cell viability (LDH) values were obtained for both dilution levels (Table 3). Moreover, cell viability was equivalent to that obtained for cell incubation with a plain culture medium. For both cell lines, no difference in cytotoxicity was observed for water samples collected before and after treatment (Table 3, MTT assay). Cell viability was also maintained after treatment (Table 3, LDH assay), providing similar or even higher values than those obtained for plain culture media or tap water. Additionally, samples analysed right after ozonation (i.e., before UF) rendered percentages of $91 \pm 6\%$ and $23 \pm 8\%$ in the MTT and LDH assays for L929 cells, respectively, indicating that no cytotoxic compounds were produced during this step.

Table 3. Results (percentage) from MTT [a] and LDH [b] assays obtained for urban wastewater before (WW) and after treatment (TWW$_0$).

Cell Line	MTT Assay		LDH Assay [c]	
	WW	TWW$_0$	WW	TWW$_0$
L929	102 ± 13	112 ± 15	20.7 ± 2.0 (28.3 ± 3.2)	19.7 ± 1.6 (32.6 ± 4.5)
Caco-2	116 ± 8	96 ± 9	58.6 ± 4.4 (59.1 ± 6.8)	53.2 ± 7.7 (59.5 ± 5.6)

[a] Values for culture media were 100% (Relative Standard Deviation—RSD < 20%) and between 1 and 9% for Triton X-100 (total disruption of cells). Samples were diluted 5 times in culture media before incubation with cells. [b] Values for Triton X-100 (total disruption of cells) were 100% (RSD < 10%). Values for tap water were 111 ± 5 for L929 cells and 110 ± 12 for Caco-2 cells. [c] Values between brackets correspond to blank values obtained in culture media only (intact cells). Values for tap water were 26.3 ± 5.6 for L929 cells and 55.9 ± 4.7 for Caco-2 cells.

The presence of estrogenic activity was also evaluated using the YES assay for WW and TWW$_0$ samples. Yeast growth inhibition was not observed for any of the tested samples. Induction ratios (IR) were 1.02 ± 0.09 for WW, and 0.74 ± 0.02 for TWW$_0$. These values were below the kit threshold value IR10 (corresponding to 10% of the maximum IR, value of 2.82, obtained for E2 standards), which indicated no estrogenic activity.

Work on toxicity assessment of effluents treated by ozonation has provided contradictory evidence. The biological toxicity of the influent of sewage treatment plants was significantly decreased after applying different advanced treatment processes, including ozone combined with UV, using *Daphnia magna*, zebrafish (*Danio rerio*), and *Vibrio fischeri* [63] as target organisms. However, when ozone and hydrogen peroxide were used together, a slight acute toxicity was perceived for *V. fischeri* while acute toxicity was observed for *D. magna* [64]. Other work, also applying the algae *Desmodesmus quadricauda*, indicated that the toxicity class of treated wastewater may change from completely non-toxic to very high hazard category, with a clear relationship between the time of ozonation and the increase in ecotoxicity [65]. This compound-dependent behaviour was also observed in a study with zebrafish embryos where different pharmaceutical compounds were tested [66]. Therefore, our results with cell lines are in agreement with previous works, where no toxic effect was observed after treatment, particularly when low doses of ozone are applied.

4. Conclusions

The results of this study indicate that UF performed after ozonation can be a suitable approach to allow the safe reuse of urban wastewater for irrigation. The combined process resulted in an effective treatment, especially against micropollutants detected in the UWWTP secondary effluent, and in the reduction in the microbial load. Treated wastewater stored for seven days maintained the quality required for irrigation, with the physico-chemical parameters, and enterobacteria and nematode egg counts below the maximum values recommended in water quality standards. In addition, no harmful biological effects were detected concerning the viability and estrogenicity tests. However, the fact that total bacterial cells, total cultivable heterotrophs as well as the *intI*1 gene reactivated to values close to those observed for untreated wastewater, shows that there is still room for additional improvement of this process.

Author Contributions: Conceptualisation, A.M.T.S., O.C.N., C.M.M., M.A.S., S.M.C.-S.; methodology, A.M.T.S., O.C.N., C.M.M., M.A.S., S.M.C.-S.; investigation, C.A.L.G., S.R.-S., J.A.-S., I.I.R., A.R.R.; writing—original draft preparation, C.A.L.G., S.R.-S., J.A.-S., I.I.R.; writing—review and editing, A.M.T.S., O.C.N., C.M.M., M.A.S., A.R.R.; supervision, A.M.T.S., O.C.N., C.M.M., M.A.S.; project administration, A.M.T.S., O.C.N., C.M.M., M.A.S., S.M.C.-S.; funding acquisition, A.M.T.S., O.C.N., C.M.M., M.A.S., S.M.C.-S. All authors have read and agreed to the published version of the manuscript.

Acknowledgments: We would like to thank the scientific collaboration under Base Funding-UIDB/50020/2020 of the Associate Laboratory LSRE-LCM and Base Funding-UIDB/00511/2020 of the Laboratory for Process Engineering, Environment, Biotechnology and Energy—LEPABE, both funded by national funds through the FCT/MCTES (PIDDAC), and FCT project UID/Multi/50016/2013 (Associate Laboratory CBQF) and UIDB/50006/2020 (LAQV, REQUIMTE).

References

1. UN General Assembly. Transforming Our World: The 2030 Agenda for Sustainable Development, 21 October 2015, A/RES/70/1. Available online: https://www.refworld.org/docid/57b6e3e44.html (accessed on 22 April 2020).
2. Jimenez, B.; Asano, T. *Water Reuse: An International Survey of Current Practice, Issues and Needs (Scientific and Technical Report)*; IWA Publishing: London, UK, 2018.
3. Bixio, D.; De Heyder, B.; Cicurel, H.; Muston, M.; Miska, V.; Joksimovic, D.; Schäfer, A.I.; Ravazzini, A.; Aharoni, A.; Savic, D.; et al. Municipal wastewater reclamation: Where do we stand? An overview of treatment technology and management practice. *Water Sci. Tech. W. Sup.* **2005**, *5*, 77–85. [CrossRef]
4. Pedrero, F.; Kalavrouziotis, I.; Alarcón, J.J.; Koukoulakis, P.; Asano, T. Use of treated municipal wastewater in irrigated agriculture-Review of some practices in Spain and Greece. *Agric. Water Manag.* **2010**, *97*, 1233–1241. [CrossRef]
5. Vaz-Moreira, I.; Nunes, O.C.; Manaia, C.M. Bacterial diversity and antibiotic resistance in water habitats: Searching the links with the human microbiome. *FEMS Microbiol. Rev.* **2014**, *38*, 761–778. [CrossRef] [PubMed]
6. Barbosa, M.; Moreira, N.F.F.; Ribeiro, A.R.; Pereira, M.F.R.; Silva, A.M.T. Occurrence and removal of organic micropollutants: An overview of the watch list of EU Decision 2015/495. *Water Res.* **2016**, *94*, 257–279. [CrossRef] [PubMed]
7. EUR-Lex. Regulation (EU) 2020/741 of the European Parliament and of the Council of 25 May 2020 on minimum requirements for water reuse. *Off. J. Eur. Union L* **2020**, *177*, 32–55.
8. Michael-Kordatou, I.; Karaolia, P.; Fatta-Kassinos, D. The role of operating parameters and oxidative damage mechanisms of advanced chemical oxidation processes in the combat against antibiotic-resistant bacteria and resistance genes present in urban. *Water Res.* **2018**, *129*, 208–230. [CrossRef]
9. Rizzo, L.; Malato, S.; Antakyali, D.; Beretsou, V.G.; Đolić, M.B.; Gernjak, W.; Heath, E.; Ivancev-Tumbas, I.; Karaolia, P.; Ribeiro, A.R.L.; et al. Consolidated vs new advanced treatment methods for the removal of contaminants of emerging concern from urban wastewater. *Sci. Total Environ.* **2019**, *655*, 986–1008. [CrossRef]
10. Wei, C.; Zhang, F.; Hu, Y.; Feng, C.; Wu, H. Ozonation in water treatment: The generation, basic properties of ozone and its practical application. *Rev. Chem. Eng.* **2016**, *33*, 49–90. [CrossRef]
11. Von Gunten, U. Oxidation processes in water treatment: Are we on track? *Environ. Sci. Technol.* **2018**, *52*, 5062–5075. [CrossRef]
12. Von Gunten, U. Ozonation of drinking water: Part, I. Oxidation kinetics and product formation. *Water Res.* **2003**, *37*, 1443–1467. [CrossRef]
13. Ikehata, K.; Jodeiri Naghashkar, N.; Gamal El-Din, M. Degradation of aqueous pharmaceuticals by ozonation and advanced oxidation processes: A review. *Ozone Sci. Eng.* **2006**, *28*, 353–414. [CrossRef]
14. Moreira, N.F.F.; Sousa, J.M.; Macedo, G.; Ribeiro, A.R.; Barreiros, L.; Pedrosa, M.; Faria, J.L.; Pereira, M.F.R.; Castro-Silva, S.; Segundo, M.A.; et al. Photocatalytic ozonation of urban wastewater and surface water using immobilized TiO2 with LEDs: Micropollutants, antibiotic resistance genes and estrogenic activity. *Water Res.* **2016**, *94*, 10–22. [CrossRef] [PubMed]
15. Sousa, J.M.; Macedo, G.; Pedrosa, M.; Becerra-Castro, C.; Castro-silva, S.; Pereira, M.F.R.; Silva, A.M.T.; Nunes, O.C.; Manaia, C.M. Ozonation and UV254 nm radiation for the removal of microorganisms and antibiotic resistance genes from urban wastewater. *J. Hazard. Mater.* **2017**, *323*, 434–441. [CrossRef]
16. Iakovides, I.C.; Michael-Kordatou, I.; Moreira, N.F.F.; Ribeiro, A.R.; Fernandes, T. Continuous ozonation of

urban wastewater: Removal of antibiotics, antibiotic-resistant Escherichia coli and antibiotic resistance genes and phytotoxicity. *Water Res.* **2019**, *159*, 333–347. [CrossRef] [PubMed]

17. Rizzo, L.; Manaia, C.; Merlin, C.; Schwartz, T.; Dagot, C.; Ploy, M.C.; Michael, I.; Fatta-Kassinos, D. Urban wastewater treatment plants as hotspots for antibiotic resistant bacteria and genes spread into the environment: A review. *Sci. Total Environ.* **2013**, *447*, 345–360. [CrossRef] [PubMed]

18. Hembach, N.; Alexander, J.; Hiller, C.; Wieland, A.; Schwartz, T. Dissemination prevention of antibiotic resistant and facultative pathogenic bacteria by ultrafiltration and ozone treatment at an urban wastewater treatment plant. *Sci. Rep.* **2019**, *9*, 12843. [CrossRef]

19. Czekalski, N.; Imminger, S.; Salhi, E.; Veljkovic, M.; Kleffel, K.; Drissner, D.; Von Gunten, U. Inactivation of antibiotic resistant bacteria and resistance genes by ozone: From laboratory experiments to full-scale wastewater treatment. *Environ. Sci. Technol.* **2016**, *50*, 11862–11871. [CrossRef] [PubMed]

20. Rizzo, L.; Gernjak, W.; Krzeminski, P.; Malato, S.; McArdell, C.S.; Sanchez Perez, J.A.; Schaar, H.; Fatta-Kassinos, D. Best available technologies and treatment trains to address current challenges in urban wastewater reuse for irrigation of crops in EU countries. *Sci. Tot. Env.* **2020**, *710*, 136312. [CrossRef]

21. EUR-Lex. Directive 2013/39/EU of the European Parliament and of the Council of 12 August 2013 amending directives 2000/60/EC and 2008/105/EC as regards priority substances in the field of water policy. *Off. J. Eur. Union L* **2013**, *226*, 1–17.

22. EU Decision 495/2015. Commission implementing Decision (EU) 2015/495 of 20 March 2015 establishing a watch list of substances for Union-wide monitoring in the field of water policy pursuant to directive 2008/105/EC of the European Parliament and of the Council. *Off. J. Eur. Union L* **2015**, *78*, 40–42.

23. Spencer, P.; Domingos, S.; Edwards, B.; Howes, D.; Shorney-Darby, H.; Scheerman, H.; Milton, G.; . Clement, J. Ozone enhanced ceramic membrane filtration for wastewater recycling. *Water Pract. Technol* **2019**, *14*, 331–340. [CrossRef]

24. Si, X.; Hu, Z.; Huang, S. Combined process of ozone oxidation and ultrafiltration as an effective treatment technology for the removal of endocrine-disrupting chemicals. *Appl. Sci.* **2018**, *8*, 1240. [CrossRef]

25. Wang, H.; Park, M.; Liang, H.; Wu, S.; Lopez, I.J.; Ji, W.; Li, G.; Snyder, S.A. Reducing ultrafiltration membrane fouling during potable water reuse using pre-ozonation. *Water Res.* **2017**, *125*, 42–51. [CrossRef] [PubMed]

26. Acero, J.L.; Benitez, F.J.; Real, F.J.; Rodriguez, E. Elimination of selected emerging contaminants by the combination of membrane filtration and chemical oxidation processes. *Water Air Soil Pollut.* **2015**, *226*, 139. [CrossRef]

27. FAOLEX. *Decree-Law 236/98 Establishing Water Quality Standards*; Republic Diary No. 176/1998, Series I-A of 1998-08-01; Portuguese Presidency of the Council of Ministers: Lisboa, Portugal, 1998; pp. 3676–3716.

28. FAOLEX. *Decree-Law No. 119/2019 Establishing the Legal Scheme of the Production of Water for Reuse*; Republic Diary No. 159/2019, Series I of 2019-08-21; Portuguese Presidency of the Council of Ministers: Lisboa, Portugal, 2019; pp. 21–44.

29. WHO. *A Compendium for Standards for Wastewater Reuse in the Eastern Mediterranean Region*; World Health Organisation (WHO): Cairo, Egypt, 2006.

30. Becerra-Castro, C.; Rita, A.; Vaz-Moreira, I.; Silva, E.F.; Manaia, C.M.; Nunes, O.C. Wastewater reuse in irrigation: A microbiological perspective on implications in soil fertility and human and environmental health. *Environ. Int.* **2015**, *75*, 117–135. [CrossRef]

31. Graça, C.A.L.; Lima, R.B.; Pereira, M.F.R.; Silva, A.M.T.; Ferreira, A. Intensification of the ozone-water mass transfer in an oscillatory flow reactor with innovative design of periodic constrictions: Optimization and application in ozonation water treatment. *Chem. Eng. J.* **2020**, *389*, 124412.

32. Marchese, J.; Ochoa, N.A.; Pagliero, C.; Almandoz, C. Pilot-scale ultrafiltration of an emulsified oil wastewater. *Environ. Sci. Technol.* **2000**, *34*, 2990–2996. [CrossRef]

33. Mansas, C.; Mendret, J.; Brosillon, S.; Ayral, A. Coupling catalytic ozonation and membrane separation: A review. *Sep. Purif. Technol.* **2020**, *236*, 1161221. [CrossRef]

34. You, S.-H.; Tseng, D.-H.; Hsu, W.-C. Effect and mechanism of ultrafiltration membrane fouling removal by ozonation. *Desalination* **2007**, *202*, 224–230. [CrossRef]

35. Rice, E.W.; Baird, R.B.; Eaton, A.D. *Standard Methods for the Examination of Water and Wastewater*; American Public Health Association, American Water Works Association, Water Environment Federation: Washington, DC, USA, 2017.

36. Barbosa, M.O.; Ribeiro, A.R.; Ratola, N.; Hain, E.; Homem, V.; Pereira, M.F.R.; Blaney, L.; Silva, A.M.T. Spatial

and seasonal occurrence of micropollutants in four Portuguese rivers and a case study for fluorescence excitation-emission matrices. *Sci. Total Environ.* **2018**, *644*, 1128–1140. [CrossRef]

37. Ribeiro, A.R.; Pedrosa, M.; Moreira, N.F.F.; Pereira, M.F.R.; Silva, A.M.T. Environmental friendly method for urban wastewater monitoring of micropollutants defined in the Directive 2013/39/EU and Decision 2015/495/EU. *J. Chromatogr. A* **2015**, *1418*, 140–149. [CrossRef] [PubMed]

38. Rocha, J.; Manaia, C.M. Cell-based internal standard for qPCR determinations of antibiotic resistance indicators in environmental water samples. *Ecol. Indic.* **2020**, *113*, 106194. [CrossRef]

39. Narciso-da-Rocha, C.; Rocha, J.; Vaz-Moreira, I.; Lira, F.; Tamames, J.; Henriques, J.L.; Manaia, C.M. Bacterial lineages putatively associated with the dissemination of antibiotic resistance genes in a full-scale urban wastewater treatment plant. *Environ. Int.* **2018**, *118*, 179–188. [CrossRef] [PubMed]

40. Pärnänen, K.M.M.; Narciso-da-Rocha, C.; Kneis, D.; Berendonk, T.U.; Cacace, D.; Do, T.T.; Elpers, C.; Fatta-Kassinos, D.; Henriques, I.; Jaeger, T.; et al. Antibiotic resistance in European wastewater treatment plants mirrors the pattern of clinical antibiotic resistance prevalence. *Sci. Adv.* **2019**, *5*, eaau9124.

41. Denman, S.E.; McSweeney, C.S. Development of a real-time PCR assay for monitoring anaerobic fungal and cellulolytic bacterial populations within the rumen. *FEMS Microbiol. Ecol.* **2006**, *58*, 572–582. [CrossRef] [PubMed]

42. Barraud, O.; Baclet, M.C.; Denis, F.; Ploy, M.C. Quantitative multiplex real-time PCR for detecting class 1, 2 and 3 integrons. *J. Antimicrob. Chemother.* **2010**, *65*, 1642–1645. [CrossRef] [PubMed]

43. Brankatschk, R.; Bodenhausen, N.; Zeyer, J.; Bürgmann, H. Simple absolute quantification method correcting for quantitative PCR efficiency variations for microbial community samples. *Appl. Environ. Microbiol.* **2012**, *78*, 4481–4489. [CrossRef]

44. Rocha, J.; Cacace, D.; Kampouris, I.; Guilloteau, H.; Jäger, T.; Marano, R.B.M.; Karaolia, P.; Manaia, C.M.; Merlin, C.; Fatta-Kassinos, D.; et al. Inter-laboratory calibration of quantitative analyses of antibiotic resistance genes. *J. Environ. Chem. Eng.* **2018**, *8*, 102214. [CrossRef]

45. Ferreira, M.; Chaves, L.L.; Lima, S.A.C.; Reis, S. Optimization of nanostructured lipid carriers loaded with methotrexate: A tool for inflammatory and cancer therapy. *Int. J. Pharm.* **2015**, *492*, 65–72. [CrossRef]

46. Dorais, M.; Alsanius, B.W.; Voogt, W.; Pepin, S.; Tüzel, H.; Tüzel, Y.; Möller, K. *Impact of Water Quality and Irrigation Management on Organic Greenhouse Horticulture*; BioGreenhouse COST Action FA1105: Bleiswijk, Netherlands, 2020; ISBN 978-94-6257-538-7.

47. Water Salinity and Plant Irrigation. Available online: https://www.agric.wa.gov.au/water-management/water-salinity-and-plant-irrigation (accessed on 8 October 2020).

48. Khuntia, S.; Majumder, S.K.; Ghosh, P. Removal of Ammonia from Water by Ozone Microbubbles. *Ind. Eng. Chem. Res.* **2013**, *52*, 318–326. [CrossRef]

49. Ballabio, C.; Panagos, P.; Lugato, E.; Huang, J.-H.; Orgiazzi, A.; Jones, A.; Fernández-Ugalde, O.; Borrelli, P.; Montanarella, L. Copper distribution in European topsoils: An assessment based on LUCAS soil survey. *Sci. Total Environ.* **2018**, *636*, 282–298. [CrossRef] [PubMed]

50. Lucheta, A.R.; Lambais, M.R. Sulfur in agriculture. *Rev. Bras. Ciênc. Solo* **2012**, *36*, 1369–1379. [CrossRef]

51. Mousel, D.; Palmowski, L.; Pinnekamp, J. Energy demand for elimination of organic micropollutants in municipal wastewater treatment plants. *Sci. Total Environ.* **2017**, *575*, 1139–1149. [CrossRef] [PubMed]

52. Li, Y.; Zhang, S.; Zhang, W.; Xiong, W.; Ye, Q.; Hou, X.; Wang, C.; Wang, P. Life cycle assessment of advanced wastewater treatment processes: Involving 126 pharmaceuticals and personal care products in life cycle inventory. *J. Environ. Manag.* **2019**, *238*, 442–450. [CrossRef] [PubMed]

53. Arzate, S.; Pfister, S.; Oberschelp, C.; Sánchez-Pérez, J.A. Environmental impacts of an advanced oxidation process as tertiary treatment in a wastewater treatment plant. *Sci. Total Environ.* **2019**, *694*, 133572. [CrossRef] [PubMed]

54. Pesqueira, J.F.J.R.; Pereira, M.F.R.; Silva, A.M.T.S. Environmental impact assessment of advanced urban wastewater treatment technologies for the removal of priority substances and contaminants of emerging concern: A review. *J. Clean. Prod.* **2020**, *261*, 121078. [CrossRef]

55. Spuhler, D.; Andrés Rengifo-Herrera, J.; Pulgarin, C. The effect of Fe^{2+}, Fe^{3+}, H$_2$O$_2$ and the photo-Fenton reagent at near neutral pH on the solar disinfection (SODIS) at low temperatures of water containing Escherichia coli K12. *Appl. Catal. B Environ.* **2010**, *96*, 126–141. [CrossRef]

54. Pesqueira, J.F.J.R.; Pereira, M.F.R.; Silva, A.M.T.S. Environmental impact assessment of advanced urban wastewater treatment technologies for the removal of priority substances and contaminants of emerging concern: A review. *J. Clean. Prod.* **2020**, *261*, 121078. [CrossRef]

55. Spuhler, D.; Andrés Rengifo-Herrera, J.; Pulgarin, C. The effect of Fe^{2+}, Fe^{3+}, H_2O_2 and the photo-Fenton reagent at near neutral pH on the solar disinfection (SODIS) at low temperatures of water containing Escherichia coli K12. *Appl. Catal. B Environ.* **2010**, *96*, 126–141. [CrossRef]

56. Zhao, X.; Hu, H.-Y.; Yu, T.; Su, C.; Jiang, H.; Liu, S. Effect of different molecular weight organic components on the increase of microbial growth potential of secondary effluent by ozonation. *J. Environ. Sci.* **2014**, *26*, 2190–2197. [CrossRef]

57. Giannakis, S.; Merino Gamo, A.I.; Darakas, E.; Escalas-Cañellas, A.; Pulgarin, C. Monitoring the post-irradiation E. coli survival patterns in environmental water matrices: Implications in handling solar disinfected wastewater. *Chem. Eng. J.* **2014**, *253*, 366–376. [CrossRef]

58. Ubomba-Jaswa, E.; Navntoft, C.; Polo-López, M.I.; Fernandez-Ibáñez, P.; McGuigan, K.G. Solar disinfection of drinking water (SODIS): An investigation of the effect of UV-A dose on inactivation efficiency. *Photochem. Photobiol. Sci.* **2009**, *8*, 587–595. [CrossRef]

59. Clancy, S. DNA damage & repair: Mechanisms for maintaining DNA integrity. *Nat. Educ.* **2008**, *1*, 103.

60. EPA. *Guidelines for Water Reuse*; Environmental Protection Agency (EPA): Wasghinton, DC, USA, 2012; (EPA/600/R-12/618).

61. Alcalde-Sanz, L.; Gawlik, B.M. *Minimum Quality Requirements for Water Reuse in Agricultural Irrigation and Aquifer Recharge—Towards A Water Reuse Regulatory Instrument at EU Level, EUR 28962 EN*; Publications Office of the European Union: Luxembourg, 2017.

62. Trintinaglia, L.; Bianchi, E.; Silva, L.; Nascimento, C.; Spilki, F.; Ziulkoski, A. Cytotoxicity assays as tools to assess water quality in the Sinos River basin. *Braz. J. Biol.* **2015**, *75*, 75–80. [CrossRef] [PubMed]

63. Zhang, Y.; Yuan, Y.X.; Wang, Y.F.; Li, C.; Zhu, J.; Li, R.F.; Wu, Y.H. Comprehensive evaluation on the bio-toxicity of three advanced wastewater treatment processes. *Water Air Soil Pollut.* **2020**, *231*, 110. [CrossRef]

64. Nahim-Granados, S.; Rivas-Ibanez, G.; Perez, J.A.S.; Oller, I.; Malato, S.; Polo-Lopez, M.I. Synthetic fresh-cut wastewater disinfection and decontamination by ozonation at pilot scale. *Water Res.* **2020**, *170*, 115304. [CrossRef] [PubMed]

65. Affek, K.; Muszynski, A.; Zaleska-Radziwill, M.; Doskocz, N.; Zietkowska, A.; Widomski, M. Evaluation of ecotoxicity and inactivation of bacteria during ozonation of treated wastewater. *Desalin. Water Treat.* **2020**, *192*, 176–184. [CrossRef]

66. Pohl, J.; Ahrens, L.; Carlsson, G.; Golovko, O.; Norrgren, L.; Weiss, J.; Orn, S. Embryotoxicity of ozonated diclofenac, carbamazepine, and oxazepam in zebrafish (*Danio rerio*). *Chemosphere* **2019**, *225*, 191–199. [CrossRef]

12

A Comparison of the Mechanism of TOC and COD Degradation in Rhodamine B Wastewater by a Recycling-Flow Two- and Three-dimensional Electro-Reactor System

Jin Ni [1], Huimin Shi [1], Yuansheng Xu [1] and Qunhui Wang [1,2,]*

[1] Department of Environmental Engineering, School of Energy and Environmental Engineering, University of Science and Technology Beijing, 30 Xueyuan Road, Haidian District, Beijing 10083, China; jolinxiaopang@163.com (J.N.); shihuimin99@hotmail.com (H.S.); xys519828120@163.com (Y.X.)

[2] Beijing Key Laboratory on Resource-oriented Treatment of Industrial Pollutants, University of Science and Technology Beijing, 30 Xueyuan Road, Beijing 10083, China

* Correspondence: wangqh59@sina.com

Abstract: Dye wastewater, as a kind of refractory wastewater (with a ratio of biochemical oxygen demand (BOD) and chemical oxygen demand (COD) of less than 0.3), still needs advanced treatments in order to reach the discharge standard. In this work, the recycling-flow three-dimensional (3D) electro-reactor system was designed for degrading synthetic rhodamine B (RhB) wastewater as dye wastewater (100 mg/L). After 180 min of degradation, the removal of total organic carbon (TOC) and chemical oxygen demand (COD) of RhB wastewater were both approximately double the corresponding values in the recycling-flow two-dimensional (2D) electro-reactor system. Columnar granular activated carbon (CGAC), as micro-electrodes packed between anodic and cathodic electrodes in the recycling-flow 3D electro-reactor system, generated an obviously characteristic peak of anodic catalytic oxidation, increased the mass transfer rate and electrochemically active surface area (EASA) by 40%, and rapidly produced 1.52 times more hydroxyl radicals (·OH) on the surface of CGAC electrodes, in comparison to the recycling-flow 2D electro-reactor system. Additionally, the recycling-flow 3D electro-reactor system can maintain higher current efficiency (CE) and lower energy consumption (Es).

Keywords: recycling-flow; three-dimensional electro-reactor system; two-dimensional electro-reactor system; rhodamine B; wastewater treatment

1. Introduction

Nowadays, the unsafe disposal of dye wastewater, which still contains lots of complex pollutants and toxic matter, such as aromatic, chloric, and azo compounds, is seriously threating environmental and ecological systems and human health [1–5]. The reason is that dye wastewater, as a kind of refractory wastewater (with a ratio of biochemical oxygen demand (BOD) and chemical oxygen demand (COD) of less than 0.3), is severely difficult to degrade in order to fall under discharge standards in the activated sludge process using traditional and biological wastewater treatment methods [6]. Therefore, there is a definite urgent need for methods which efficiently degrade dye wastewater after biological treatment [7–10].

In comparison to the two-dimensional (2D) electro-reactor, consisting of an anode electrode, cathode electrode, and electrolyzer, the three-dimensional (3D) electro-reactor contains a certain number of small granular substances, such as activated carbon particles and diatomite particles, charged by an electric field to form micro-electrodes and then acquires an electrochemically oxidative ability with a third electrode, which is placed between the anode and the cathode electrodes [11–14]. Simultaneously, the degradation of dye pollutants from wastewater can occur on the surface of these small granular electrodes and anode and cathode electrodes in the 3D electro-reactor [15,16]. Hence, the 3D electro-reactor theoretically shows more brilliant promise in the advanced treatment of dye wastewater as effluent than the 2D electro-reactor.

Currently, the number of researchers [17–19] who pay attention to the study of 3D electrode technology is growing dramatically, but they are always focusing on the novel methods of granular electrode modification in order to improve the oxidative degradation ability of the 3D electro-reactor instead of the design and amendment of the 3D electro-reactor system. However, the processes of granular electrode modification normally require quite serious and extreme conditions, such as high temperature and pressure, and special materials, such as noble gases and metal [20,21].

As is well known, the fixed bed 3D electro-reactor system and the fluid bed 3D electro-reactor system are usually used to treat dye wastewater. The former system has the main disadvantage of low treatment efficiency due to extremely a high hydraulic retention time (HRT) and, meanwhile, the main disadvantage of the latter system is effluent's high total organic carbon (TOC) and COD above the discharged standard due to a low HRT [22–24]. Therefore, a recycling-flow 3D electro-reactor system, taking advantages of the strong oxidative degradation ability, high treatment efficiency, no secondary pollution, and being operated under normal temperature and atmosphere pressure, is designed for degrading dye wastewater in our work.

Choosing RhB wastewater as one kind of dye wastewater, this paper is going to analyze the mechanism of the TOC and COD degradation of RhB wastewater in the aspect of mass transfer, the electrochemically active surface area (EASA) of electrodes, the instant concentration of hydroxyl radicals (·OH), current efficiency (CE), and energy consumption (Es) in the recycling-flow 3D electro-reactor system, compared with the recycling-flow 2D electro-reactor system. Additionally, it carries on, finding out the mechanism of TOC and COD degradation with different voltages, with different electrolytes, and at different HRTs in the recycling-flow 3D electro-reactor system.

2. Materials and Methods

2.1. Materials

RhB as a dye has a molecular formula of $C_{28}H_{31}ClN_2O_3$ and a molecular weight of 479.01. Columnar granular activated carbon (CGAC) from coconut shells (average size: 1.50 mm) was purchased from Henan Lianhua Carbon Manufacturing Co. Two $Ti/RuO_2/TiO_2$ board electrodes (size: 60 mm × 100 mm × 2 mm) were provided by the Second Research Institute of the China Aerospace Science and Industry Group. Anhydrous sodium sulfate (Na_2SO_4), sodium chloride (NaCl), mercury(II) sulfate ($HgSO_4$), phosphoric acid (H_3PO_4), potassium dichromate ($K_2Cr_2O_7$), sulfuric acid (H_2SO_4), and potassium ferrocyanide ($K_4Fe(CN)_6$) were of analytical grade and used without any further purification. Silver sulfate (Ag_2SO_4) was not less than 99.7%. Ammonium iron(II) sulfate (($NH_4)_2Fe(SO_4)_2$ was not less than 99.5%. The ferroin indicator solution standard is Q/12NK4019-2011.

2.2. Experimental Setup and Procedure

A virtual diagram of the experimental setup is shown in Figure 1. The electro-reactor was a plexiglass rectangular tank (Organic Glass Factory, Beijing, China) with two $Ti/RuO_2/TiO_2$ board electrodes as the anode and cathode in the 2D electro-reactor. The anode and cathode were positioned vertically and parallel to each other with an inter-electrode gap of 30 mm. The CGAC electrodes were packed between the anode and cathode up to a height of 80 mm in the 3D electro-reactor; the liquid level equaled the height of the packed bed.

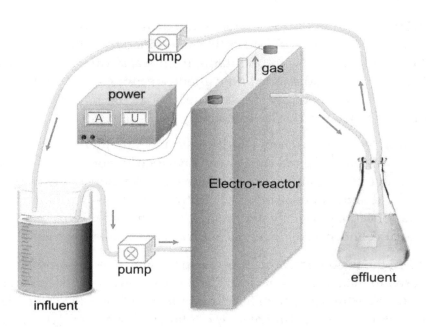

Figure 1. The virtual diagram of the recycling-flow 3D electro-reactor system.

In this recycling system, one peristaltic pump (Youji Keyi ZQ000S, Baoding, Hebei, China) was used to pump the influent RhB wastewater from a beaker into the bottom of the electro-reactor. Then the RhB wastewater flowed out from the upper outlet of the electro-reactor and was collected in an Erlenmeyer flask. Simultaneously, the other peristaltic pump was used to pump the effluent RhB wastewater from the Erlenmeyer flask back to the beaker.

Prior to commencing the electrochemical oxidation treatment, CGAC electrodes adsorbed RhB solution until becoming saturated in order to minimize the effect of adsorption on TOC and COD removal. All experiments used a digital DC power supply (DC 30 V/5 A; DH1716-6D). After completing the experiment, all treated samples were collected and filtered through 0.45 mm filters. The filtrate was then analyzed, as described in the next sub-sections.

2.3. Analytical Methods

TOC was measured by a Vario TOC analysis device. According to the instructions of the Vario TOC analysis device, when carbonaceous compounds are burned in an oxygen-rich environment, the carbon is completely converted into CO_2, and then the non-scattering infrared detector (NDIR) detects the amount of CO_2 and converts it into total carbon (TC) in the sample. After the sample is acidified by phosphoric acid (H_3PO_4) (1% v/v) and pH decreasing, the carbonate and bicarbonate in the sample are converted into CO_2, which is blown out and enters the NDIR, and then the detected amount of CO_2 is converted into total inorganic carbon (TIC). The value of TOC is TC minus TIC.

COD was measured by a microwave digestion method. According to the instructions of the Kedibo microwave digestion device, mercury(II) sulfate ($HgSO_4$) as a masking agent, 0.05 mol/L potassium dichromate ($K_2Cr_2O_7$) as a digestion solution, a mix of 10 g silver sulfate (Ag_2SO_4) and 1 L sulfuric acid (H_2SO_4) as a catalyst, ferroin solution as an indicator, and potassium ferrocyanide ($K_4Fe(CN)_6$) as a standard solution were used.

Electrochemical measurements were performed using a conventional three-electrode cell and a CHI 660E electrochemical workstation (CHI, Beijing, China). Ag/AgCl and $Ti/RuO_2/TiO_2$ board electrodes served as the reference and counter electrodes, respectively.

CE (%) is the rate of the efficient current and total current in a period and Es (KW·h/kg TOC) is the electricity consumption of 1 kg TOC degradation. They were calculated according to the following equations [25–27]:

$$CE = \frac{TOC_0 - TOC_t}{480I}FQ \qquad (1)$$

$$Es = \frac{UI}{60(TOC_0 - TOC_t)Q} \qquad (2)$$

where TOC_0 (g/L) and TOC_t (g/L) correspond to the total organic carbon at $t = 0$ min and $t = t$ min, respectively. I is the average current (A), F is the Faraday constant (96,485 C/mol), Q is the flow rate of water (L/min), and U is the applied electric voltage (V).

3. Result and Discussion

3.1. Electrochemical Properties of TOC and COD Degradation in the Recycling-Flow Elecro-Reactor System

As shown in Figure 2, the TOC removal of RhB wastewater in the recycling-flow 3D electro-reactor system was always higher than that in the recycling-flow 2D electro-reactor system, and the highest TOC removal was 72.0%, which was 1.98 times that in the recycling-flow 2D electro-reactor system (36.3%). From the perspective of COD, the COD removal of RhB wastewater in the recycling-flow 3D electro-reactor system was much higher than that in the recycling-flow 2D electro-reactor system and the highest COD removal was up to 86.9%. By the 30th minute, COD removal in the recycling-flow 3D electro-reactor system had already reached 63.4%, meanwhile, the recycling-flow 2D electro-reactor system just reached 41.9% by the 180th minute.

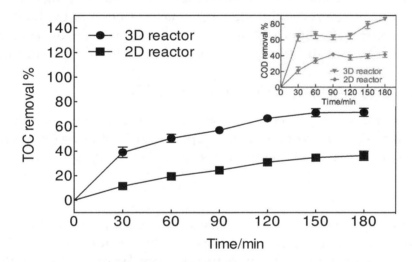

Figure 2. Total organic carbon (TOC) and chemical oxygen demand (COD) removal of rhodamine B (RhB) wastewater in the recycling-flow 3D and 2D electro-reactor systems (RhB wastewater initial concentration is 100 mg/L; volume is 500 mL; initial pH is 7; Na$_2$SO$_4$ as an electrolyte, initial concentration is 2 g/L; voltage is 5 V; hydraulic retention time (HRT) is 20 min).

It is normally considered that the mass transfer and EASA of the electrodes play major roles. Therefore, the higher TOC and COD removal in the recycling-flow 3D electro-reactor system is due to the presence of CGAC as conductive particles packed in the 3D electro-reactor and constitute a number of microelectrodes, which dramatically increase the area of the reaction electrode and benefit organic matter degradation easily and quickly in oxidative processes [28,29].

In order to obtain the mass transfer rate and EASA of the electrodes in the 2D and 3D electro-reactors, respectively, cyclic voltammograms (CVs) of the 2D and 3D electro-reactors were measured in a 0.05 mol/L K$_4$Fe(CN)$_6$ + 0.45 mol/L Na$_2$SO$_4$ solution at different scan rates from 0.015 V/s to 0.1 V/s by a CHI 660E electrochemical workstation (CHI, China). The 3D electro-reactor had a characteristic peak of anodic catalytic oxidation, as illustrated in Figure 3. Taking a sweep rate of 0.1 V/s (red line) as an example, when the potential was 0.7 V, it should be 0.2 A if the potential and the corresponding current were in a linear relationship (demonstrated in Figure 3b). However, the Figure 3a curve shows that the corresponding current is as high as 0.35 A at the potential of 0.7 V. This was because the corresponding current incurred a mutation at 0.7 V, which is called a characteristic peak of anodic catalytic oxidation.

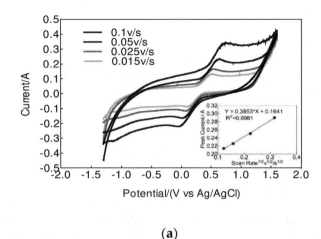

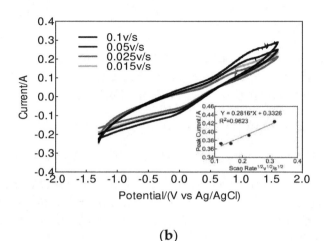

(a)   (b)

Figure 3. Cyclic voltammograms of the 3D and 2D electro-reactors in a 50 mmol/L $K_4Fe(CN)_6$ + 0.45 mol/L Na_2SO_4 solution at different scan rates. (**a**) 3D electro-reactor, (**b**) 2D electro-reactor. Insets show the plots of the peak current vs. the square root of the scan rate.

It was found that both the anodic peak current and the cathodic peak current increased as the scan rate increased, indicating a reversible electrochemical reaction of the $[Fe(CN)_6]^{4-}/[Fe(CN)_6]^{3-}$ redox couple. At the same time, Figure 3 (insets) shows a brilliant linear relationship between the oxidation peak current (I_p) and the square root of the scan rate ($v^{1/2}$), according to the following equation [21,30,31]:

$$I_p = \left(2.69 \times 10^5\right)n^{2/3}AD_R^{1/2}C_R v^{1/2} \tag{3}$$

where n is the number of transferred electrons, A is the EASA (cm^2), D_R is the diffusion coefficient of the reduced species (cm^2/s), C_R is the bulk reduced species concentration (mmol/L), and v is the scan rate (mV/s). The above equation can be simplified into the following equation:

$$I_p = kv^{1/2} \tag{4}$$

where k is a coefficient only relevant to A and D_R because n and C_R are constant in this study.

As shown in the insets of Figure 3a,b, the slopes of the linear relationship between I_p and $v^{1/2}$, named as the value of k, representing the mass transfer rate, were obtained according to the linear fitting of the plots. As expected, the k value of the 3D electro-reactor (0.3953) was larger than the corresponding value of the 2D electro-reactor (0.2816). This demonstrates that the 3D electro-reactor had greater mass transfer properties than the 2D electro-reactor [32,33].

In addition, the EASA can also be derived from the k value by assuming that the D_R value of $[Fe(CN)_6]^{4-}$ is constant in this study. The obtained EASA value of the 3D electro-reactor is also higher than that of the 2D electro-reactor. In particular, it was 1.40 times the corresponding value of the 2D electro-reactor. The higher EASA of the 3D electro-reactor means that the 3D electro-reactor will provide much more electrochemically active sites for RhB oxidation, and thus will be beneficial in improving the oxidation of organics on the electrode surface [34].

Based on the mass transfer rate and EASA result and discussion above, the 3D electro-reactor has been demonstrated to possess a much higher electro-catalytic activity for degrading organic matter (RhB) than the 2D electro-reactor.

It is well known that the hydroxyl radical (·OH), of which the oxidation potential (2.8 eV) is the second highest and regarded as a powerful oxidizing chemical in nature, plays an important role in the electrochemical oxidation of RhB wastewater [35]. Therefore, the production of ·OH was detected by using the electron paramagnetic resonance (EPR) technique by adding the ·OH scavenger 5,5-dimenthyl,1-pyrroline-N-oxide (DMPO) to further reveal the underlying mechanism of the TOC and COD degradation of RhB wastewater. As expected, typical EPR spectra of the DMPO–·OH adduct with a 1:2:2:1 quartet were acquired in the 3D and 2D electro-reactors when current and voltage were applied, as shown in Figure 4. It is worth noting that the 3D electro-reactor achieved the higher EPR intensity, which was nearly 2.52 times higher than the corresponding value of the 2D electro-reactor and demonstrated that the electro-generation of ·OH occurred on the CGAC electrodes and the usage of the CGAC electrodes could enhance ·OH generation effectively in electrochemical oxidation processes [36].

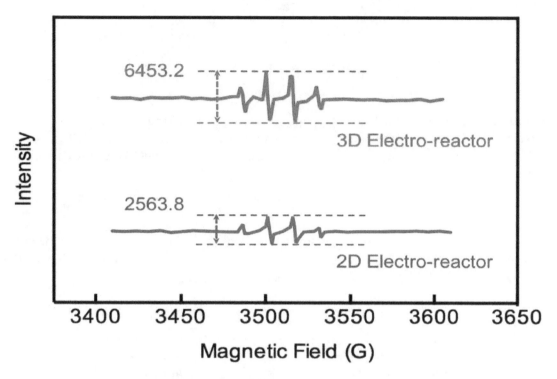

Figure 4. Dimenthyl,1-pyrroline-N-oxide (DMPO) spin trapping the electron paramagnetic resonance (EPR) spectra of hydroxyl radicals (OH) in the 3D and 2D electro-reactors.

Overall, the degradation of organic matter (RhB) by the 3D electro-reactor depends on more ·OH being adsorbed on the surface of CGAC electrodes, forming more electrochemically active sites for RhB oxidation. However, the 2D electro-reactor can just undergo less ·OH adsorption and the degradation strength is also reduced due to the absence of CGAC as conductive particles.

The CE and Es of the recycling-flow 3D and 2D electro-reactor systems are shown in Figure 5. Overall, the CE of the recycling-flow 3D and 2D electro-reactor systems both went up from 8.3% and 2.5% to 14.4% and 7.6%, respectively, and the Es of the recycling-flow 3D and 2D electro-reactor systems both decreased from 202.6 KW·h/kg TOC and 658.1 KW·h/kg TOC to 116.3 KW·h/kg TOC and 219.9 KW·h/kg TOC, respectively, during the electrochemical processes. The CE of the recycling-flow 3D electro-reactor system was always nearly twice as much as the corresponding value of the recycling-flow 2D electro-reactor system in the treatment period. Additionally, the lowest CE value of the recycling-flow 3D electro-reactor system, by the 30th minute, had reached 8.3%, which approximately equaled the highest CE value of the recycling-flow 2D electro-reactor system (7.6%).

From the perspective of Es, by the 30th minute, the Es of the recycling-flow 3D electro-reactor system was 202.6 KW·h/kg TOC, which was less than one third the corresponding value of the recycling-flow 2D electro-reactor system (658.1 KW·h/kg TOC). Although the Es of the recycling-flow 2D electro-reactor system reduced slightly, the minimum Es was still as high as 219.9 KW·h/kg TOC, which was almost double the corresponding value of the recycling-flow 3D electro-reactor system (116.3 KW·h/kg TOC).

This verified that the existence of CGAC electrodes in the recycling-flow 3D electro-reactor system increased the mass transfer rate and EASA, rapidly generated much more ·OH on the surface of CGAC electrodes, and improved CE and reduced Es.

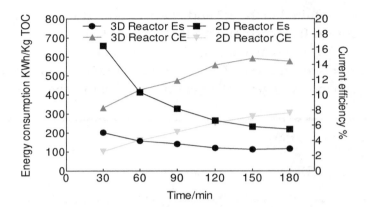

Figure 5. Energy consumption (Es) and current efficiency (CE) of the recycling-flow 3D and 2D electro-reactor systems (RhB wastewater initial concentration is 100 mg/L; volume is 500 mL; initial pH is 7, Na_2SO_4 as an electrolyte, initial concentration is 2 g/L; voltage is 5 V; HRT is 20 min; current density is 60 mA/cm^2).

3.2. Mechanism of TOC and COD Degradation with Different Electrolytes in the Recycling-Flow Electro-Reactor System

An electrolyte is frequently added to wastewater to enhance the conductivity of the solution and reduce impedance in an electrochemical reaction. It is actually necessary to research the effect of the electrolyte on the degradation of RhB wastewater, as shown in Figure 6.

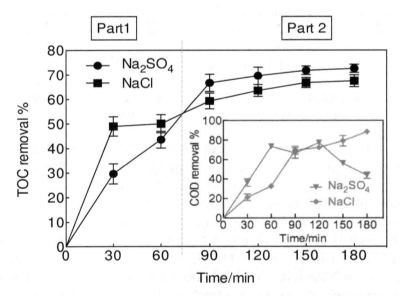

Figure 6. TOC and COD removal of RhB wastewater in the recycling-flow 3D electro-reactor system (NaCl and Na_2SO_4 as electrolytes, initial concentration is 2 g/L; RhB wastewater initial concentration is 100 mg/L; volume is 500 mL; initial pH is 7; voltage is 7 V; HRT is 20 min).

From the beginning to the 75th min as Part 1, the TOC removal of RhB wastewater (NaCl) was higher than that of RhB wastewater (Na_2SO_4). The reason was that Cl$^-$ in the RhB wastewater (NaCl) dramatically generated lots of active chlorine (Cl), which reacted with ·OH in a synergistic process to increase TOC removal through electrochemical reaction processes. However, from the 75th min to the 180th min as Part 2, the TOC removal of RhB wastewater (Na_2SO_4) started to go beyond that of RhB wastewater (NaCl). The former and the latter peak rates reached 72.8% and 67.3%, respectively. As some active chlorine (Cl) was converted into chlorine (Cl_2) in the RhB wastewater (NaCl), Cl_2 partially escaped into the air. The electrons lost during the formation of Cl_2 could no longer be used due to the balance of electron gain and loss. Therefore, RhB wastewater (NaCl) had a final TOC removal lower than that of RhB wastewater (Na_2SO_4) [37].

Additionally, the COD removal of RhB wastewater (NaCl) increased gradually and then reached 89.3%. Meanwhile, the COD removal of RhB wastewater (Na_2SO_4) first increased and then dropped to 43.9%. The highest value could just reach 77.6%. RhB decomposed into small molecules which could not be oxidized by potassium dichromate ($K_2Cr_2O_7$) in the solution (Na_2SO_4) from the 60th min, whilst a side reaction occurred in the solution (NaCl) to generate ClO^- which had strong oxidizing properties, and enhanced the ability to degrade COD [38].

Obviously, the CE of the recycling-flow 3D electro-reactor system with NaCl was higher than the corresponding value of the recycling-flow 3D electro-reactor system with Na_2SO_4 before the 75th minute, but the CE of the recycling-flow 3D electro-reactor system with Na_2SO_4 was higher than the corresponding value of the recycling-flow 3D electro-reactor system with NaCl up to the 180th min, as illustrated in Figure 7. Overall, the CE of the recycling-flow 3D electro-reactor systems with NaCl and Na_2SO_4 both went up smoothly and then got to the highest value (12.2% and 11.3%, respectively).

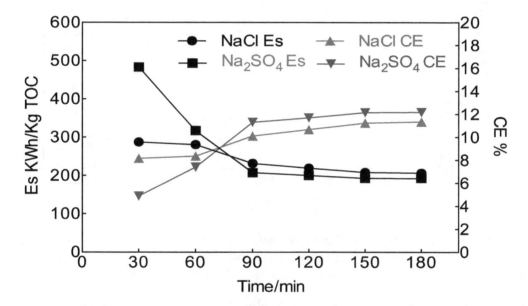

Figure 7. Es and CE of the recycling-flow 3D electro-reactor system with NaCl and Na_2SO_4 as electrolytes (NaCl and Na_2SO_4 initial concentration is 2 g/L; RhB wastewater initial concentration is 100 mg/L; volume is 500 mL; initial pH is 7; voltage is 7 V; HRT is 20 min; current density is 60 mA/cm^2).

Meanwhile, the Es of the recycling-flow 3D electro-reactor system with Na_2SO_4 was higher than the corresponding value of the recycling-flow 3D electro-reactor system with NaCl up to the 80th min. In particular, the Es of the recycling-flow 3D electro-reactor system with Na_2SO_4 was 483.5 KW·h/kg TOC, which was 68.3% higher than the corresponding value of the recycling-flow 3D electro-reactor system with NaCl (287.2 KW·h/kg TOC) by the 30th min. However, the Es of the recycling-flow 3D electro-reactor system with Na_2SO_4 started to be lower than the corresponding value of the recycling-flow 3D electro-reactor system with NaCl from the 80th min. In addition, the Es of the recycling-flow 3D electro-reactor systems with NaCl and Na_2SO_4 both declined to the lowest value (192.7 KW·h/kg TOC and 206.7 KW·h/kg TOC, respectively) in this period.

These again indicate that NaCl, as an electrolyte, had a high conductive efficiency in the former period due to active chlorine (Cl) generation and then changed to low conductive efficiency due to Cl_2 escaping into the air in the latter period from CE and Es [22].

3.3. Mechanism of TOC and COD Degradation in Different Voltages in the Recycling-Flow Electro-Reactor System

As seen in Figure 8, the COD removal curve shows a smooth increase and then reaches 30.1% with the voltage of 3 V. In addition, the COD removal of RhB wastewater rose dramatically with the voltages of 5 V and 7 V up to the 60th min and then kept nearly flat between the 60th min and 120th min. Finally, COD removal with 5 V carried on increasing to 86.9%, whilst the corresponding value with 7 V started decreasing to 43.9%. This could be explained by the equation below:

$$COD\ removal(\%) = \frac{COD_0 - COD_t}{COD_0} \times 100\% = 1 - \frac{COD_t}{COD_0} \times 100\% \tag{5}$$

where COD_0 is the initial COD of RhB wastewater and COD_t is the COD of RhB wastewater after treating t minutes.

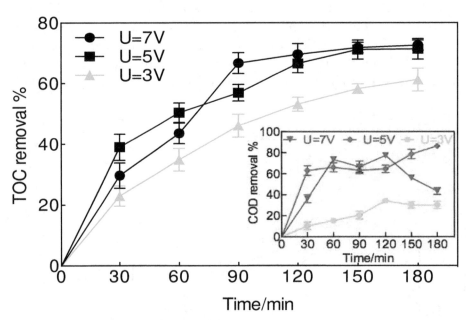

Figure 8. TOC and COD removal of RhB wastewater with different voltages in the recycling-flow 3D electro-reactor system (RhB wastewater initial concentration is 100 mg/L; volume is 500 mL; initial pH is 7; Na_2SO_4 as an electrolyte, initial concentration is 2 g/L; HRT is 20 min).

The COD_0 value was smaller than the actual value due to some macromolecular substances being unable to be oxidized by $K_2Cr_2O_7$ in the initial RhB wastewater, and then more macromolecules could be oxidized into smaller molecules with the voltage of 7 V set in this experiment, compared with the voltage of 5 V, which were easily oxidized by $K_2Cr_2O_7$, resulting in a COD_t value and COD removal that are greater simultaneously [39].

From the perspective of TOC, as shown in Figure 8, the TOC removal of RhB wastewater grew the most slowly and the final TOC removal just arrived at 61.0% when the voltage was 3 V. TOC removal was always higher with the voltages of 7 V and 5 V, and the highest values were basically equivalent (72.2% and 72.0%, respectively), which were nearly 1.18 times the corresponding value with the voltage of 3V.

Interestingly, from the 0th min to the 70th min, TOC removal with the voltage of 5 V was higher than that with the voltage of 7 V. TOC degradation processes are illustrated in Figure 9. First of all, TOC was converted into TIC and then TIC was converted into CO_2 and H_2O. The concentration of ·OH in the RhB wastewater was higher, so that oxidation was stronger between the board electrodes with 7 V. TIC was quickly oxidized into CO_2 and H_2O and TOC was converted into TIC as main processes. With the voltage of 5 V, the oxidation was weaker [40–42]. Electron transfer processes were normally used to convert TOC into TIC whilst only some of TIC was converted into CO_2 and H_2O. Hence, TOC removal was higher with 5 V. After the 70th min, TOC removal with the voltage of 7 V was beyond the corresponding value with the voltage of 5 V. TIC from TOC was basically converted into CO_2 and H_2O with the voltage of 7 V, while TIC just started being rapidly converted into CO_2 and H_2O as main processes with the voltage of 5 V. Hence, TOC removal was higher with 7 V. The final TOC removal with 7 V was slightly higher than the corresponding value with 5 V since the potential of 7 V was higher than 5 V up to the 180th min [43,44].

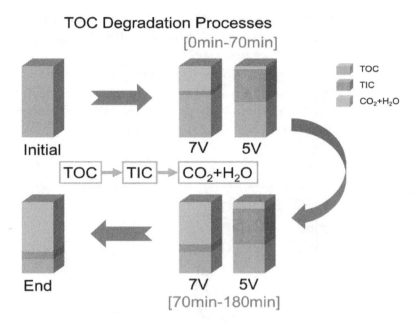

Figure 9. The diagram of TOC degradation processes in RhB wastewater with voltages of 7 V and 5 V.

CE with the voltage of 3 V rose the fastest from 8.0% by the 30th minute to 21.8% by the 180th minute, which increased by nearly two times, meanwhile, the CEs with 5 V and 7 V both showed gradually increasing trends to 14.1% and 12.2%, respectively, as indicated in Figure 10. Plus, the Es with different voltages (3 V, 5 V, and 7 V) gradually decreased by 63.1%, 42.6%, and 60.2%, respectively, from the 30th min up to the 180th min. In the whole electrolysis process, the higher the voltage was, the higher the Es was. At the 180th min, Es with the voltage of 3 V was 46.1 KW·h/kg TOC, which was 39.6% of the corresponding value with 5 V (116.3 KW·h/kg TOC) and 23.9% of the corresponding value with 7 V (192.7 KW·h/kg TOC).

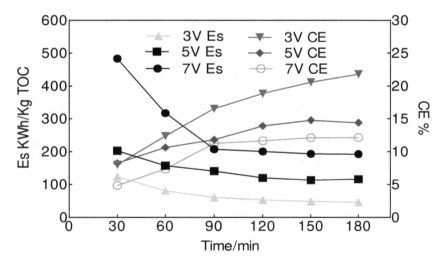

Figure 10. Es and CE of the recycling-flow 3D electro-reactor system with different voltages (RhB wastewater initial concentration is 100 mg/L; volume is 500 mL; initial pH is 7; Na_2SO_4 as an electrolyte, initial concentration is 2 g/L; HRT is 20 min; current density is 60 mA/cm^2).

3.4. Mechanism of TOC and COD Degradation at Different HRTs in the Recycling-Flow Electro-Reactor System

HRT refers to the residence time of wastewater in the reactor, which can be calculated according to the following equation [45]:

$$HRT = \frac{V}{Q} \tag{6}$$

where V is the reactor volume or pool capacity and Q is the influent flow rate.

In this experiment, the flow rate was controlled by operating the pumps in order to study the effect of HRT on the degradation of RhB wastewater.

COD removal curves were almost growing coincidently at different HRTs (20 min, 40 min, and 60 min), as illustrated in Figure 11. Finally, COD removal (86.9%) at HRT = 20 min was slightly higher than the corresponding values at HRT = 40 min (80.1%) and HRT = 60 min (83.4%). This indicated that HRT had little effect on the degradation of COD in the RhB wastewater.

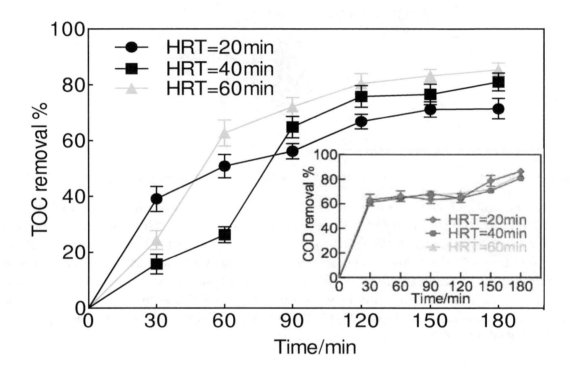

Figure 11. TOC and COD removal of RhB wastewater at different HRTs in the recycling-flow 3D electro-reactor system (RhB wastewater initial concentration is 100 mg/L; volume is 500 mL; initial pH is 7; Na_2SO_4 as an electrolyte, initial concentration is 2 g/L; voltage is 5 V).

TOC removal at different HRTs (20 min, 40 min, and 60 min) gradually increased and then reached the maximum value at the 180th min. The highest TOC removal (85.0%) at HRT = 60 min was 13.8% and 6.8% higher than the corresponding value (71.2% and 78.2%) at HRT = 20 min and HRT = 40 min, respectively. In the comparisons of TOC removal, it can be concluded that HRT = 60 min had a better removal effect.

However, HRT is equal to V/Q and the pool capacity of HRT = 60 min is three times that of HRT = 20 min, which means that the construction cost must be more, as the influent flow rate is constant. In addition, the initial concentration of RhB wastewater was 100 mg/L and the corresponding COD was 215 mg/L. The final COD removal (86.9%) at HRT = 20 min was that of the COD of the effluent, which was 28.2 mg/L after treatment, which reached the Grade A standard (< 50 mg/L) of the Pollutant Discharge Standard for Urban Sewage Treatment Plants [46]. In summary, HRT = 20 min, with less construction cost, is the optimal HRT.

The CE at HRT = 20 min (8.3–14.8%) was always higher than that at HRT = 40 min (CE: 2.4–12.4%) and HRT = 60 min (CE: 2.5–8.7%), as shown in Figure 12. The maximum CE (14.8%) at HRT = 20 min was 1.2 times and 1.7 times the corresponding values at HRT = 40 min (12.4%) and HRT = 60 min (8.7%), respectively. From the perspective of Es, Es at HRT = 20 min was always lower than that at HRT = 40 min and HRT = 60 min, and the minimum Es at HRT = 20 min, 40 min, and 60 min were 116.3 KW·h/kg TOC, 135.3 KW·h/kg TOC, and 192.4 KW·h/kg TOC, respectively.

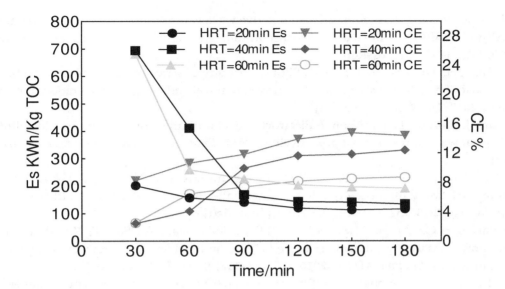

Figure 12. Es and CE of the recycling-flow 3D electro-reactor system at different HRTs (RhB wastewater initial concentration is 100 mg/L; volume is 500 mL; initial pH is 7; Na_2SO_4 as an electrolyte, initial concentration is 2 g/L; voltage is 5 V; current density is 60 mA/cm^2).

4. Conclusions

In conclusion, CGAC, as micro-electrodes between anodic and cathodic electrodes in the recycling-flow 3D electro-reactor system, generated an obviously characteristic peak of anodic catalytic oxidation, increased the mass transfer rate and EASA by 40%, and rapidly produced 1.52 times more ·OH on the surface of CGAC electrodes so that the TOC and COD removal of RhB wastewater were both approximately double the corresponding values in the recycling-flow 2D electro-reactor system after 3 h of treatment in the same experimental conditions, with higher a CE and lower Es.

Treating RhB wastewater in the recycling-flow 3D electro-reactor system, Na_2SO_4 as an electrolyte was more beneficial for TOC degradation, and got higher CE and lower Es, in long electrolyzing times (more than 75 min and less than 180 min). On the contrary, NaCl as an electrolyte could improve COD removal more, and get higher a CE and lower Es in short electrolyzing times (less than 75 min). Plus, TOC and COD removal were the best in the proper voltage (5 V), not the highest one or lowest one. Normally, the higher the voltage, the lower the CE and the more the Es. The HRT condition had little effect on COD removal but high HRT was good for TOC degradation.

Author Contributions: Data curation, J.N., H.S. and Y.X.; Writing—original draft, J.N.; Writing—review and editing, Supervision, Resources, Q.W. All authors have read and agreed to the published version of the manuscript.

Acknowledgments: The authors appreciate the constructive suggestions from reviewers and editors that helped improve this paper and the support from National Environmental and Energy Base for International Science and Technology Cooperation.

References

1.   Shannon, M.A.; Bohn, P.W.; Elimelech, M.; Georgiadis, J.G.; Marinas, B.J.; Mayes, A.M. Science and technology for water purification in the coming decades. *Nature* **2008**, *452*, 301–310. [CrossRef]

2.   Holkar, C.R.; Jadhav, A.J.; Pinjari, D.V.; Mahamuni, N.M.; Pandit, A.B. A critical review on textile wastewater treatments: Possible approaches. *J. Environ. Manag.* **2016**, *182*, 351–366. [CrossRef]

3.   Massoud, M.A.; Tarhini, A.; Nasr, J.A. Decentralized approaches to wastewater treatment and management: Applicability in developing countries. *J. Environ. Manag.* **2009**, *90*, 652–659. [CrossRef]

4.   Naumczyka, J.H.; Kucharskab, M.A.; Ładyńskab, J.A.; Wojewódkaa, D. Electrochemical oxidation process in application to raw and biologically pre-treated tannery wastewater. *Desalin. Water Treat.* **2019**, *162*, 166–175. [CrossRef]

5. Cui, M.-H.; Gao, J.; Wang, A.-J.; Sangeetha, T. Azo dye wastewater treatment in a bioelectrochemical-aerobic integrated system: Effect of initial azo dye concentration and aerobic sludge concentration. *Desalin. Water Treat.* **2019**, *165*, 314–320. [CrossRef]

6. Zhao, R.; Zhao, H.; Dimassimo, R.; Xu, G. Pilot scale study of sequencing batch reactor (sbr) retrofit with integrated fixed film activated sludge (ifas): Nitrogen removal and design consideration. *Environ. Sci. Water Res. Technol.* **2018**, *4*, 569–581. [CrossRef]

7. Kim, S.; Kim, J.; Kim, S.; Lee, J.; Yoon, J. Electrochemical lithium recovery and organic pollutant removal from industrial wastewater of a battery recycling plant. *Environ. Sci. Water Res. Technol.* **2018**, *4*, 175–182. [CrossRef]

8. Soares, P.A.; Batalha, M.; Souza, S.M.A.G.U.; Boaventura, R.A.R.; Vilar, V.J.P. Enhancement of a solar photo-fenton reaction with ferric-organic ligands for the treatment of acrylic-textile dyeing wastewater. *J. Environ. Manag.* **2015**, *152*, 120–131. [CrossRef] [PubMed]

9. Jorfi, S.; Barzegar, G.; Ahmadi, M.; Soltani, R.D.C.; Takdastan, A.; Saeedi, R.; Abtahi, M. Enhanced coagulation-photocatalytic treatment of acid red 73 dye and real textile wastewater using uva/synthesized mgo nanoparticles. *J. Environ. Manag.* **2016**, *177*, 111–118. [CrossRef] [PubMed]

10. Le Luua, T.; Tiena, T.T.; Duongb, N.B.; Phuongb, N.T.T. Study of the treatment of tannery wastewater after biological pretreatment by using electrochemical oxidation on bdd/ti anode. *Desalin. Water Treat.* **2019**, *137*, 194–201. [CrossRef]

11. Li, X.; Zhu, W.; Wang, C.; Zhang, L.; Qian, Y.; Xue, F.; Wu, Y. The electrochemical oxidation of biologically treated citric acid wastewater in a continuous-flow three-dimensional electrode reactor (ctder). *Chem. Eng. J.* **2013**, *232*, 495–502. [CrossRef]

12. Zhang, C.; Jiang, Y.; Li, Y.; Hu, Z.; Zhou, L.; Zhou, M. Three-dimensional electrochemical process for wastewater treatment: A general review. *Chem. Eng. J.* **2013**, *228*, 455–467. [CrossRef]

13. Feng, Y.; Yang, L.; Liu, J.; Logan, B.E. Electrochemical technologies for wastewater treatment and resource reclamation. *Environ. Sci. Water Res. Technol.* **2016**, *2*, 800–831. [CrossRef]

14. Yousefi, Z.; Zafarzadeh, A.; Mohammadpour, R.A.; Zarei, E.; Mengelizadeh, N.; Ghezel, A. Electrochemical removal of acid red 18 dye from synthetic wastewater using a three-dimensional electrochemical reactor. *Desalin. Water Treat.* **2019**, *165*, 352–361. [CrossRef]

15. Liu, Y.; Yu, Z.; Hou, Y.; Peng, Z.; Wang, L.; Gong, Z.; Zhu, J.; Su, D. Highly efficient pd-fe/ni foam as heterogeneous fenton catalysts for the three-dimensional electrode system. *Catal. Commun.* **2016**, *86*, 63–66. [CrossRef]

16. Qiying, L.; Hongyu, S.; Xibo, L.; Junwu, X.; Fei, X.; Limin, L.; Jun, L.; Shuai, W. Ultrahigh capacitive performance of three-dimensional electrode nanomaterials based on α-mno 2 nanocrystallines induced by doping au through Å-scale channels. *Nano Energy* **2016**, *21*, 39–50.

17. Yu, X.; Hua, T.; Liu, X.; Yan, Z.; Xu, P.; Du, P. Nickel-based thin film on multiwalled carbon nanotubes as an efficient bifunctional electrocatalyst for water splitting. *Acs Appl. Mater. Interfaces* **2014**, *6*, 15395–15402. [CrossRef]

18. Yu, X.; Sun, Z.; Yan, Z.; Xiang, B.; Liu, X.; Du, P. Direct growth of porous crystalline nico2o4 nanowire arrays on a conductive electrode for high-performance electrocatalytic water oxidation. *J. Mater. Chem. A* **2014**, *2*, 20823–20831. [CrossRef]

19. Yu, X.; Xu, P.; Hua, T.; Han, A.; Liu, X.; Wu, H.; Du, P. Multi-walled carbon nanotubes supported porous nickel oxide as noble metal-free electrocatalysts for efficient water oxidation. *Int. J. Hydrog. Energy* **2014**, *39*, 10467–10475. [CrossRef]

20. Wu, W.; Huang, Z.H.; Lim, T.T. Enhanced electrochemical oxidation of phenol using hydrophobic tio2-nts/sno2-sb-ptfe electrode prepared by pulse electrodeposition. *RSC Adv.* **2015**, *5*, 32245–32255. [CrossRef]

21. Li, X.; Wu, Y.; Zhu, W.; Xue, F.; Qian, Y.; Wang, C. Enhanced electrochemical oxidation of synthetic dyeing wastewater using sno 2 -sb-doped tio 2 -coated granular activated carbon electrodes with high hydroxyl radical yields. *Electrochim. Acta* **2016**, *220*, 276–284. [CrossRef]

22. Liu, W.; Ai, Z.; Zhang, L. Design of a neutral three-dimensional electro-fenton system with foam nickel as particle electrodes for wastewater treatment. *J. Hazard. Mater.* **2012**, *243*, 257–264. [CrossRef] [PubMed]

23. Chen, J.-y.; Li, N.; Zhao, L. Three-dimensional electrode microbial fuel cell for hydrogen peroxide synthesis coupled to wastewater treatment. *J. Power Sources* **2014**, *254*, 316–322. [CrossRef]

24. Hao, R.; Li, S.; Li, J.; Meng, C. Denitrification of simulated municipal wastewater treatment plant effluent using a three-dimensional biofilm-electrode reactor: Operating performance and bacterial community. *Bioresour. Technol.* **2013**, *143*, 178–186. [CrossRef]

25. Neti, N.R.; Misra, R. Efficient degradation of reactive blue 4 in carbon bed electrochemical reactor. *Chem. Eng. J.* **2012**, *184*, 23–32. [CrossRef]

26. Pang, T.; Wang, Y.; Yang, H.; Wang, T.; Cai, W. Dynamic model of organic pollutant degradation in three dimensional packed bed electrode reactor. *Chemosphere* **2018**, *206*, 107–114. [CrossRef]

27. Liu, Z.; Wang, F.; Li, Y.; Xu, T.; Zhu, S. Continuous electrochemical oxidation of methyl orange waste water using a three-dimensional electrode reactor. *J. Environ. Sci.* **2011**, *23*, S70–S73. [CrossRef]

28. Zheng, T.; Wang, Q.; Shi, Z.; Fang, Y.; Shi, S.; Wang, J.; Wu, C. Advanced treatment of wet-spun acrylic fiber manufacturing wastewater using three-dimensional electrochemical oxidation. *J. Environ. Sci.* **2016**, *50*, 21–31. [CrossRef]

29. Zhao, H.Z.; Sun, Y.; Xu, L.N.; Ni, J.R. Removal of acid orange 7 in simulated wastewater using a three-dimensional electrode reactor: Removal mechanisms and dye degradation pathway. *Chemosphere* **2010**, *78*, 46–51. [CrossRef]

30. Wei, L.; Guo, S.; Yan, G.; Chen, C.; Jiang, X. Electrochemical pretreatment of heavy oil refinery wastewater using a three-dimensional electrode reactor. *Electrochim. Acta* **2010**, *55*, 8615–8620. [CrossRef]

31. Jung, K.-W.; Hwang, M.-J.; Park, D.-S.; Ahn, K.-H. Performance evaluation and optimization of a fluidized three-dimensional electrode reactor combining pre-exposed granular activated carbon as a moving particle electrode for greywater treatment. *Sep. Purif. Technol.* **2015**, *156*, 414–423. [CrossRef]

32. Chi, Z.; Wang, Z.; Liu, Y.; Yang, G. Preparation of organosolv lignin-stabilized nano zero-valent iron and its application as granular electrode in the tertiary treatment of pulp and paper wastewater. *Chem. Eng. J.* **2018**, *331*, 317–325. [CrossRef]

33. Can, W.; Yao-Kun, H.; Qing, Z.; Min, J. Treatment of secondary effluent using a three-dimensional electrode system: Cod removal, biotoxicity assessment, and disinfection effects. *Chem. Eng. J.* **2014**, *243*, 1–6. [CrossRef]

34. Li, X.-Y.; Xu, J.; Cheng, J.-P.; Feng, L.; Shi, Y.-F.; Ji, J. Tio2-sio2/gac particles for enhanced electrocatalytic removal of acid orange 7 (ao7) dyeing wastewater in a three-dimensional electrochemical reactor. *Sep. Purif. Technol.* **2017**, *187*, 303–310. [CrossRef]

35. Zhang, B.; Hou, Y.; Yu, Z.; Liu, Y.; Huang, J.; Qian, L.; Xiong, J. Three-dimensional electro-fenton degradation of rhodamine b with efficient fe-cu/kaolin particle electrodes: Electrodes optimization, kinetics, influencing factors and mechanism. *Sep. Purif. Technol.* **2019**, *210*, 60–68. [CrossRef]

36. Chen, H.; Feng, Y.; Suo, N.; Long, Y.; Li, X.; Shi, Y.; Yu, Y. Preparation of particle electrodes from manganese slag and its degradation performance for salicylic acid in the three-dimensional electrode reactor (tde). *Chemosphere* **2019**, *216*, 281–288. [CrossRef]

37. He, W.; Ma, Q.; Wang, J.; Yu, J.; Bao, W.; Ma, H.; Amrane, A. Preparation of novel kaolin-based particle electrodes for treating methyl orange wastewater. *Appl. Clay Sci.* **2014**, *99*, 178–186. [CrossRef]

38. Zhan, J.; Li, Z.; Yu, G.; Pan, X.; Wang, J.; Zhu, W.; Han, X.; Wang, Y. Enhanced treatment of pharmaceutical wastewater by combining three-dimensional electrochemical process with ozonation to in situ regenerate granular activated carbon particle electrodes. *Sep. Purif. Technol.* **2019**, *208*, 12–18. [CrossRef]

39. Nidheesh, P.V.; Gandhimathi, R. Trends in electro-fenton process for water and wastewater treatment: An overview. *Desalination* **2012**, *299*, 1–15. [CrossRef]

40. Yu, X.; Zhou, M.; Ren, G.; Ma, L. A novel dual gas diffusion electrodes system for efficient hydrogen peroxide generation used in electro-fenton. *Chem. Eng. J.* **2015**, *263*, 92–100. [CrossRef]

41. Wang, C.-T.; Chou, W.-L.; Chung, M.-H.; Kuo, Y.-M. Cod removal from real dyeing wastewater by electro-fenton technology using an activated carbon fiber cathode. *Desalination* **2010**, *253*, 129–134. [CrossRef]

42. Wang, C.T.; Hu, J.L.; Chou, W.L.; Kuo, Y.M. Removal of color from real dyeing wastewater by electro-fenton technology using a three-dimensional graphite cathode. *J. Hazard. Mater.* **2008**, *152*, 601–606. [CrossRef]

43. Lei, H.; Li, H.; Li, Z.; Li, Z.; Chen, K.; Zhang, X.; Wang, H. Electro-fenton degradation of cationic red x-grl using an activated carbon fiber cathode. *Process Saf. Environ. Prot.* **2010**, *88*, 431–438. [CrossRef]

44. Pérez, J.F.; Sabatino, S.; Galia, A.; Rodrigo, M.A.; Llanos, J.; Sáez, C.; Scialdone, O. Effect of air pressure on the electro-fenton process at carbon felt electrodes. *Electrochim. Acta* **2018**, *273*, 447–453. [CrossRef]

45.   Huang, Z.; Ong, S.L.; Ng, H.Y. Submerged anaerobic membrane bioreactor for low-strength wastewater treatment: Effect of hrt and srt on treatment performance and membrane fouling. *Water Res.* **2011**, *45*, 705–713. [CrossRef] [PubMed]
46.   Zhang, Q.H.; Yang, W.N.; Ngo, H.H.; Guo, W.S.; Jin, P.K.; Dzakpasu, M.; Yang, S.J.; Wang, Q.; Wang, X.C.; Ao, D. Current status of urban wastewater treatment plants in china. *Environ. Int.* **2016**, *92*, 11–22. [CrossRef]

13

Degradation of Hexacyanoferrate (III) from Gold Mining Wastewaters via UV-A/LED Photocatalysis using Modified TiO $_2$ P25

Augusto Arce-Sarria [1,2], **Kevin Mauricio Aldana-Villegas** [1], **Luis Andres Betancourt-Buitrago** [1], **Jose Ángel Colina-Márquez** [3], **Fiderman Machuca-Martínez** [1] and **Miguel Angel Mueses** [3,*]

[1] Escuela de Ingeniería Química, Universidad del Valle, Cali 760032, Colombia; augusto.arce@correounivalle.edu.co (A.A.-S.); kevin.aldana@correounivalle.edu.co (K.M.A.-V.); betancourt.luis@correounivalle.edu.co (L.A.B.-B.); fiderman.machuca@correounivalle.edu.co (F.M.-M.)

[2] Tecnoparque Nodo Cali, GIDEMP Materials and Products Research Group, ASTIN Center, SENA Regional Valle, Cali 760003, Colombia

[3] Modeling & Application of Advanced Oxidation Processes, Photocatalysis & Solar Photoreactors Engineering, Chemical Engineering Program, Universidad de Cartagena, Cartagena 130001, Colombia; jcolinam@unicartagena.edu.co

* Correspondence: mmueses@unicartagena.edu.co

Abstract: The photocatalytic degradation of potassium hexacyanoferrate (III) was assessed in a bench-scale compound parabolic collectors (CPC) reactor assisted with a light-emitting diode (LED) UV-A source emitting at 365 nm, and using a modified TiO_2 as a catalyst via the hydrothermal treatment of commercial Aeroxide P25. The experiments were performed under oxic and anoxic conditions in order to observe a possible reduction of the iron. The modified TiO_2 showed a specific surface area 2.5 times greater than the original Aeroxide P25 and its isotherm and hysteresis indicated that the modified catalyst is mesoporous. The bandgap energy (E_g) of the modified TiO_2 increased (3.34 eV) compared to the P25 TiO_2 band gap (3.20 eV). A specific reaction rate constant of 0.1977 min^{-1} and an electrical oxidation efficiency of 7.77 kWh/m^3 were obtained in the photocatalytic degradation. Although the TiO_2 P25 yields a photocatalytic degradation 9.5% higher than that obtained one with the modified catalyst (hydrothermal), this catalyst showed better performance in terms of free cyanide release. This last aspect is a significant benefit since this can help to avoid the pollution of fresh water by reusing the treated wastewater for gold extraction. A photocatalytic degradation of the cyanocomplex of 93% was achieved when the process occurred under oxic conditions, which favored the removal. Summarizing, the hydrothermal method could be a promising treatment to obtain TiO_2-based catalysts with larger specific areas.

Keywords: photocatalysis; UV-LED; TiO_2; hexacyanoferrate; mining; hydrothermal method

1. Introduction

Small and medium industries of gold extraction use the leaching process with sodium cyanide for mining the gold contained in the extracted ore, before precipitation of the metallic gold in the presence of zinc. During the process, the cyanide extracts undesired metals and thus forms several types of cyano complexes. The produced wastewater is rich in metallic complexes that are formed when the free cyanide interacts with the different metals present in the ores such as Ni, Fe, Co, Au, Ag, etc. These cyano complexes are very stable and recalcitrant compounds, which are hard to remove by natural remediation, resulting in the pollution of rivers, lakes and groundwater sources. Besides, solar photolysis releases free cyanide, which is highly harmful to ecosystems [1]. Advanced oxidation processes (AOPs), such as ozone-based treatments, alkaline chlorination, hydrogen peroxide-based processes, biological and photocatalytic processes, can be used as alternative treatment technologies for these mining wastewaters [2].

The heterogeneous photocatalysis is an AOP where a solid semiconductor, assisted by UV radiation, promotes the generation of free hydroxyl radicals ($\bullet OH$) and the degradation of diverse pollutants. The most commonly used semiconductor is the titanium dioxide (TiO_2), and it can be used as a base oxide for the synthesis of other photoactive catalysts as well. The TiO_2 is preferred because of its low cost, easy handling, and low toxicity. In general, when the photocatalyst is irradiated with photons with energy greater than the bandgap (Eg) of the semiconductor, the excited electrons are promoted from the valence band to the conduction band of the semiconductor, leading to the formation of electron–hole pairs. The strong oxidative potential of the holes (h^+) oxidizes the hydroxyl anions of water for generating $\bullet OH$, whereas the electrons of the conduction band can react with oxygen for generating superoxide ions ($O_2^{\bullet-}$) or promote other reduction reactions. Those radicals are the main species responsible for the oxidation reactions in the photocatalytic process [3,4].

To improve the semiconductors' $\bullet OH$-generating performance, several studies have been focused on the preparation of semiconductors with enhanced radiation absorption. Different methods of preparation have been reported, namely hydrothermal [5], sol-gel [6], anodic oxidation, template method, and chemical vapor deposition (CVD) [7,8]. TiO_2 catalysts doped with rare earth and transition metals have been modified to improve their $\bullet OH$ electron transfer properties. Some modifications on their morphology have also been made to produce structures such as nanorods, nanotubes, nanospheres, nanoflowers, among others [9–12].

The hydrothermal method has been widely used for the nanomaterial synthesis of TiO_2 with diverse morphologies. This methodology is controlled by different variables, namely the precursors used, pH, temperature and reaction time [13]. Nowadays, TiO_2-based nanowires, TiO_2 nanotubes [14], carbon nanotubes [15], nanofibers, nanoflowers, and others have been successfully modified by hydrothermal treatment [16]. This method has become a very important tool for obtaining advanced materials due to its advantages, such as low cost, low operating temperatures, energy saving and lower impact to the environment (according to the principles of green chemistry) [10,12,17], in comparison to anodic oxidation and CVD methods. The hydrothermal treatment has been applied to the synthesis of nitrogen and carbon co-doped TiO_2 [18], Sn-doped TiO_2 nanoparticles composites [19], silica-titania combination of sol-gel-hydrothermal TiO_2 nanoparticles [20], and both anatase and rutile TiO_2 [21]. Moreover, several applications of TiO_2 nanoparticles synthesized by the hydrothermal method have been reported such as hydrogen production via CO_2 reduction, degradation of emergent pollutants and selective oxidation [22].

Huang and Chien [23] showed that the degradation of methylene blue increases from 65% to 95% with titania nanotubes compared to the powder. Camposeco et al. [24] compared the degradation between nanotubes and Evonik P25, showing that the catalytic activity was improved from 54 to 93% for methylene blue degradation and from 37% to 60% for the elimination of methylene orange. However, there is a lack of specific information about the use of titania modified via hydrothermal process for treating gold mining wastewater under UV/LED radiation.

In this work, the degradation of potassium hexacyanoferrate, which is a complex occurring as a by-product in gold mining wastewaters, via photocatalysis with hydrothermally treated TiO_2, was studied. The mechanism proposed by Grieken et al. 2005 [25] or the hexacyanoferrate (III) reduction to hexacyanoferrate (II) and the subsequent degradation by heterogeneous photocatalysis is depicted in Figure 1. After the progressive abatement of the CN^- groups in the molecule, the free cyanide can remain stable in solution due to the high pH of treatment or to produce cyanate by photocatalytic degradation, which is less toxic than the free cyanide. Nonetheless, the free cyanide is an advantage if the treated wastewater can be reused for the gold extraction. This would reduce the fresh water and cyanide consumptions and a consequent diminution of cyanide presence in water bodies.

The mechanism of free cyanide release is congruent with the reported literature [26–29]. The oxic conditions were analyzed in order to compare these results with the obtained ones in our previous work [27]. A further contribution respect to the reported literature is the use of the modified P25 via hydrothermal treatment and its potential improvement for the potassium hexacyanoferrate removal.

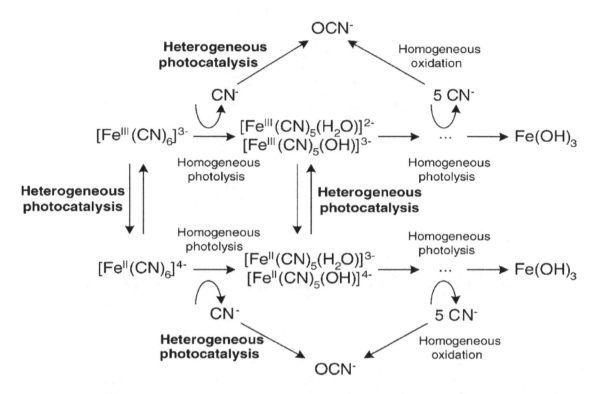

Figure 1. Mechanism of the heterogeneous photocatalytic degradation of Hexacyanoferrate [25]. Reprinted from Applied Catalysis B: Environmental, 55, Rafael van Grieken *, José Aguado, María-José López-Muñoz, Javier Marugán, Photocatalytic degradation of iron–cyanocomplexes by TiO_2 based catalysts, 201–211.

The photocatalytic performance of the obtained titania was evaluated by analyzing the effect of the catalyst load on the overall efficiency of the photodegradation under both oxic and anoxic conditions. In addition, the impact of the variation of the power supplied by the UV source and of the initial concentration of the cyanocomplex, was assessed. All the experiments were carried out in a bench-scale compound parabolic collector (CPC) photoreactor with artificial UV/LED radiation.

2. Materials and Methods

2.1. Catalyst Treatment

The catalyst was modified by using the hydrothermal treatment [30–34]. Six grames of Aeroxide P25 (Evonik®, Essen, Germany) were mixed with 100 mL of a 10-M solution of NaOH (Merck, Darmstadt, Germany). The solution was stirred to avoid the formation of agglomerates and then it was decanted into a 120-mL beaker. Subsequently, it was transferred to a stainless-steel sealed reactor. The reactor temperature increased up to 120 or 180 °C during 24 or 72 h, according to the 2^3 experimental design described in Table 1. The white precipitate was washed with a 0.1-M HCl (Merck, Darmstadt, Germany) solution under stirring. The solid was recovered by centrifugation followed by a series of washing cycles with deionized water until the pH of the supernatant was 7.4. After drying the solid at 100 °C for 24 h, it was calcinated at 400 or 500 °C during four hours, with a heating gradient of 10 °C/min. Figure 2 shows the detailed procedure for the synthesis of photocatalysts.

Table 1 shows the different conditions of reaction time, reaction temperature and calcination temperature used to prepare each of the eight catalysts. For the statistical analysis, an analysis of variance (ANOVA) was carried out, considering a significance level of 0.05.

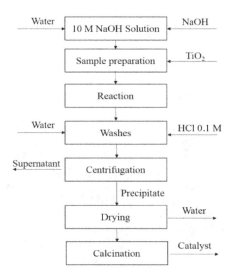

Figure 2. Schematic diagram of the hydrothermal synthesis method used.

Table 1. Experimental runs of the 2^3 factorial design.

Labels	Reaction Temperature (°C)	Reaction Time (h)	Calcination Temperature (°C)
SL400	120 (S)	24 (L)	400
SL500	120 (S)	24 (L)	500
SH400	120 (S)	72 (H)	400
SH500	120 (S)	72 (H)	500
LL400	180 (L)	24 (L)	400
LL500	180 (L)	24 (L)	500
LH400	180 (L)	72 (H)	400
LH500	180 (L)	72 (H)	500

2.2. Catalyst Evaluation

The evaluation of the performance of the modified catalysts was carried out in a bench-scale CPC reactor assisted by a UV/LED radiation source [27,35]. The reactor consisted of four Pyrex tubes with an outside diameter of 2 cm and a length of 11 cm, which were connected to a 750-mL container through a centrifugal pump. The input power of the centrifugal pump was 50 W. The container was sealed at the top with a stopper, which had openings for sampling and oxygen/nitrogen inlet to the gas diffuser [27].

Four 30 W LEDs (TaoYuan Electron Ltd. TY-365 nm, Hong Kong, China) connected in parallel, were used as the artificial light source. The light output was set up with a tilt angle of 115–125° and of 900–1200 mW of radiation intensity per LED [27,35]. Each LED (model GW GPS-3030D, GWINSTEK, Veldhoven, Netherlands) was equipped with a cooling system consisting of heat sinks and a 12-V fan. The UVA radiation intensity was measured with a UV radiometer (DELTA OHM model HD2102.2, Deltha Ohm S.r.l., Padova, Italy) and it was varied by adjusting the current intensity supplied to the LEDs at a constant voltage of 30 V. The reactor had a reactive volume and a total irradiated area of 138.23 cm^3 and 276.4 cm^2, respectively. The ratio of the illuminated volume to the total volume was 0.23. This ratio is useful to characterize the reactive system volume used with respect to those used by other authors and thus be able to compare its performance.

Once the system was loaded with the matrix to be degraded, the LEDs were placed above the tubes at approximately 3 cm of height, whereas the parabolic collectors were placed below the reactor. The use of these reflective surfaces provides a more homogeneous distribution of the radiation reflected to reactor walls since the bottom of the tubes could be illuminated evenly [36].

The hexacyanoferrate III (K$_3$[Fe(CN)$_6$], CAS 13746-66-2, (Panreac AppliChem, Darmstadt, Germany) was selected as the model cyanocomplex of the gold mining wastewaters. The control experiments (physical adsorption, i.e., without light; or photolysis, i.e., without catalyst) were carried out with 60 mL of solutions of 100 ppm of the pollutant. For the physical adsorption experiment, the solution was kept under continuous stirring in a 500-mL beaker, under darkness conditions. For the photolysis experiment, the power of the UV-LEDs was set at 30 W that supplies the maximum intensity of UV radiation. For both experiments, an aliquot of 5 mL was taken every 10 min during two hours (time set for the reaction).

The results obtained for the removal were estimated with the Equation (1):

$$\%Degradation = \left(1 - \frac{C}{C_0}\right) \times 100 \tag{1}$$

where C is the final concentration and C_0 the initial concentration

For each optimization step, 500 mL of a solution of 100 ppm of hexacyanoferrate was prepared. For keeping the solution pH above 12, 1 mL of a 10 M solution of NaOH was previously added to 500 mL of hexacyanoferrate solution. After an adsorption stage carried out under darkness conditions for 20 min, the LEDs were turned on to perform the photocatalytic runs. The experiments were carried out at room temperature (20 °C) and 10 mL aliquots (less than 10% of the total volume) were taken at different time intervals. For oxic and anoxic experiments, air or nitrogen was sparged, according to the case, into the solution at a constant flow rate of 0.5 L/min. The optimization study was executed in four stages:

(1)   Variation of the catalyst dose (0.1, 0.3, 0.5 and 0.7 g/L) to determine the best performing catalyst dose, at oxic conditions for an hour.

(2)   Comparison of reactions (during two hours) under anoxic and oxic conditions, using the best performing catalyst dose selected in the previous stage to select the best conditions for the following experiments: oxic (air) or anoxic (nitrogen).

(3)   Variation of the radiation intensity, by testing the power supplied by the LEDs (10, 20 and 30 W) during 3 h of reaction.

(4)   Variation of the initial concentration of the contaminant (50–100 ppm) during three hours of reaction.

The hexacyanoferrate (III) concentration was followed by UV-VIS (JASCO V-730 spectrophotometer, Easton, MD, USA) at 303 nm, corresponding to its maximum absorbance wavelength in the UV spectrum. The measurement of total dissolved iron was performed using atomic absorption spectrometry (Thermo Scientific iCE 3000, Waltham, MA, USA) and the measurement of CN^- by titration with $AgNO_3$ according to the Standard Methods 4500 [37].

A kinetic law with a two-step reaction was used to describe the degradation of hexacyanoferrate (III). The first step (faster) corresponds to the adsorption of $Fe(CN)_6^{3-}$ onto the surface of TiO_2 and degradation of the iron modified, whereas the second step (slower) corresponds to the reduction of the iron present in the cyano-metallic complex (that corresponds to the removal of dissolved iron) [38].

For the kinetic analysis of the photo reductive process of the iron cyanocomplex, a pseudo first-order reaction rate equation was proposed (Equations (2) and (3)), as suggested by previous studies [39–41]:

$$-\frac{dC}{dt} = k'C \tag{2}$$

$$ln(C_0/C) = k't \tag{3}$$

where k' is the pseudo first-order rate constant (min^{-1}), C_0 and C are the initial and final concentrations of the iron complex in solution, respectively. The $ln\,(C_0/C)$ was plotted versus time for obtaining the k' value, which is the slope of the equation of the line.

2.3. Characterization

The crystalline phases of the resulting solid from the hydrothermal synthesis were characterized using X-ray diffraction (XRD) on a X'per PRO-PANalytical diffractometer with CuKα radiation

(0.1542 nm) with a 2θ sweep between 0° and 90°. The surface area was determined by the Brunauer–Emmett–Teller method (BET) by adsorption–desorption of nitrogen (N_2) at 77 K and the volume and size of the pore were determined by the Barrett–Joyner–Halenda method (BJH) in a Micromeritics equipment ASAP 2020 V4.01 (Micromeritics, Norcross, GA, USA).

The morphology was analyzed by scanning electron microscopy (SEM) and X-ray energy dispersion spectrometry (EDS) was used for the analysis of elemental composition of the catalyst in a JEOL JSM 6490 LV brand equipment. The semiconductor bandgap (E_g) was estimated by measuring the material transmittance with UV-vis diffuse reflectance spectroscopy (UV DRS) in a Thermo Scientific Evolution 300 PC series EVOP068001 spectrophotometer. Finally, the Fourier-transform infrared spectroscopy (FT-IR) was used to identify the functional groups of the inorganic and organic substances (FT/IR-4100 type-A).

2.4. Estimation of the Electric Oxidation Efficiency (E_{Eo})

The IUPAC has proposed methods to calculate the electrical consumption of an AOP, depending on the type of reactor and the amount of contaminant to be treated. For low concentrations, it is proposed to use the electric energy per order (E_{Eo}). This parameter consists of the electrical energy (kWh) required to remove the pollutant up to 90% of its initial concentration per volume unit. The E_{Eo} can be calculated using the Equation (4), following the methodology proposed by Shirzad-Siboni et al. [41] and Daneshvar et al. [40]:

$$E_{E_o} = \frac{1000\, P\, t}{60\, V\, log\left(C_0/C_f\right)} \tag{4}$$

where P is the power supplied to the system (kW) and it is defined as the product of electric potential and the current intensity (A); V is the total reactive volume (L), and t is time (h). From Equations (3) and (4), the E_{Eo} can be calculated as follows:

$$E_{E_o} = \frac{38.4P}{Vk'} \tag{5}$$

3. Results and Discussion

3.1. Photolysis and Adsorption

The control tests in 3-h experiments showed that the photolysis contributes moderately to the removal of contaminants and the release of free cyanide. A 17% of photolytic removal of hexacyanoferrate and a 12% of cyanide release were achieved, which is in agreement with the results reported in this literature review [26]. On the contrary, the adsorption had a minor effect both in the elimination of contaminants and in the release of cyanide, respectively, 10% and less than 5% after three hours of experimentation. It was observed that 8% of the initial hexacyanoferrate concentration was adsorbed during the first 20 min of the experiment and therefore the dark period for the photocatalytic runs was set at 20 min.

3.2. Evaluation of Synthesized Materials

3.2.1. Catalyst Load

This behavior observed in the Figure 3 is explained by the lower flow of photons into the reactive system resulting from the higher turbidity (catalyst loads higher than 0.5 g/L) of the slurry to treat [42]. This screening effect limits the effectiveness of the treatment by decreasing the local volumetric rate of photon absorption for tubular photoreactors, which has been analyzed by Colina-Marquez et al. in 2010 [43] and Mueses et al. in 2013 [44]. Those studies reported an optimal catalyst load of 0.3 g/L for CPC reactors, approximately. In turn, Osathaphan et al. [45] used catalyst loads between 0.1 and 4 g/L without affecting the reductive treatment considerably. Given the best results when using 0.5 g/L of both SL400 and SL500, both catalysts were promising to degrade the cyanocomplex. To select the best performing catalysts modified, photocatalytic experiments were performed using 0.5 g/L of each catalyst to degrade the pollutant during 2 h of reaction. For the further experiments, 0.5 g/L of SL400 was selected, due to the better performance and also in order to save energy in the calcination process.

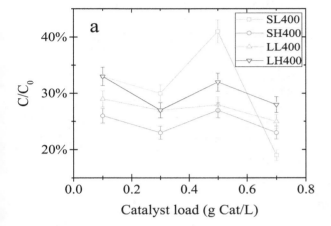
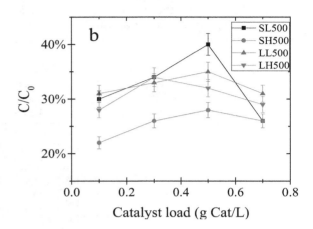

Figure 3. Evaluation of the best catalyst load to remove the cyanocomplex (hexacyanoferrate III) after 1 h of reaction (**a**) Catalysts calcined at 400 °C; (**b**) Catalysts calcined at 500 °C. Operating conditions: Initial pollutant concentration of 100 ppm, 20 min of adsorption, LED power supply of 20 W, air flow of 0.5 L/min.

Table 2 shows the results of the degradation obtained at different synthesis temperatures, calcination temperature and synthesis times.

Table 2. Degradation percentage of $K_3[Fe(CN)_6]$ after 2 h of reaction, using a catalyst load of 0.5 g/L.

Calcination Temperature	400 °C		500 °C	
Synth. Time / Synthesis Temp.	24 (L)	72 (H)	24 (L)	72 (H)
120 °C (S)	SL400	SH400	SL500	SH500
	53 51	50 50	49 48	47 49
180 °C (L)	LL400	LH400	LL500	LH500
	50 48	48 50	43 44	51 50

Operating conditions: initial pollutant concentration of 100 ppm, 20 min of adsorption, LED nominal power of 20 W, air flow of 0.5 L/min. Each experiment was done in duplicate.

A statistical analysis (see Table 3) of the information reported in Table 2 was carried out by using Statgraphics® Centurion XVI (version 16.2.04, Statpoint Technologies Inc., The Plains, VA, USA) and it was found that the calcination temperature was the most significant effect on the response variable within the evaluated intervals (see Figure 4), obtaining better results with 400 °C. The second most significant effect was the synthesis temperature and the best results were obtained at 120 °C; however, it is not statistically significant. Comparing the information of the table with the Pareto chart (Figure 4), it can be observed that the calcination temperature has a negative effect; that means that an increase of this variable represents a degradation decrease. This behavior can be attributed to the reduction of the surface area of the catalyst or material sintering at higher temperatures [6].

Table 3. ANOVA for degradation percentage of $K_3[Fe(CN)_6]$.

Source of Variation	SS	df	MS	F	p-Value
A: Calcination Temperature	225.625	1	225.625	10.62	0.0116
B: Synthesis time	50.625	1	50.625	2.38	0.1613
C: Synthesis temperature	105.625	1	105.625	4.97	0.0563
AB	180.625	1	180.625	8.50	0.0194
AC	0.5625	1	0.5625	0.26	0.6208
BC	225.625	1	225.625	10.62	0.0116
Blocks	0.0625	1	0.0625	0.03	0.8681
Total error	17.0	8	2.125		
Total (corr.)	964.375	15			

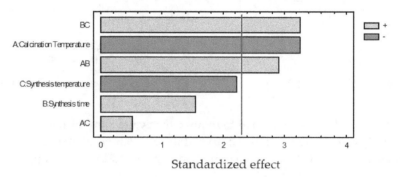

Figure 4. Standardized Pareto Chart for degradation percentage of $K_3[Fe(CN)_6]$.

On the other hand, although the synthesis time was not significant, its interactions with the other variables were meaningful and synergistic. This behavior is interesting because it means that a simultaneous increase of the calcination and synthesis temperatures with the synthesis time represents an improvement on the pollutant removal. In fact, the interaction between the synthesis time and the synthesis temperature (BC) is as significant as the effect of the calcination temperature. In addition, it was found that the best results for the degradation of the cyanocomplex were obtained for the catalyst modified at 24 h—120 °C to 400 °C (SL400). Considering all these facts, the following stages were carried out using SL400.

3.2.2. Tests Under Oxic and Anoxic Conditions

The degradation of the cyanocomplex by photocatalysis using SL400 was evaluated under oxic and anoxic conditions, to evaluate the importance of the presence of oxygen (Figure 5).

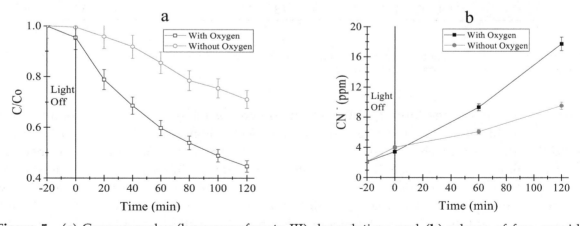

Figure 5. (**a**) Cyanocomplex (hexacyanoferrate III) degradation; and (**b**) release of free cyanide. Operating conditions: Initial concentration of $K_3[Fe(CN)_6]$ of 100 ppm, catalyst load of 0.5 g/L, 20 min of adsorption, Power supply: 20 W, air or nitrogen flow of 0.5 L/min, reaction time of 2 h.

After two hours of reaction, 56% of the cyanocomplex was degraded in the presence of oxygen, whereas it was only 29% when air was replaced by nitrogen. In turn, the cyanide release was two times higher when air containing oxygen was used (18 ppm in the presence of oxygen and 9 ppm using nitrogen). Finally, for the total removal of iron, a removal of 40% was achieved in the presence of oxygen and only 15% under an inert atmosphere. The higher degradation of the cyanocomplex and release of free cyanide in the presence of oxygen can be ascribed to the, electrons directly reducing iron and the oxidation of the complex by holes, hydroxyl radicals and superoxide anions. In contrast to our results that showed that the presence of oxygen during the reaction increases the degradation of the complex, Yang et al. [46] and Ku and Jung [47] reported a better performance of the P25 TiO_2 for the removal of the studied contaminants under anoxic conditions. In these reports, the authors observed that the presence of oxygen did not have a significant effect on the contaminant removal, whereas a higher reduction was showed with nitrogen.

3.2.3. Effect of the Radiation Intensity

The availability of UV photons directly affects the generation of electron–hole pairs. By comparing the results obtained at 10, 20 and 30 W (Figure 6a), it can be observed that the radiation intensity higher effect when increasing from 10 to 20 W than after a further increase to 30 W. Regarding to the degradation of the cyanocomplex, removals of 55, 73 and 79% were obtained with 10, 20 and 30 W, respectively. Additionally, iron removals of 30, 48 and 60% were achieved for 10, 20, and 30 W, respectively. The dissolved iron concentration was analyzed to corroborate its removal from the solution and its deposition onto the catalyst surface (Figure 6b).

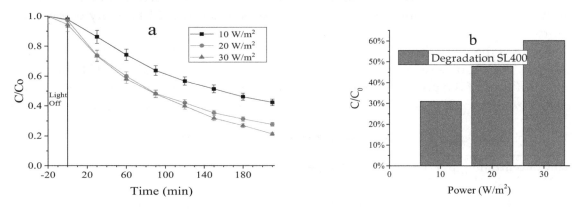

Figure 6. (**a**) Degradation of the cyanocomplex (hexacyanoferrate III) at 10 W, 20 W and 30 W; (**b**) Total removal of dissolved iron. Operating conditions: Initial concentration of $K_3[Fe(CN)_6]$ of 100 ppm, catalyst load of 0.5 g/L, 20 min of adsorption, air flow of 0.5 L/min, reaction time of 3.5 h.

The degradation values obtained with 20 and 30 W exhibited similar behaviors. An energy increase of 33% (20 to 30 W) yielded just an increase of 8.12% for the cyanocomplex degradation. This means that this energy increase is not enough to significantly affect the degradation performance. Therefore, the radiation intensity of 20 W was selected as the best condition due to the less energy consumption. Similar results were obtained by Rodriguez and Ossa [27], reporting a better but not significant performance when working at 30 than 20 W, and thus the selection of an inferior power supply to avoid an additional electrical consumption.

3.2.4. Comparison between Modified TiO_2 and the Raw P25

By comparing the raw and treated TiO_2, the degradation efficiency obtained with SL400 was 70%, whereas TiO_2 P25 led to a photocatalytic removal of 80% (Figure 7a). In turn, 20 ppm of cyanide are released by SL400 and less 10% is observed for TiO_2 P25, with 18 ppm of cyanide released (Figure 7b). Although for the complex degradation, the TiO_2 P25 showed better results; regarding to the free cyanide release, the SL400 showed a performance 10% higher. As the initial concentration of contaminant increases, the degradation decreases, as it was documented in the studies of Yang et al. [46] and Samarghandi et al. [39]. The cyanide release can be beneficial since it can be reused in the mining processes where such cyanide can be returned for the mineral (gold) re-extraction process. This feature would make the use of the synthesized material economically and environmentally attractive and also attenuate its weakness against P25 in terms of degradation of the hexacyanoferrate complex.

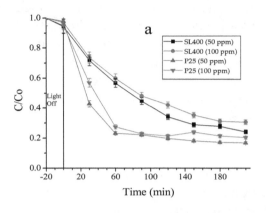 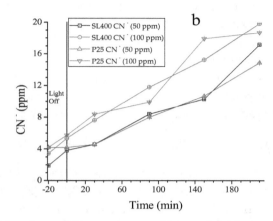

Figure 7. (**a**) Cyanocomplex degradation; (**b**) Free cyanide released. Operating conditions: 0.5 L/min of air bubbled, 0.5 g cat/L solution, 3.5 h of reaction with 30 min of adsorption and power supply of 20 W.

Van Grieken et al. [25] reported that the oxidative degradation of hexacyanoferrate (100 ppm of initial concentration) releases around 20 ppm of CN^- in 240 min of irradiation by using mercury lamps. In this study, the same amount of cyanide ion was released in 210 min by using a UVA/LED photon source.

Table 4 shows the values found for the pseudo first-order speed constant (min^{-1}) for a reaction time of 210 min. As it can be seen, the P25 TiO_2 rate constants are higher than the SL400 ones for both initial concentrations of the pollutant. This can be explained because of the differences in superficial area, particle size distribution, semiconductor purity and other features in electronic properties.

Table 4. Pseudo first order rate constants.

Initial Concentration (ppm)	Catalyst Type	Apparent Reaction Rate Constant k' (min^{-1})
100	P25	0.1924
100	SL400	0.1679
50	P25	0.211
50	SL400	0.1977

3.2.5. Electric Oxidation Efficiency

Table 5 shows the E_{Eo} values obtained for the P25 and the SL400 sample with two different concentrations of hexacyanoferrate.

Table 5. Electrical oxidation efficiency for the catalysts used.

Catalyst Type	Voltage (V)/Amperage (A)	Initial Concentration of Contaminant	E_{Eo} (kWh/m^3)
P25	30/0.8	100	7.98
SL400	30/0.8	100	9.15
P25	30/0.8	50	7.28
SL400	30/0.8	50	7.77

The P25 still exhibits better performance regarding to the energy consumption. This behavior is related to the higher activity of the commercial standard, which was discussed previously. The obtained results are similar to the reported ones by Daneshvar et al. [40], which did not exceed 10 kWh/m^3.

On the other hand, when the value obtained is compared with the study of Rodriguez and Ossa [27], it was found that the E_{Eo} is 40 and 20 times lower, respectively, than the presented ones in Table 5. In these works, it was reported the same concentration of $Fe(CN)_6$ but with the use of different catalysts.

3.3. Characterization of the Photocatalyst

3.3.1. Fourier-Transform Infrared Spectroscopy (FT-IR)

The Figure 8 shows the IR spectra of the SL400 before and after usage in the photocatalytic experiments. Four bands are highlighted that are common in both spectra. As described by Thennarasu et al. [48], the peaks observed around 3300–3400 cm^{-1} correspond to the stretching vibrations (stress) of the •OH and around 1600 cm^{-1} arises from the water bending mode that can be associated with water absorbed by the catalyst due to the presence of moisture in the materials by contact with air. The main bands below 1000 cm^{-1} were attributed to the Ti-O and Ti-O-Ti bending vibrations. The band around 1300 cm^{-1} is attributed to the C-H bending vibrations.

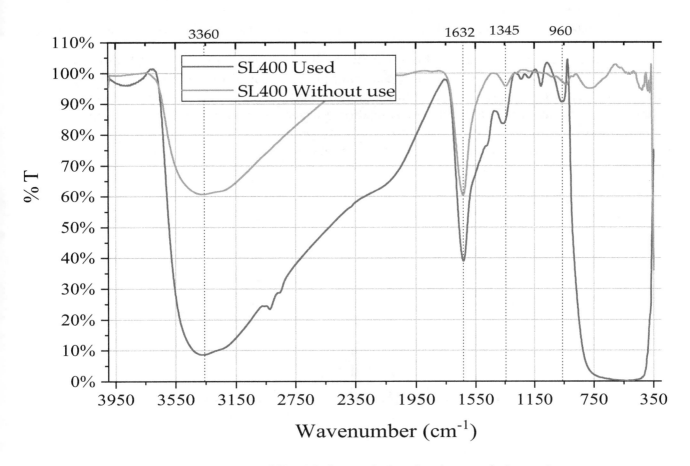

Figure 8. FTIR spectra of SL400 before and after the photocatalytic reaction.

3.3.2. XRD Results

According to Mozia et al. [33], the peaks found at 2θ of 24°, 28° and 48° as those observed for SL400 (Figure 9) correspond to titanates of the form $A_2Ti_2O_5 \cdot H_2O$ and $A_2Ti_3O_7$. The sodium titanates ($Ti_{12}O_{36}Na_4$ or Ti_3O_9Na) exhibit peaks at 10°, 24°, 28°, 48° and 62°, which evidence the presence of the anatase phase of TiO_2 at 25°, 62°, and 82°. The analysis showed no significant amount of rutile since may be found at calcination temperatures over 600 °C.

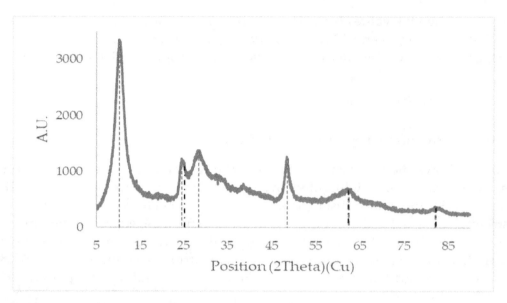

Figure 9. XRD pattern for SL400 without use. The blue dotted lines represent the peaks associated with titanates and the black dotted line represents the peaks associated with the anatase phase of TiO_2.

The most significant difference between the SL400 diffractogram (Figure 9) and that of P25 (Figure 10) without modifications [49], is the sharper peaks obtained by XRD for the commercial P25. This means a more crystalline structure for the unmodified P25 and some amorphous characteristics for the modified material (SL400). This modification affected the overall performance of the modified material regarding to the activity and, therefore, the pollutant removal. In addition, the XRD of SL400 does not have characteristic peaks of rutile phase as P25, which are known to improve the photocatalytic activity thanks to its synergistic effect with the anatase.

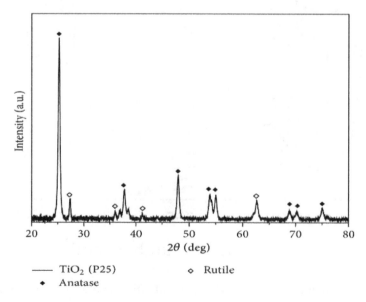

Figure 10. XRD pattern for TiO_2 P25 [49].

3.3.3. EDS Results

The Figure 11 shows a micrograph obtained from SL400. Additionally, an energy-dispersive X-ray spectroscopy analysis (EDS) was performed for elemental detection of the modified catalyst (see Figure 11). This analysis shows the type of elements present in different analyzed areas of the catalyst, where the presence of Carbon (C), Oxygen (O), Sodium (Na) and Titanium (Ti) were exhibited, with their respective composition, as shown in Table 6.

Figure 11. EDS analysis of the modified catalyst SL400.

Table 6. EDS results in % weight in the modified catalyst.

Spectra	C (%)	O (%)	Na (%)	Ti (%)	Total (%)
1	-	34.87	5.13	60.00	100
2	4.62	44.52	7.37	43.49	100
3	4.56	43.86	6.22	45.36	100
4	3.18	47.63	7.69	41.50	100
5	5.67	46.00	7.02	41.31	100
Max	5.67	47.63	7.69	60.00	
Min	3.18	34.87	5.13	41.31	
Average	3.61	43.38	6.69	46.33	

According to the EDS results, the presence of carbon in the material (3–6%), probably from impurities in the precursors used for the synthesis, can affect negatively the photocatalytic performance because of the number of active sites on the semiconductor surface decreases as the carbon occupies them.

3.3.4. Surface Area Results

The surface area was 127.84 m^2/g, which is greater than the surface area of the precursor material (50 m^2/g). The pore volume of the total amount absorbed was 0.197 cm^3/g and the pore size distribution analyzed by the BJH method was approximately 58 Å (5.8 nm) for an average particle size of 469 Å (46.9 nm). An isotherm of type IV was observed (Figure 12) with a hysteresis type III, which suggests that this catalyst is a mesoporous solid (2–50 nm).

Although the sample SL400 has a surface area higher than the P25's one, the number of active sites could not exceed the amount of sites of the TiO_2 P25, since the modified catalyst did not exceed the photocatalytic activity of the precursor. In addition, the absence of rutile phase affects the overall activity of the TiO_2, since this phase in the P25 acts synergistically with the anatase to improve the activity of the catalyst. The surface area is similar to those obtained by Turki et al. [50], Sikhwivhilu et al. [51] and Fen et al. [52]. On the other hand, some studies have obtained values higher than 200 m^2/g as is the case of Thennarasu et al. [48] and Camposeco et al. [53] with important photocatalytic activity.

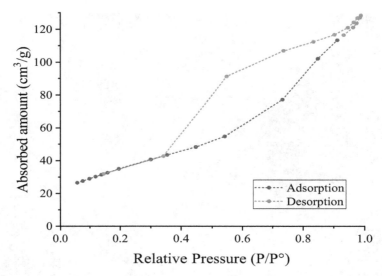

Figure 12. Isotherm absorption-desorption of the modified catalyst SL400.

3.3.5. Bandgap Energy Estimation by DRS

The bandgap energy (E_g) is one of the most important parameters in the photocatalytic activity of TiO$_2$ since it determines the effective wavelength interval for photon absorption. This parameter was estimated with the Kubelka-Munk theory according to the methodology reported by López and Gómez [54] (see Figure 13). It has to be considered that the crystal size, the particle size, the aggregation state of the particles, and the impurities present in the solid and the method of synthesis, can significantly affect the E_g.

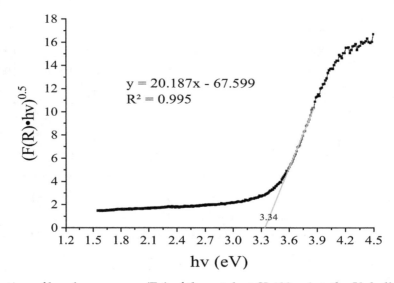

Figure 13. Estimation of bandgap energy (E_g) of the catalyst SL400 using the Kubelka-Munk function, the red points represent those used to obtain the slope of the line and obtain the intercept on the x axis.

The energy of the bandgap obtained was 3.34 eV and the wavelength (estimated with Equation (6)) at which the catalyst is activated is 370 nm.

$$\lambda = \frac{h \times c}{h_v} \qquad (6)$$

If these values are compared with those reported for TiO$_2$ P25 (E_g = 3.20 eV; λ = 385 nm), it is expected that the modified catalyst underperform respect to the commercial standard, regarding the UV photons absorption. This can be a significant drawback when it is intended to use a wide spectrum photons source.

4. Conclusions

The modified TiO_2 P25, via the hydrothermal method, did not improve the $Fe(CN)_6$ removal with respect to the obtained one with the original P25. This could be attributed to the loss of both the rutile phase and the material crystallinity. In addition, the increase of the bandgap energy for the modified P25 is another drawback since it affects the photon absorption by the semiconductor. Although the higher free cyanide release achieved with the modified material can be considered as a shortcoming regarding to the environmental potential of this material, in this particular case, this can be beneficial since this free cyanide could be reused for the gold extraction process and so, obtain a closed cycle for the water use. Furthermore, the increase of the specific surface area can be a promising result, in terms of physical adsorption of the studied pollutant or metallic cations.

While at a first sight the hydrothermal method did not improve the activity of the P25, further studies should be carried out to obtain more information about the structural modifications of the catalyst and potential advantages for photocatalytic applications.

Author Contributions: Conceptualization, A.A.-S. and L.A.B.-B.; methodology, A.A.-S. and L.A.B.-B.; software, L.A.B.-B. and A.A.-S.; validation, A.A.-S., L.A.B.-B., F.M.-M. and J.Á.C.-M.; formal analysis, K.M.A.-V. and A.A.-S.; investigation, K.M.A.-V., A.A.-S. and L.A.B.-B.; resources, F.M.-M.; data curation, K.M.A.-V. and L.A.B.-B.; writing—original draft preparation, K.M.A.-V.; writing—review and editing, M.A.M. and J.Á.C.-M.; visualization, M.A.M.; supervision, A.A.-S.; project administration, A.A.-S.; funding acquisition, F.M.-M. and M.A.M. All authors have read and agreed to the published version of the manuscript.

Acknowledgments: The authors are grateful to Universidad del Valle and the COLCIENCIAS for the Ph.D. scholarship 567-2012. Also, the authors thank the Biotechnology and Nanotechnology Laboratory of Tecnoparque (Nodo Cali) for the support with the analytical techniques. Colina-Márquez and Mueses thank the University of Cartagena. All authors send thanks to Ana Rita Lado Ribeiro from the University of Porto for the much-appreciated writing revisions of this manuscript.

References

1. Moran, R. El Cianuro en la Minería: Algunas Observaciones sobre la Química, Toxicidad y Análisis de las Aguas Asociadas con la Minería. *Ecología* **1999**, *99*, 23.
2. Kuyucak, N.; Akcil, A. Cyanide and Removal Options from Effluents in Gold Mining and Metallurgical Processes. *Miner. Eng.* **2013**, *50–51*, 13–29. [CrossRef]
3. López-Vásquez, A.F.; Colina-Márquez, J.A.; Machuca-Martínez, F. Multivariable analysis of 2,4-d herbicide photocatalytic degradation. *DYNA* **2011**, *78*, 119–125.
4. Pichat, P. A brief survey of the practicality of using photocatalysis to purify the ambient air (indoors or outdoors) or air effluents. *Appl. Catal. B* **2019**, *245*, 770–776. [CrossRef]
5. Lozano-Morales, S.; Morales, G.; Lopez Zavala, M.; Arce, A.; Machuca, F. Photocatalytic Treatment of Paracetamol Using TiO_2 Nanotubes: Effect of pH. *Processes* **2019**, *7*, 319. [CrossRef]
6. Arce-Sarria, A.; Machuca-Martínez, F.; Bustillo-Lecompte, C.; Hernández-Ramírez, A.; Colina-Márquez, J. Degradation and Loss of Antibacterial Activity of Commercial Amoxicillin with TiO2/WO3-Assisted Solar Photocatalysis. *Catalysts* **2018**, *8*, 222. [CrossRef]
7. Sulka, G.D.; Brzózka, A.; Liu, L. Fabrication of diameter-modulated and ultrathin porous nanowires in anodic aluminum oxide templates. *Electrochim. Acta* **2011**, *56*, 4972–4979. [CrossRef]
8. Soto Rodriguez, P.E.D.; Olivares, F.; Gómez-Ruiz, S.; Cabrera, G.; Villalonga, R.; Segura del Río, R. Functionalized carbon nanotubes decorated with fluorine-doped titanium dioxide nanoparticles on silicon substrate as template for titanium dioxide film photo-anode grown by chemical vapour deposition. *Thin Solid Film.* **2018**, *656*, 30–36. [CrossRef]
9. Wang, H.; Guo, Z.; Wang, S.; Liu, W. One-dimensional titania nanostructures: Synthesis and applications in dye-sensitized solar cells. *Thin Solid Film.* **2014**, *558*, 1–19. [CrossRef]
10. Byrappa, K.; Adschiri, T. Hydrothermal technology for nanotechnology. *Prog. Cryst. Growth Charact. Mater.* **2007**, *53*, 117–166. [CrossRef]

11. Wang, Y.; He, Y.; Lai, Q.; Fan, M. Review of the progress in preparing nano TiO$_2$: An important environmental engineering material. *J. Environ. Sci.* **2014**, *26*, 2139–2177. [CrossRef] [PubMed]

12. Pang, Y.L.; Lim, S.; Ong, H.C.; Chong, W.T. A critical review on the recent progress of synthesizing techniques and fabrication of TiO$_2$-based nanotubes photocatalysts. *Appl. Catal. A* **2014**, *481*, 127–142. [CrossRef]

13. Kasuga, T.; Hiramatsu, M.; Hoson, A.; Sekino, T.; Niihara, K. Titania Nanotubes Prepared by Chemical Processing. *Adv. Mater.* **1999**, *11*, 1307–1311. [CrossRef]

14. Suzuki, Y.; Yoshikawa, S. Synthesis and Thermal Analyses of TiO$_2$-Derived Nanotubes Prepared by the Hydrothermal Method. *J. Mater. Res.* **2004**, *19*, 982–985. [CrossRef]

15. Gogotsi, Y.; Libera, J.A.; Yoshimura, M. Hydrothermal synthesis of multiwall carbon nanotubes. *J. Mater. Res.* **2000**, *15*, 2591–2594. [CrossRef]

16. Nguyen Phan, T.-D.; Pham, H.-D.; Viet Cuong, T.; Jung Kim, E.; Kim, S.; Woo Shin, E. A simple hydrothermal preparation of TiO$_2$ nanomaterials using concentrated hydrochloric acid. *J. Cryst. Growth* **2009**, *312*, 79–85. [CrossRef]

17. Yoshimura, M.; Byrappa, K. Hydrothermal Processing of Materials: Past, Present and Future. *J. Mater. Sci.* **2007**, *43*, 2085–2103. [CrossRef]

18. Dolat, D.; Quici, N.; Kusiak-Nejman, E.; Morawski, A.W.; Li Puma, G. One-step, hydrothermal synthesis of nitrogen, carbon co-doped titanium dioxide (N,CTiO$_2$) photocatalysts. Effect of alcohol degree and chain length as carbon dopant precursors on photocatalytic activity and catalyst deactivation. *Appl. Catal. B* **2012**, *115–116*, 81–89. [CrossRef]

19. Nguyen-Phan, T.-D.; Pham, V.H.; Chung, J.S.; Chhowalla, M.; Asefa, T.; Kim, W.-J.; Shin, E.W. Photocatalytic performance of Sn-doped TiO$_2$/reduced graphene oxide composite materials. *Appl. Catal. A* **2014**, *473*, 21–30. [CrossRef]

20. Li, Z.; Hou, B.; Xu, Y.; Wu, D.; Sun, Y.; Hu, W.; Deng, F. Comparative study of sol–gel-hydrothermal and sol–gel synthesis of titania–silica composite nanoparticles. *J. Solid State Chem.* **2005**, *178*, 1395–1405. [CrossRef]

21. Andersson, M.; Österlund, L.; Ljungström, S.; Palmqvist, A. Preparation of Nanosize Anatase and Rutile TiO2 by Hydrothermal Treatment of Microemulsions and Their Activity for Photocatalytic Wet Oxidation of Phenol. *J. Phys. Chem. B* **2002**, *106*, 10674–10679. [CrossRef]

22. Liu, N.; Chen, X.; Zhang, J.; Schwank, J.W. A review on TiO$_2$-based nanotubes synthesized via hydrothermal method: Formation mechanism, structure modification, and photocatalytic applications. *Catal. Today* **2014**, *225*, 34–51. [CrossRef]

23. Huang, K.-C.; Chien, S.-H. Improved visible-light-driven photocatalytic activity of rutile/titania-nanotube composites prepared by microwave-assisted hydrothermal process. *Appl. Catal. B* **2013**, *140–141*, 283–288. [CrossRef]

24. Camposeco, R.; Castillo, S.; Mejía-Centeno, I.; Navarrete, J.; Gomez, R. Effect of the Ti/Na molar ratio on the acidity and the structure of TiO$_2$ nanostructures: Nanotubes, nanofibers and nanowires. *Mater. Charact.* **2014**, *90*, 113–120. [CrossRef]

25. Van Grieken, R.; Aguado, J.; López-Muñoz, M.-J.; Marugán, J. Photocatalytic degradation of iron–cyanocomplexes by TiO$_2$ based catalysts. *Appl. Catal. B* **2005**, *55*, 201–211. [CrossRef]

26. Betancourt-Buitrago, L.A.; Hernandez-Ramirez, A.; Colina-Marquez, J.A.; Bustillo-Lecompte, C.F.; Rehmann, L.; Machuca-Martinez, F. Recent Developments in the Photocatalytic Treatment of Cyanide Wastewater: An Approach to Remediation and Recovery of Metals. *Processes* **2019**, *7*, 225. [CrossRef]

27. Betancourt-Buitrago, L.A.; Ossa-Echeverry, O.E.; Rodriguez-Vallejo, J.C.; Barraza, J.M.; Marriaga, N.; Machuca-Martínez, F. Anoxic photocatalytic treatment of synthetic mining wastewater using TiO2 and scavengers for complexed cyanide recovery. *Photochem. Photobiol. Sci.* **2019**, *18*, 853–862. [CrossRef]

28. Caicedo, D.F.; Brum, I.A.S.; Buitrago, L.A.B. Photocatalytic degradation of ferricyanide as synthetic gold mining wastewater using TiO$_2$ assisted by H$_2$O$_2$. *REM-Int. Eng. J.* **2020**, *73*, 99–107. [CrossRef]

29. Kim, S.H.; Lee, S.W.; Lee, G.M.; Lee, B.-T.; Yun, S.-T.; Kim, S.-O. Monitoring of TiO$_2$-catalytic UV-LED photo-oxidation of cyanide contained in mine wastewater and leachate. *Chemosphere* **2016**, *143*, 106–114. [CrossRef]

30. Liu, C.; Zhang, L.; Liu, R.; Gao, Z.; Yang, X.; Tu, Z.; Yang, F.; Ye, Z.; Cui, L.; Xu, C.; et al. Hydrothermal synthesis of N-doped TiO$_2$ nanowires and N-doped graphene heterostructures with enhanced photocatalytic properties. *J. Alloy. Compd.* **2016**, *656*, 24–32. [CrossRef]

31. Procek, M.; Stolarczyk, A.; Pustelny, T.; Maciak, E. A Study of a QCM Sensor Based on TiO_2 Nanostructures for the Detection of NO_2 and Explosives Vapours in Air. *Sensors* **2015**, *15*, 9563–9581. [CrossRef] [PubMed]

32. Chang, G.; Cheng, Z.; Warren, R.; Song, G.; Shen, J.; Lin, L. Highly Efficient Photocatalysts for Surface Hybridization of TiO_2 Nanofibers with Carbon Films. *ChemPlusChem* **2015**, *80*, 827–831. [CrossRef] [PubMed]

33. Mozia, S.; Borowiak-Paleń, E.; Przepiórski, J.; Grzmil, B.; Tsumura, T.; Toyoda, M.; Grzechulska-Damszel, J.; Morawski, A.W. Physico-chemical properties and possible photocatalytic applications of titanate nanotubes synthesized via hydrothermal method. *J. Phys. Chem. Solids* **2010**, *71*, 263–272. [CrossRef]

34. Mozia, S. Application of temperature modified titanate nanotubes for removal of an azo dye from water in a hybrid photocatalysis-MD process. *Catal. Today* **2010**, *156*, 198–207. [CrossRef]

35. Betancourt-Buitrago, L.A.; Vásquez, C.; Veitia, L.; Ossa-Echeverry, O.; Rodriguez-Vallejo, J.; Barraza-Burgos, J.; Marriaga-Cabrales, N.; Machuca-Martínez, F. An approach to utilize the artificial high power LED UV-A radiation in photoreactors for the degradation of methylene blue. *Photochem. Photobiol. Sci.* **2017**, *16*, 79–85. [CrossRef]

36. Blanco, J.; Malato, S.; Peral, J.; Sánchez, B.; Cardona, I. Diseño de reactores para fotocatálisis: Evaluación comparativa de las distintas opciones. In *Eliminación de Contaminantes por Fotocatálisis Heterogénea*; Blesa, M., Ed.; CYTED: Buenos Aires, Argentina, 2001; p. 253.

37. Rice, E.W.; Bridgewater, L.; American Public Health; American Water Works; Water Environment. *Standard Methods for the Examination of Water and Wastewater*; American Public Health Association: Washington, DC, USA, 2012.

38. Osathaphan, K.; Ruengruehan, K.; Yngard, R.; Sharma, V. Photocatalytic degradation of Ni(II)-Cyano and Co(III)-cyano complexes. *Water Air Soil Pollut.* **2013**, *224*, 1647. [CrossRef]

39. Samarghandi, M.; Yang, J.-K.; Giahi, O.; Shirzad-Siboni, M. Photocatalytic reduction of hexavalent chromium with illuminated amorphous FeOOH. *Environ. Technol.* **2014**, *36*, 1–30. [CrossRef]

40. Daneshvar, N.; Aleboyeh, A.; Khataee, A.R. The evaluation of electrical energy per order (EEo) for photooxidative decolorization of four textile dye solutions by the kinetic model. *Chemosphere* **2005**, *59*, 761–767. [CrossRef]

41. Shirzad-Siboni, M.; Farrokhi, M.; Darvishi Cheshmeh Soltani, R.; Khataee, A.; Tajassosi, S. Photocatalytic Reduction of Hexavalent Chromium over ZnO Nanorods Immobilized on Kaolin. *Ind. Eng. Chem. Res.* **2014**, *53*, 1079–1087. [CrossRef]

42. Barakat, M.; Chen, Y.T.; Huang, C.P. Removal of toxic cyanide and Cu(II) Ions from water by illuminated TiO. *Appl. Catal. B* **2004**, *53*, 13–20. [CrossRef]

43. Colina-Márquez, J.; Machuca, F.; Li Puma, G. Radiation Absorption and Optimization of Solar Photocatalytic Reactors for Environmental Applications. *Environ. Sci. Technol.* **2010**, *44*, 5112–5120. [CrossRef] [PubMed]

44. Mueses, M.A.; Machuca-Martinez, F.; Li Puma, G. Effective quantum yield and reaction rate model for evaluation of photocatalytic degradation of water contaminants in heterogeneous pilot-scale solar photoreactors. *Chem. Eng. J.* **2013**, *215–216*, 937–947. [CrossRef]

45. Osathaphan, K.; Chucherdwatanasak, B.; Rachdawong, P.; Sharma, V.K. Photocatalytic oxidation of cyanide in aqueous titanium dioxide suspensions: Effect of ethylenediaminetetraacetate. *Sol. Energy* **2008**, *82*, 1031–1036. [CrossRef]

46. Yang, J.-K.; Lee, S.-M.; Farrokhi, M.; Giahi, O.; Shirzad-Siboni, M. Photocatalytic removal of Cr(VI) with illuminated TiO2. *Desalin. Water Treat.* **2012**, *46*, 1–6. [CrossRef]

47. Ku, Y.; Jung, I.-L. Photocatalytic reduction of Cr(VI) in aqueous solutions by UV irradiation with the presence of titanium dioxide. *Water Res.* **2001**, *35*, 135–142. [CrossRef]

48. Thennarasu, S.; Rajasekar, K.; Balkis Ameen, K. Hydrothermal temperature as a morphological control factor: Preparation, characterization and photocatalytic activity of titanate nanotubes and nanoribbons. *J. Mol. Struct.* **2013**, *1049*, 446–457. [CrossRef]

49. White, L.; Koo, Y.; Yun, Y.; Sankar, J. TiO_2 Deposition on AZ31 Magnesium Alloy Using Plasma Electrolytic Oxidation. *J. Nanomater.* **2013**, *2013*, 319437. [CrossRef]

50. Turki, A.; Kochkar, H.; Guillard, C.; Berhault, G.; Ghorbel, A. Effect of Na content and thermal treatment of titanate nanotubes on the photocatalytic degradation of formic acid. *Appl. Catal. B* **2013**, *138–139*, 401–415. [CrossRef]

51. Sikhwivhilu, L.; Sinha Ray, S.; Coville, N. Influence of bases on hydrothermal synthesis of titanate nanostructures. *Appl. Phys. A* **2009**, *94*, 963–973. [CrossRef]

52.   Fen, L.B.; Han, T.K.; Nee, N.M.; Ang, B.C.; Johan, M.R. Physico-chemical properties of titania nanotubes synthesized via hydrothermal and annealing treatment. *Appl. Surf. Sci.* **2011**, *258*, 431–435. [CrossRef]

53.   Camposeco, R.; Castillo, S.; Navarrete, J.; Gomez, R. Synthesis, characterization and photocatalytic activity of TiO2 nanostructures: Nanotubes, nanofibers, nanowires and nanoparticles. *Catal. Today* **2016**, *266*, 90–101. [CrossRef]

54.   López, R.; Gomez, R. Band-Gap Energy Estimation from Diffuse Reflectance Measurements on Sol–Gel and Commercial TiO2: A Comparative Study. *J. Sol-Gel Sci. Technol.* **2011**, *61*, 1–7. [CrossRef]

14

Effect of Zr Impregnation on Clay-Based Materials for H$_2$O$_2$-Assisted Photocatalytic Wet Oxidation of Winery Wastewater

Vanessa Guimarães *[ID], Ana R. Teixeira, Marco S. Lucas[ID] and José A. Peres[ID]

Vila Real Chemistry Center (CQVR), University of Trás-os-Montes and Alto Douro (UTAD), Quinta de Prados, 5000-801 Vila Real, Portugal; ritamourateixeira@gmail.com (A.R.T.); mlucas@utad.pt (M.S.L.); jperes@utad.pt (J.A.P.)
* Correspondence: guimavs@gmail.com

Abstract: UV-activated Zr-doped composites were successfully produced through the impregnation of Zr on the crystal lattice of different clay materials by a one-step route. Fixing the amount of Zr available for dopage (4%), the influence of different supports, submitted to different chemical treatments, on the photocatalytic activity of the resulting Zr-doped pillared clay materials (PILC) was assessed. Both chemical characterization and structural characterization suggest that the immobilization of Zr on montmorillonite and PILC structures occurred through isomorphic substitution between Si and Zr in the tetrahedral sheet of the clay material. This structural change was demonstrated by significant modifications on Si-OH stretching vibrations (1016 cm^{-1}, 1100 cm^{-1} and 1150 cm^{-1}), and resulted in improved textural properties, with an increase in surface area from 8 m^2/g (natural montmorillonite) to 107 m^2/g after the pillaring process, and to 118 m^2/g after the pillaring and Zr-doping processes ((Zr)Al-Cu-PILC). These materials were tested in the UV-photodegradation of agro-industrial wastewater (AIW), characterized by high concentrations of recalcitrant contaminants. After Zr-dopage on AlCu-PILC heterogeneous catalyst, the total organic carbon (*TOC*) removals of 8.9% and 10.4% were obtained through adsorption and 77% and 86% by photocatalytic oxidation, at pH 4 and 7, respectively. These results suggest a synergetic effect deriving from the combination of Zr and Cu on the photocatalytic degradation process.

Keywords: Zr-doped materials; pillared clays; advanced oxidation processes; photocatalysis; agro-industrial wastewater

1. Introduction

Agro-industrial activities are one of the main sources of wastewater pollution and its impact on the environment has received special attention in recent years [1,2]. Winery wastewater (WW) is characterized by high load of recalcitrant organic compounds [1,3], and its unregulated discharge represents a great threat to aquatic ecosystems and human health [4]. In this regard, the development of effective and low cost methods for the treatment of WW is now imperative.

Currently, different techniques have been developed to treat this type of effluent, including adsorption [5], coagulation [6] and biological processes [7]. However, some of the drawbacks include the limited adsorption capacity and the formation of a potential second pollution source, since these processes only transfer contaminants from one phase to another instead of destroying them [8]. Biological degradation is the most common process applied, however, the microbial activity can be inhibited by the recalcitrant character and toxicity of the organic contaminants [9]. To overcome these problems, advanced oxidation processes (AOPs) have been proposed as effective, fast and non-expensive technologies for the degradation of recalcitrant contaminants [10–12]. Different homogeneous AOPs have already been applied in the treatment of agro-industrial wastewaters, particularly ozonation [13], Fenton [14] and photo-Fenton (solar and UV-A LEDs) processes [15]. Nonetheless, despite the interesting results obtained, Fenton processes have important limitations, namely the acidic conditions needed to improve the degradation efficiency, the additional procedure to remove the homogeneous catalyst from treated effluent, and the neutralization of the treated effluent to meet the legal discharge limits (pH 6.0–9.0) [16]. In order to overcome these drawbacks, heterogeneous AOPs have been the main focus of research interest in the last years, due to the substantial reduction in the effective costs associated with the sludge treatment, as well as the easy catalyst recovery and potential reuse [17,18].

The Catalytic Wet Peroxide Oxidation (CWPO) process is one of the most efficient, economical and environmental-friendly advanced oxidation processes for the treatment of non-biodegradable pollutants under milder conditions, and was successfully applied in the treatment of several organic contaminants, using different types of supports [19–22]. Considering the recalcitrant character of some type of effluents and the fairly poor results that have been obtained so far with the conventional Fenton process, the combination of UV light irradiation in the oxidation processes was proposed with a significant improvement in degradation efficiency [22–26].

The application of CWPO process in the treatment of WW is quite limited. Among our previous research studies, where different clay-based supported catalysts were applied for the first time in the heterogeneous UV/H_2O_2-assisted treatment of a real winery wastewater [27,28], only few heterogeneous catalysts including Fe-graphite [29] and natural clay [30] were applied to improve the efficiency of CWPO in the treatment of a winery wastewater. The results obtained by other studies are very interesting, with significant *TOC* removals, 80% and 55%, respectively. However, the authors did not explore the influence of crucial operational conditions, namely the variation of pH conditions, which may affect the efficiency of the photo-catalytic process.

According to our previous studies, the application of AlCu pillared clay (PILC) as heterogeneous catalyst revealed great stability along the treatment process and a high performance at neutral pH conditions, reaching a *TOC* removal of 83% ($[H_2O_2]_0$ = 98 mM; catalyst dosage = 3.00 g/L). This is particularly important, once it allows the catalyst reuse and eliminates the cost of effluent neutralization before its discharge. Thus, in this work, a novel Zr-doped AlCu-pillared clay ((Zr)AlCu-PILC) was prepared, attending to the ZrO_2 excellent electrical, mechanical, chemical and photocatalytic properties. Accordingly, AlCu-PILC was chosen for this purpose owing to their low cost, environmental stability, high surface area and adsorption capacity, as well as great photo-catalytic activity, which combined with Zr may be significantly enhanced [20,21]. Zr-nanocomposites have been prepared by quite a few methods, including the sol–gel process [31,32], combustion [33], the hydrothermal method [34], microwave irradiation [35], etc. However, it continues to be a challenge to find a simple, efficient and low cost methodology to prepare these nanocomposites.

This work intends to develop a one-step route to incorporate Zr onto clay lattice, promoting great stability and improved photocatalytic activity. The resulting photocatalyst will be tested in the photodegradation of a real WW under UV-C irradiation, and the influence of Zr immobilization on the properties and photoactivity of the heterogeneous catalysts will be discussed.

Different models have been employed to describe the kinetics of catalytic processes involving a heterogeneous liquid–solid system [36–38]. Reaction control models, such as pseudo-first-order and pseudo-second-order models, were considered unsuitable to describe the kinetics of heterogeneous photocatalytic processes, because the two separated linear regression analyses obtained did not take into account relevant factors, namely, the transient period between each linear region, the non-linear behavior during the induction period, and the objective determination of each region, which is subjective when applying two separated linear regressions. The Fermi's model provides a single fit to experimental results showing a transition between the induction period (slow degradation) and the subsequent rapid degradation step of an organic compound (inverted S-shaped transient curve) [39–41]. Considering that the degradation process does not have to follow any particular kinetics or reaction order, it is worth noting that Fermis based model was specifically developed to describe the kinetics of complex systems, involving mixtures of unknown pollutants and several reaction intermediates formed during the photocatalytic process. Therefore, it includes lumped analytical parameters, such as TOC, that can be derived in groups of compounds with different reactivity [42].

In a previous work [42], a lumped kinetic model based on Fermi's equation was developed to describe the TOC histories for the degradation of a dye by catalytic wet peroxide oxidation, as shown in Equation (1)

$$\frac{TOC}{TOC_0} = \frac{1 - x_{TOC}}{1 + \exp\left[k_{TOC}(t - t^*_{TOC})\right]} + x_{TOC} \tag{1}$$

where k_{TOC} corresponds to the apparent reaction rate constant; t^*_{TOC} represents the transition time related to the TOC content curve's inflection point, and x_{TOC} corresponds to the fraction of non-oxidazable compounds that are formed during the reaction.

The Lumped kinetic model based on Fermi's equation has successfully described the kinetics of our previous experiments using pillared clays in the H_2O_2-assisted photocatalytic wet oxidation of WW and, therefore, it is intended to apply this method in order to describe the kinetics of the WW degradation process using the new proposed materials as heterogeneous catalysts.

2. Materials and Methods

2.1. Reagents and Winery Wastewater Sampling

$ZrOCl_2.8H_2O$ (99%) was supplied by Alfa-Aesar, $CuCl_2.2H_2O$ (99%) by Panreac, H_2O_2 (30% w/v) by Sigma-Aldrich. NaOH and H_2SO_4 (95%) were both obtained from Analar NORMAPUR. Deionized water was used to prepare the respective solutions. The agro-industrial wastewater (AIW) was collected from a Portuguese winery cellar located in the Douro region (Northeast of Portugal). The main chemical parameters measured are shown in Table 1. Prior to the oxidation process, the wastewater was submitted to a primary treatment, where the suspended solids were removed from the effluent.

Table 1. Agro-industrial wastewater characterization.

Parameter	Value
pH	3.8 ± 0.1
Chemical Oxygen Demand (mg O_2/L)	1420 ± 45
Biochemical Oxygen Demand (mg O_2/L)	610 ± 15
Total Organic Carbon (mg C/L)	500 ± 12
Total Polyphenols (mg gallic acid/L)	105 ± 3
Phosphates (mg P_2O_5/L)	2.7 ± 0.2
Sulphates (mg SO_4^{2-}/L)	17.8 ± 1.0
Total Iron (mg Fe/L)	0.45 ± 0.02
Aluminium (µg Al/L)	17.5 ± 0.9
Cadmium (µg Cd/L)	2.1 ± 0.1
Copper (µg Cu/L)	400 ± 18
Chromium (µg Cr/L)	0.05 ± 0.003
Manganese (µg Mn/L)	29 ± 1.4
Zinc (µg Zn/L)	4200 ± 200

2.2. Clay Mineral

Natural montmorillonite (MT) was purchased from Fluka, Alfa-Aesar. The chemical composition and main surface properties of natural clay mineral are listed in Tables 2 and 3, respectively. The chemical data was determined by energy dispersive X-ray spectroscopy (EDS/EDAX, FEI QUANTA–400). The total iron expressed as Fe_2O_3 content in raw-montmorillonite was found to be 4.28%. The cation exchange capacity (CEC) of the mineral fractions was measured following the ammonium acetate method proposed by Chapman [43].

Table 2. Main chemical compositions of raw montmorillonite and its derived catalysts, obtained by EDS/EDAX (wt.%).

Sample	SiO$_2$ (%)	Al$_2$O$_3$ (%)	Fe$_2$O$_3$ (%)	MgO (%)	Na$_2$O (%)	CaO (%)	K$_2$O (%)	CuO (%)	ZrO$_2$ (%)	Al/Si	Zr/Si	CEC (meq/g)
MT	68.80	21.97	1.58	3.13	2.54	0.95	0.31	-	-	0.32	-	0.61
Zr-MT	63.35	20.09	1.50	2.90	2.90	0.85	0.31	-	5.54	0.32	11.44	0.22
AlCu-PILC	64.15	26.36	1.36	2.53	0.97	0.26	0.37	1.32	-	0.41	-	0.23
(Zr)Al-PILC	60.66	25.62	1.69	2.15	1.17	0.56	0.6	-	5.14	0.42	11.80	0.21
(Zr)AlCu-PILC	60.46	25.64	1.03	1.97	1.02	0.24	0.23	1.30	5.41	0.42	11.18	0.22

Table 3. Specific surface areas and pore characteristics of MT and their respective catalysts.

Sample	S$_{BET}$ (m^2/g)	V$_{total\ pore}$ (cm^3/g)
MT	8.5	0.047
Zr-MT	65	0.109
Al-Cu-PILC	107	0.202
(Zr)Al-PILC	81	0.146
(Zr)Al-Cu-PILC	118	0.217

2.3. Analytical Techniques

Several physical-chemical parameters were measured in order to characterize the agro-industrial wastewater, namely the chemical oxygen demand (COD), the biological oxygen demand (BOD$_5$), the total organic carbon (TOC) and the total polyphenols (mg gallic acid/L) presented in Table 1. The COD and BOD$_5$ were determined according to Standard Methods (5220D; 5210D; respectively) [44]. COD analysis was carried out in a COD reactor from HACH Co. and a HACH DR 2400 spectrophotometer was used for colorimetric measurement. Biochemical oxygen demand (BOD$_5$) was determined using a respirometric OxiTop system. pH evolution was followed by means of a pH-meter (HANNA Instruments, Rhode Island, USA). The TOC content (mg C/L) was determined using a Shimadzu TOC-L CSH analyzer (Tokyo, Japan). Total polyphenols were evaluated following the Folin–Ciocalteu method [45].

2.4. Catalysts Preparation

The preparation of the pillared clays was carried out following a conventional procedure described in detail by Molina, et al. [46]. AlCu-PILC was prepared through the intercalation between montmorillonite fractions and poly(hydroxy)aluminium (Al$_3$(OH)$_4$$^{5+}$) and copper Cu$_3(OH)_4$$^{2+}$ species. The pillaring solution was prepared by slow addition of a 0.2 M NaOH solution to a mixture of 0.1 M AlCl$_3$ and 0.1 M CuCl$_2$ (Cu/(Al+Cu) = 0.1), under constant stirring until the molar ratio OH/Al = 2.5 was reached. The resulting solution was adjusted to pH 6 and was further aged for 8 h at 298 K. The intercalation process was initiated by the addition of a suspension of 0.1 wt.% montmorillonite in deionized water to the pillaring solution, applying the stoichiometry of 10 mmol Al/g clay. The cationic exchange process was carried out at room temperature for 12 h under constant stirring. The resulting suspension was washed by centrifugation with deionized water in order to reach ionic conductivity values lower than 10 µS. After air-drying, the resulted material was calcinated for 2 h at 400 °C.

The Al-Cu oligomeric solution was adjusted to pH 6 in order to achieve the higher proportion of oligomeric species: 100% of both Al$_3$(OH)$_4$$^{5+}$ and species. The aqueous speciation was calculated by Visual MINTEQ, version 3.0. After the pillaring process, Cu^{2+} oligomeric species were converted to the respective metal oxide clusters by dehydration and dehydroxylation along the calcination process. The Al-PILC was prepared following the same procedure adopted to AlCu-PILC, but only using poly(hydroxy)aluminium (Al$_3$(OH)$_4$$^{5+}$) species.

The Zr-doped catalysts (Zr-MT, (Zr)Al-PILC and (Zr)AlCu-PILC)) were prepared by the incipient wetness impregnation method. The precursor solution was prepared with ZrOCl$_2$.8H$_2$O in order to obtain a zirconium load of 4 wt.%. After impregnation process, the doped catalysts were dried at 100 °C overnight and calcinated for 3 h at 400 °C. The results obtained from the chemical characterization (Table 2) confirm that 4 wt.% of zirconium were successfully immobilized on different heterogeneous catalysts.

2.5. Catalysts Characterization

The FTIR spectra were obtained by mixing 1 mg natural montmorillonite with 200 mg KBr. The powder mixtures were then inserted into molds and pressed at 10 ton/cm^2 to obtain the transparent pellets. The samples were analyzed with a Bruker Tensor 27 spectrometer and the infrared spectra in transmission mode were recorded in the 4000–400 cm^{-1} frequency region. The microstructural characterization was carried out by scanning electron microscopy (SEM/ESEM FEI QUANTA 400) and the chemical composition of the different catalysts was estimated (Table 2) using energy dispersive X-ray spectroscopy (EDS/EDAX).

The textural parameters of samples were obtained from N$_2$ adsorption–desorption isotherms at 77 K using a Micromeritics ASAP 2020 apparatus (Norcross, Georgia, USA). The samples were degassed at 150 °C up to 10^{-4} Torr before analysis. The specific surface area (S_{BET}) was determined by applying the Gurevitsch's rule at a relative pressure p/p$_0$ = 0.30 and according to the Brunauer, Emmet, Teller (BET) method from the linear part of the nitrogen adsorption isotherms. Different pore volumes were determined by the Barrett, Joyner, Halenda model (BJH model).

2.6. Adsorption Tests

Different adsorption tests were carried out in order to predict the amount of organic carbon removed through adsorption. The adsorption batch experiments were carried out at different pH conditions (pH 4.0 and pH 7.0) by adding 3.00 g/L of each heterogeneous catalyst into 500 mL of WW (500 mg C/L). The temperature was kept constant throughout the experiments. After the adsorption runs, the samples were centrifuged and the TOC content of the supernatant solution was measured. The percentage of organic carbon removed through adsorption was calculated according to Equation (2) [47,48]:

$$TOC_{rem.}(\%) = \frac{TOC_0 - TOC_t}{TOC_0} \times 100 \qquad (2)$$

where TOC_0 is the initial TOC content (mg C/L) and TOC_t is TOC value at instant t (mg C/L).

2.7. Photocatalytic Experiments and Kinetic Modelling

The photocatalytic experiments were performed in a batch cylindrical photoreactor (600 cm^3) equipped with a UV-C low pressure mercury vapour lamp (TNN 15/32)—working power = 15 W (795.8 W/m^2) and λ_{max} = 254 nm (Heraeus, Germany). The UV absorption spectrum of the AIW reveals a maximum at ca. 275 nm (with and without the catalysts) and a high absorption at the wavelength where the UV-C lamp emits. In a typical run, 3.0 g/L of catalyst was mixed with 500 mL of the AIW (TOC = 500 mg C/L) for 15 min. After this, a specific amount of H$_2$O$_2$ (98 mM) was added to the suspension and the UV light was turned on at the same time. The initial pH varied from 4.0 to 7.0, and was adjusted by adding 1 M of H$_2$SO$_4$ or 1 M of NaOH. After the reaction has started, 20 mL of solution was withdrawn for TOC measurements at different reaction times, completing a total period of 240 min. The samples were centrifuged and the Zr and Cu concentrations were analyzed by atomic absorption spectroscopy (AAS) using a Thermo Scientific iCE 3000 SERIES. All experiments were performed in triplicate and the observed standard deviation was always less than 5% of the reported values.

A kinetic modelling based on a lumped kinetic model traduced by Fermi's equation was carried out in order to describe the WW degradation process. The experiments were conducted at different pH conditions (pH 4 and pH 7), where temperature, effluent volume, contaminant concentration, H$_2$O$_2$ concentration and catalyst dosage were kept constant. A nonlinear least squares regression, based on the Levenberg–Marquardt (LM) algorithm, was applied using the OriginPro 8.5 "Sigmoidal Fit Tool". As a result, a unique semi-empirical function is applied to simultaneously describe the initial low *TOC* conversion (induction period) and subsequent rapid degradation step. Therefore, both the initial transition period and pseudo-first order kinetic period can be expressed with the proposed model [49].

3. Results and Discussion

3.1. Catalysts Characterization

The X-ray diffractograms corresponding to natural montmorillonite (MT) and to Zr-doped and undoped PILCs, are shown in Figure 1. The hkl reflections associated with MT diffraction pattern are characteristic of a montmorillonite clay mineral with mixed interlayer composition including different proportions of Na^+ and Ca^{2+} ions (13.08 Å). This assumption is in agreement with the chemical characterization data, which shows proportions of 0.95% and 2.54% of CaO and Na_2O (Table 2), respectively. The MT samples modified with previously synthesized oligomeric species ($Cu_3(OH)_4^{2+}$ and/or $Al_3(OH)_4^{5+}$) show a shift of the basal reflection d001 from 13.08 Å (MT) to 18.02 Å and to 17.01 Å for (Zr)AlCu-PILC and (Zr)Al-PILC, respectively, confirming the insertion of the oligomeric species in the interlayer region of montmorillonite and the successful pillaring process. The higher expansion observed when both Cu- and Al-oligomeric species were intercalated on montmorillonite results from the higher pillars formed, indicating that the number, charge, size and shape of the oligomeric species affect the pillar size. These results were also suggested by the textural properties obtained for these materials (Table 3), since the (Zr)AlCu-PILC has higher surface area (ABET=x) and higher number of total pore volume than (Zr)Al-PILC, suggesting an increase in contact area available for absorption due to the higher pillars formed.

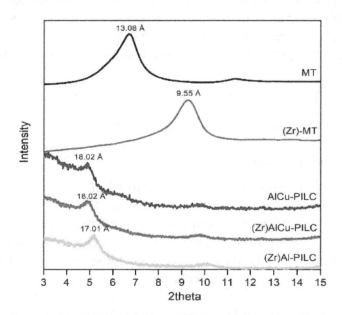

Figure 1. X-ray diffraction results obtained before (MT) and after the pillaring process (AlCu-PILC) and Zr-dopage ((Zr)-Al-PILC, (Zr)AlCu-PILC).

Comparing both (Zr)AlCu-PILC and AlCu-PILC diffraction patterns, it is possible to observe an identical behaviour, confirming that Zr was probably incorporated into the AlCu-PILC lattice without structural modification. The chemical composition of both samples before and after the Zr-doping process is also in agreement with this previous conclusion, given that the increase in Zr amount in doped-clay minerals is accompanied by a decrease in Si proportion, suggesting the isomorphic substitution between Si and Zr in the tetrahedral sheet of the pillared clay. This mechanism is triggered by the similar ionic radii of both cations, where the new one may have identical or lower ionic charge than the replaced one. In this case, both Si and Zr have similar ionic radii and the same ionic charge (+4) and, therefore, no structural charge was developed and no significant structural changes have occurred. This is particularly important, because once the AlCu-PILC has not been structurally affected by the doping process, the adsorption capacity, which is crucial for its catalytic activity, was also not negatively affected. Moreover, enhanced catalyst stability is expected; once Zr is directly incorporated on the crystal lattice of montmorillonite, the risk of metal leaching is significantly lower.

Comparing both (Zr)-MT and MT spectra, no additional conclusions are achieved, since, after the Zr impregnation process, sample (Zr)-MT was submitted to the calcination process, which resulted in the total interlayer collapse to 9.55 Å by dehydration. Therefore, independently of the position of Zr (tetrahedral sheet or interlayer region) on the MT structure, the structural collapse will occur and avoid additional conclusion by means of X-ray diffraction (XRD).

Figure 2 depicts the FTIR spectra obtained before (MT) and after the pillaring process (AlCu-PILC), as well as before (AlCu-PILC) and after the Zr-doping process ((Zr)AlCu-PILC). The results show some structural alterations on montmorillonite after the uptake of metal poly(hydroxy)-complexes and consequent formation of pillars on its internal surface. This is traduced by the decrease in intensity and shift of peaks in the range between 800 and 950 cm^{-1}, after the pillaring process, which are assigned to Al-OH, Fe-OH and Mg-OH vibration modes, at 916 cm^{-1}, 877 cm^{-1} and 849 cm^{-1}, respectively. These structural changes were only observed for PILC samples, which, according to Zhou et al. [50], can be attributed to the interactions between Al or Al/Cu mixed poly(hydroxy) species and the alumina octahedral layers.

Significant modifications on Si-OH stretching vibrations were observed after the Zr-doping process, due to the shift of the main band from 1016 cm^{-1} to 1040 cm^{-1}, and the reduction in intensity of the additional stretching vibrations assigned to the Si-O group, at 1150 cm^{-1} and 1100 cm^{-1}, confirming the incorporation of Zr ions directly in the crystal lattice by isomorphic substitution of Si ions in the tetrahedral sheet of montmorillonite. On the other hand, no additional changes in the vibrations associated with the octahedral sheets of montmorillonite (800–950 cm^{-1}) were observed after this process.

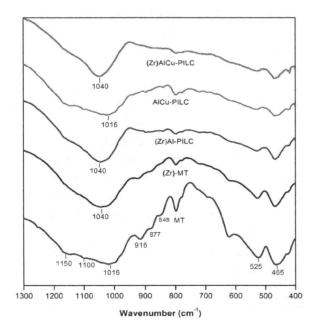

Figure 2. FTIR results obtained before (MT) and after pillaring process (AlCu-PILC) and Zr-dopage ((Zr)-Al-PILC, (Zr)AlCu-PILC), Zr-MT.

The specific surface area and total pore volume of the original montmorillonite and doped and undoped materials are shown in Table 3. These results suggest significant alterations on montmorillonite after the pillaring and Zr-doping processes, resulting in significant and progressive increases in surface area and total pore volume. Accordingly, the surface area increased from 8 m^2/g (MT) to 107 m^2/g after the pillaring process, and to 118 m^2/g after the pillaring and Zr-doping processes, whereas the total pore volume increased from 0.05 cm^3/g (MT) to 0.20 cm^3/g and to 0.22 cm^3/g for AlCu-PILC and (Zr)AlCu-PILC, respectively. The respective isotherms can be classified as type II, where unrestricted monolayer–multilayer adsorption occurs, and the behaviour of the hysteresis loops can be associated with type H3, which usually corresponds to aggregates of plate-like particles forming slit-like pores, which is in agreement with these material structures.

3.2. Adsorption vs. Reaction

Adsorption experiments were carried out using the pillared and Zr-doped catalysts to evaluate the effect of their surface chemistry on the contaminant adsorption and TOC removal. According to previous studies, adsorption plays an important role as the main mechanism involved in the initial induction period, which corresponds to the period necessary for catalyst surface activation [51]. In the present study, part of the mechanism associated with the induction period is probably associated with the adsorption of H_2O_2 and organic compounds onto the catalyst surface, producing surface complexes which promote the activation of the oxidation process through the generation of HO• radicals. Our previous research assessed, for the first time, the application of natural pillared clays (PILCs: Al-Cu-ST and Al-Fe-ST) as heterogeneous photocatalysts for the H_2O_2-assisted treatment of a real AIW [27]. The results indicated that the transition point between the induction period and surface activation and the production of HO• species was directly influenced by the amount of H_2O_2 initially dosed to the process. Accordingly, a decrease in the transition period (t *) TOC from 136 to 96 min was observed, using Al-Cu-ST as the heterogeneous catalyst (3.0 g/L), when H_2O_2 concentration increased from 29 to 98 mM, reducing the period necessary for the surface activation and, therefore, the period required to initiate the degradation process. Considering our previous conclusions, the influence of different catalysts, as well as the effect of Zr-dopage on these supports, were evaluated, taking into account the optimal experimental conditions obtained before, namely $[H_2O_2]_0$ = 98 mM and catalyst dosage = 3.0 g/L.

The evolution of TOC removal through adsorption at different pH conditions and using the different catalysts is shown in Figure 3. As expected, both pH conditions imposed and catalyst textural properties affected the catalyst adsorption capacity. The lowest contaminant adsorption was obtained for Zr-MT and (Zr)Al-PILC, at pH 4 and pH 7, respectively, which correspond to the catalysts with lower surface area (65 m^2/g and 81 m^2/g, respectively). On the contrary, the higher adsorption capacity was obtained for (Zr)AlCu-PILC, at both pH conditions, corresponding to the sample with the highest surface area (118 m^2/g) and total pore volume (0.22 cm^3/g) obtained. This behaviour has particularly impact on the TOC removal efficiency along the oxidation process (Figure 4), as it is observed that the catalyst with enhanced adsorption capacity at both pH conditions has greater activity in the degradation process. Accordingly, the TOC removals obtained using (Zr)AlCu-PILC/UV after 4 h, corresponds to 77% and 86%, at pH 4 and 7, respectively. The incorporation of Zr on montmorillonite lattice (Zr-MT) has contributed to the significant improvement of TOC removals, when compared with raw-montmorillonite, with an increase from 46% to 60% and from 37% to 61% at pH 4 and 7, respectively. The development of AlCu pillared structures had additional advantages considering the improvement of montmorillonite textural properties, resulting in additional stability and catalyst activity. This is traduced by the increase in TOC removals to 69% and 73% at pH 4 and 7, respectively, with a maximum of 33% removal in the non-catalytic UV-C/H_2O_2 experiments performed at both pH conditions. As previously observed, the Zr-dopage on AlCu-PILC has also improved its catalytic activity, promoting an increase in TOC removals to 77% and 86%, at pH 4 and pH 7, respectively, suggesting a synergetic effect of both Zr and Cu on the photocatalytic degradation process. In this case, Zr acts as a semiconductor that is excited by photons with an energy greater than its band gap (5.8–7.1 eV), generating electron–hole pairs, which migrates to the photocatalyst surface yielding radical species that can react with organic molecules upon redox reactions. The electron transfer process is enhanced by the successive redox of Cu^{2+}, a transition metal which is continuously releasing e- species, induced by the presence of a permanent irradiation source (UV-C).

During these processes, a decrease in catalyst stability, caused by the increase in Cu leaching levels from 0.0 to 1.4 mg/L, was also observed from the acidic to the neutral conditions. However, all the Cu leaching concentrations are lower than or very close to the legal discharge limit imposed by EU legislation (1.0 mg Cu/L), and only 5.0% of the Cu immobilized was released at neutral conditions, confirming that Cu immobilization on pillared clay support was successful.

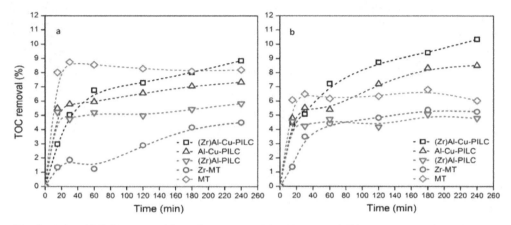

Figure 3. Evolution of *TOC* removal by adsorption: (**a**) pH 4.0 and (**b**) pH 7.0 (catalyst dosage = 3.0 g/L, $[TOC]_0$ = 500 mg C/L).

Comparing both Zr-MT and (Zr)Al-PILC performances at both pH conditions, it is possible to conclude that Zr-MT has shown increased catalytic activity, mainly from 180 min, with *TOC* removals of 60% at both pH conditions for Zr-MT, and 46% at both pH conditions for (Zr)Al-PILC. This behavior could be explained by the competition between Zr and an excess of Al for the Si tetrahedral sites on (Zr)Al-PILC, which may have hampered the Zr incorporation onto the montmorillonite crystal lattice. The lower catalytic activity observed using (Zr)Al-PILC is even more pronounced in acidic conditions, where the *TOC* conversion is very similar to MT along the treatment process. This could be explained by the increased adsorption capacity obtained by MT at pH 4.0 (Figure 3), associated with its higher CEC (Table 2), which despite the lower BET surface area of MT, when compared with (Zr)Al-PILC, has contributed to enhanced adsorption capacity and improved catalytic activity.

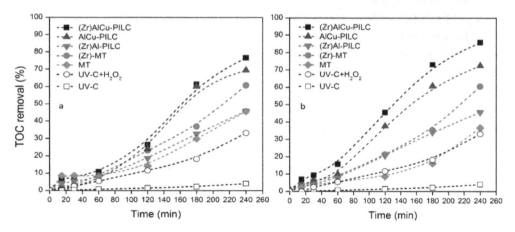

Figure 4. Evolution of *TOC* removal throughout H_2O_2-assisted photocatalytic process: (**a**) pH 4 and (**b**) pH 7 (catalyst dosage = 3.0 g/L, $[H_2O_2]_0$ = 98 mM, $[TOC]_0$ = 500 mg C/L).

3.3. Kinetic Study

In order to better understand the effect of the operational conditions on the induction period, the transition time between the induction period and the fast oxidation reaction ($t * TOC$) was obtained through the fitting of Fermi's equation based on lumped kinetic model to the experimental data. The results obtained for the different parameters are displayed in Table 4, and the fittings obtained for different catalysts are illustrated in Figure 5. The results reveal a good fitting of the kinetic model to the experimental data obtained for different catalysts, with R^2 values higher than 0.983. Considering the different parameters obtained from the modelling, it is assumed that the experimental conditions influenced the kinetic performance of our processes.

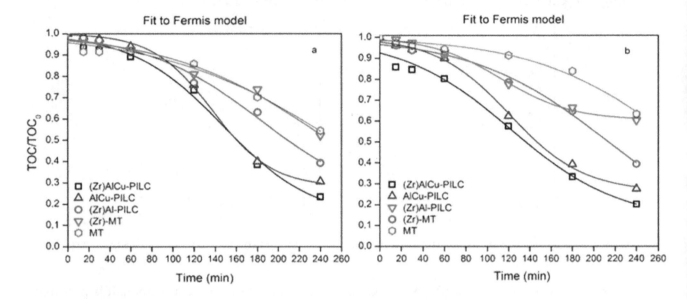

Figure 5. Normalized TOC content (TOC/TOC_0) as a function of time, using different pH conditions: (a) pH 4, (b) pH 7 (catalyst dosage = 3.0 g/L, $[H_2O_2]_0$ = 98 mM, $[TOC]_0$ = 500 mg C/L). Fit of Fermi's equation based on lumped kinetic model to the experimental data.

As observed in Figure 5, the transition point is significantly affected by the pH conditions, as well as the heterogeneous catalyst used. Accordingly, a significant decrease in $t * TOC$ was observed, from 279 to 121 min (pH 7) and from 253 to 137 min (pH 4), using raw-montmorillonite and AlCu-MT, respectively, suggesting a significant reduction in the surface activation period when the heterogeneous catalyst was applied, as well as a quicker production of $HO^\bullet$ species. This tendency was even more pronounced after Zr-dopage at neutral conditions, once the transition time has decreased from 121 to 119 min using (Zr)AlCu-PILC.

In both cases, the evolution of the H_2O_2 concentration along the photocatalytic experiments shows a decrease and subsequent increase in concentration in the first 60 min (Table 5), which suggests a possible adsorption and desorption of H_2O_2 on pillared montmorillonite. Therefore, the results suggest that the first 60 min of reaction were mainly associated with the formation of surface complexes between the H_2O_2 and catalyst surface, whereas the additional period, which completes the total induction period, may be associated with the time required for the surface activation, i.e., the time required for catalytic decomposition of the oxidant in the presence of the active phase (production of $HO^\bullet$).

Concerning the reaction rates of the different photocatalytic processes, higher reaction rates were observed using both (Zr)AlCu-PILC and AlCu-PILC catalysts when compared with the other catalysts applied (Table 4), confirming an improvement of catalytic performance during the oxidation processes. Comparing both catalysts, AlCu-PILC has the higher reaction rates at both pH conditions, 3.54×10^{-2} min^{-1} and 2.87×10^{-2} min^{-1}, at pH 4 and 7, respectively. However, (Zr)AlCu-PILC ($k_{TOC} = 2.58 \times 10^{-2}$ min^{-1}) has contributed to lower fractions of non-oxidazable compounds formed during the reaction ($x_{TOC} = 0.11$, pH 7), when compared with AlCu-PILC ($x_{TOC} = 0.26$, pH 7), which is in agreement with the higher TOC removals obtained.

Table 4. Kinetic parameters obtained by fitting Fermi's model to the experimental data (TOC/TOC_0 as function of time) using different catalysts.

Heterogeneous Catalyst	Variables	Kinetic Parameters			
		k_{TOC} (min^{-1})	t^*_{TOC} (min)	x_{TOC}	r^2
(Zr)AlCu-PILC	pH 4.0	2.35×10^{-2}	148	0.14	0.990
	pH 7.0	2.58×10^{-2}	119	0.11	0.996
AlCu-PILC	pH 4.0	3.54×10^{-2}	137	0.28	0.998
	pH 7.0	2.87×10^{-2}	121	0.26	0.997
(Zr)Al-PILC	pH 4.0	1.90×10^{-2}	158	0.45	0.992
	pH 7.0	2.16×10^{-2}	140	0.50	0.993
Zr-MT	pH 4.0	1.83×10^{-2}	183	0.19	0.983
	pH 7.0	1.56×10^{-2}	214	0.0	0.990
MT	pH 4.0	1.23×10^{-2}	253	0.0	0.960
	pH 7.0	1.50×10^{-2}	279	0.0	0.980

Table 5. H_2O_2 removal in the photocatalytic experiments using different catalysts. General conditions: pH 4.0 and pH 7.0, catalyst dosage = 3.00 g/L, $[TOC]_0$ = 500 mg C/L, UV-C irradiation.

Experiment Time (min)		H_2O_2 Removal (%)		
		Blank (H_2O_2 Only)	(Zr)AlCu-PILC	AlCu-PILC
pH 4.0	0	0.00	0.00	0.00
	15	2.03	14.03	11.18
	30	9.03	10.88	9.73
	60	20.2	54.38	35.63
	120	38.3	47.39	41.89
	180	51.1	79.04	83.58
	240	62.0	96.86	93.13
pH 7.0	0	0.00	0.0	0.00
	15	12.7	16.0	12.06
	30	8.61	12.9	10.05
	60	16.3	8.19	49.30
	120	28.5	45.6	53.29
	180	37.0	69.3	83.07
	240	46.0	88.4	94.69

3.4. Catalyst Regeneration

Considering the best performance of (Zr)AlCu-PILC, the catalyst reuse capacity was evaluated throughout three consecutive cycles of H_2O_2-assisted photocatalytic AIW treatment. The experiments were carried out at pH 7.0, using a catalyst dosage of 3.0 g/L and a H_2O_2 concentration of 98 mM. The results show that the TOC removal obtained using (Zr)AlCu-PILC corresponds to 86%, 66% and 63% after 240 min, for the first, second and third cycles, respectively. In general, a decrease in efficiency was observed from the first to the second cycle (with a loss of 20% in TOC removal). However, no additional loss of activity was observed from the second to the third cycle. All the leaching concentrations along the different cycles were very close to the legal limits imposed (1.0 mg Cu/L) and tended to decrease as the number of cycles increased, from 1.4 to 0.97 mg/L of Cu, from the first to the third cycle, revealing that the catalyst stability is not affected along the cycles.

4. Conclusions

Different catalysts submitted to different chemical treatments and/or the Zr-dopage process, were applied in the H_2O_2-assisted treatment of recalcitrant winery wastewater in order to evaluate the influence of the surface chemical properties of the doped supports on their adsorption and catalytic properties.

FTIR results show that the incorporation of Zr in the crystal lattice of montmorillonite and PILC through isomorphic substitution between Si and Zr was traduced by significant modifications on Si-OH stretching vibrations, due to the shift of the main band from 1016 cm^{-1} to 1040 cm^{-1}, and the decrease in intensity of the additional stretching vibrations assigned to the Si-O group, at, respectively, 1150 cm^{-1} and 1100 cm^{-1}.

In general, the results show that Zr-dopage on AlCu-PILC has improved its adsorption and catalytic activity, promoting an increase in *TOC* removals to 77% and 86%, with 8.85% and 10.35% of *TOC*

removed through adsorption, at pH 4 and pH 7, respectively. It suggests a synergetic effect caused by the combination of Zr and Cu on the photocatalytic degradation process, once the semiconductor electron transfer process is enhanced by the successive redox of Cu(II), induced by the presence of the UV-C irradiation source.

A significant decrease in $t * TOC$ was observed for AlCu-PILC and (Zr)AlCu-PILC, at both pH conditions, suggesting a significant reduction in the surface activation period when the heterogeneous catalyst was applied, as well as a quicker production of HO$^{\bullet}$ species. As a result, higher reaction rates were obtained using both (Zr)AlCu-PILC (2.58×10^{-2} min^{-1}) and AlCu-PILC ($k_{TOC} = 3.54 \times 10^{-2}$ min^{-1}) catalysts, confirming an improvement in catalytic performance along the oxidation processes. Comparing both catalysts, AlCu-PILC has the higher reaction rates at both pH conditions. However, (Zr)AlCu-PILC has contributed to lower fractions of non-oxidazable compounds formed during the reaction ($x_{TOC} = 0.11$, pH 7.0), making it a more efficient process.

Author Contributions: Conceptualization, V.G. and A.R.T.; methodology, V.G. and A.R.T.; validation, V.G. and A.R.T.; investigation, V.G. and A.R.T.; writing—original draft preparation, V.G. and A.R.T.; writing—review and editing, V.G., A.R.T., M.S.L. and J.A.P.; visualization, V.G., M.S.L. and J.A.P.; supervision, M.S.L. and J.A.P.; project administration, J.A.P. All authors have read and agreed to the published version of the manuscript.

Acknowledgments: The authors thank the North Regional Operational Program (NORTE 2020) and the European Regional Development Fund (ERDF), for financial support of the Project INNOVINE&WINE (BPD/UTAD/INNOVINE&WINE/WINEMAKING/754/2016).

References

1. Amor, C.; Marchão, L.; Lucas, M.S.; Peres, J.A. Application of Advanced Oxidation Processes for the Treatment of Recalcitrant Agro-Industrial Wastewater: A Review. *Water* **2019**, *11*, 205. [CrossRef]
2. Ferreira, L.C.; Fernandes, J.R.; Rodríguez-Chueca, J.; Peres, J.A.; Lucas, M.S.; Tavares, P.B. Photocatalytic degradation of an agro-industrial wastewater model compound using a UV LEDs system: Kinetic study. *J. Environ. Manag.* **2020**, *269*, 110740. [CrossRef]
3. Noukeu, N.A.; Gouado, I.; Priso, R.J.; Ndongo, D.; Taffouo, V.D.; Dibong, S.D.; Ekodeck, G.E. Characterization of effluent from food processing industries and stillage treatment trial with *Eichhornia crassipes* (Mart.) and *Panicum maximum* (Jacq.). *Water Resour. Ind.* **2016**, *16*, 1–18. [CrossRef]
4. Sousa, R.M.O.F.; Amaral, C.; Fernandes, J.M.C.; Fraga, I.; Semitela, S.; Braga, F.; Coimbra, A.M.; Dias, A.A.; Bezerra, R.M.; Sampaio, A. Hazardous impact of vinasse from distilled winemaking by-products in terrestrial plants and aquatic organisms. *Ecotoxicol. Environ. Saf.* **2019**, *183*, 109493. [CrossRef]
5. Al Bsoul, A.; Hailat, M.; Abdelhay, A.; Tawalbeh, M.; Jum'h, I.; Bani-Melhem, K. Treatment of olive mill effluent by adsorption on titanium oxide nanoparticles. *Sci. Total Environ.* **2019**, *688*, 1327–1334. [CrossRef]
6. Chen, B.; Jiang, C.; Yu, D.; Wang, Y.; Xu, T. Design of an alternative approach for synergistic removal of multiple contaminants: Water splitting coagulation. *Chem. Eng. J.* **2020**, *380*, 122531. [CrossRef]
7. Candia-Onfray, C.; Espinoza, N.; Sabino da Silva, E.B.; Toledo-Neira, C.; Espinoza, L.C.; Santander, R.;

García, V.; Salazar, R. Treatment of winery wastewater by anodic oxidation using BDD electrode. *Chemosphere* **2018**, *206*, 709–717. [CrossRef]

8. Aziz, H.A.; Abu Amr, S.S. *Advanced Oxidation Processes (AOPs) in Water and Wastewater Treatment*; IGI Global: Hershey, PA, USA, 2019. [CrossRef]

9. Trapido, M.; Tenno, T.; Goi, A.; Dulova, N.; Kattel, E.; Klauson, D.; Klein, K.; Tenno, T.; Viisimaa, M. Bio-recalcitrant pollutants removal from wastewater with combination of the Fenton treatment and biological oxidation. *J. Water Process Eng.* **2017**, *16*, 277–282. [CrossRef]

10. Amor, C.; Rodríguez-Chueca, J.; Fernandes, J.L.; Domínguez, J.R.; Lucas, M.S.; Peres, J.A. Winery wastewater treatment by sulphate radical based-advanced oxidation processes (SR-AOP): Thermally vs UV-assisted persulphate activation. *Process Saf. Environ. Prot.* **2019**, *122*, 94–101. [CrossRef]

11. Lucas, M.S.; Peres, J.A. Removal of Emerging Contaminants by Fenton and UV-Driven Advanced Oxidation Processes. *Water Air Soil Pollut.* **2015**, *226*, 273. [CrossRef]

12. Rodríguez-Chueca, J.; Amor, C.; Mota, J.; Lucas, M.S.; Peres, J.A. Oxidation of winery wastewater by sulphate radicals: Catalytic and solar photocatalytic activations. *Environ. Sci. Pollut. Res.* **2017**, *24*, 22414–22426. [CrossRef]

13. Lucas, M.S.; Peres, J.A.; Li Puma, G. Treatment of winery wastewater by ozone-based advanced oxidation processes (O$_3$, O$_3$/UV and O$_3$/UV/H$_2$O$_2$) in a pilot-scale bubble column reactor and process economics. *Sep. Purif. Technol.* **2010**, *72*, 235–241. [CrossRef]

14. Amor, C.; Lucas, M.S.; García, J.; Dominguez, J.R.; De Heredia, J.B.; Peres, J.A. Combined treatment of olive mill wastewater by Fenton's reagent and anaerobic biological process. *J. Environ. Sci. Health Part A* **2015**, *50*, 161–168. [CrossRef]

15. Rodríguez-Chueca, J.; Amor, C.; Fernandes, J.R.; Tavares, P.B.; Lucas, M.S.; Peres, J.A. Treatment of crystallized-fruit wastewater by UV-A LED photo-Fenton and coagulation–flocculation. *Chemosphere* **2016**, *145*, 351–359. [CrossRef]

16. Brink, A.; Sheridan, C.; Harding, K. Combined biological and advance oxidation processes for paper and pulp effluent treatment. *South Afr. J. Chem. Eng.* **2018**, *25*, 116–122. [CrossRef]

17. M'Arimi, M.M.; Mecha, C.A.; Kiprop, A.K.; Ramkat, R. Recent trends in applications of advanced oxidation processes (AOPs) in bioenergy production: Review. *Renew. Sustain. Energy Rev.* **2020**, *121*, 109669. [CrossRef]

18. Luo, H.; Zeng, Y.; He, D.; Pan, X. Application of iron-based materials in heterogeneous advanced oxidation processes for wastewater treatment: A review. *Chem. Eng. J.* **2020**. [CrossRef]

19. Rueda Márquez, J.J.; Levchuk, I.; Sillanpää, M. Application of Catalytic Wet Peroxide Oxidation for Industrial and Urban Wastewater Treatment: A Review. *Catalysts* **2018**, *8*, 673. [CrossRef]

20. Gil, A.; Galeano, L.A.; Vicente, M.Á. *Applications of Advanced Oxidation Processes (AOPs) in Drinking Water Treatment*; Springer International Publishing: New York, NY, USA, 2018. [CrossRef]

21. Ameta, S.C.; Ameta, R. *Advanced Oxidation Processes for Wastewater Treatment: Emerging Green Chemical Technology*; Elsevier Science: Amsterdam, The Netherlands, 2018; ISBN 9780128105252.

22. Santos Silva, A.; Seitovna Kalmakhanova, M.; Kabykenovna Massalimova, B.G.; Sgorlon, J.; Jose Luis, D.T.; Gomes, H.T. Wet Peroxide Oxidation of Paracetamol Using Acid Activated and Fe/Co-Pillared Clay Catalysts Prepared from Natural Clays. *Catalysts* **2019**, *9*, 705. [CrossRef]

23. Ormad, M.P.; Mosteo, R.; Ibarz, C.; Ovelleiro, J.L. Multivariate approach to the photo-Fenton process applied to the degradation of winery wastewaters. *Appl. Catal. B Environ.* **2006**, *66*, 58–63. [CrossRef]

24. Khare, P.; Patel, R.K.; Sharan, S.; Shankar, R. 8—Recent trends in advanced oxidation process for treatment of recalcitrant industrial effluents. In *Advanced Oxidation Processes for Effluent Treatment Plants*; Shah, M.P., Ed.; Elsevier: Amsterdam, The Netherlands, 2021; pp. 137–160. [CrossRef]

25. Jaén-Gil, A.; Buttiglieri, G.; Benito, A.; Mir-Tutusaus, J.A.; Gonzalez-Olmos, R.; Caminal, G.; Barceló, D.; Sarrà, M.; Rodriguez-Mozaz, S. Combining biological processes with UV/H$_2$O$_2$ for metoprolol and metoprolol acid removal in hospital wastewater. *Chem. Eng. J.* **2021**, *404*, 126482. [CrossRef]

26. Giannakis, S.; Jovic, M.; Gasilova, N.; Pastor Gelabert, M.; Schindelholz, S.; Furbringer, J.-M.; Girault, H.; Pulgarin, C. Iohexol degradation in wastewater and urine by UV-based Advanced Oxidation Processes (AOPs): Process modeling and by-products identification. *J. Environ. Manag.* **2017**, *195*, 174–185. [CrossRef]

27. Guimarães, V.; Teixeira, A.R.; Lucas, M.S.; Silva, A.M.T.; Peres, J.A. Pillared interlayered natural clays as heterogeneous photocatalysts for H$_2$O$_2$-assisted treatment of a winery wastewater. *Sep. Purif. Technol.* **2019**, *228*, 115768. [CrossRef]

28. Guimarães, V.; Lucas, M.S.; Peres, J.A. Combination of adsorption and heterogeneous photo-Fenton processes for the treatment of winery wastewater. *Environ. Sci. Pollut. Res.* **2019**, *26*, 31000–31013. [CrossRef]

29. Domínguez, C.M.; Quintanilla, A.; Casas, J.A.; Rodriguez, J.J. Treatment of real winery wastewater by wet oxidation at mild temperature. *Sep. Purif. Technol.* **2014**, *129*, 121–128. [CrossRef]

30. Mosteo, R.; Ormad, P.; Mozas, E.; Sarasa, J.; Ovelleiro, J.L. Factorial experimental design of winery wastewaters treatment by heterogeneous photo-Fenton process. *Water Res.* **2006**, *40*, 1561–1568. [CrossRef]

31. Tyagi, B.; Sidhpuria, K.; Shaik, B.; Jasra, R.V. Synthesis of Nanocrystalline Zirconia Using Sol–Gel and Precipitation Techniques. *Ind. Eng. Chem. Res.* **2006**, *45*, 8643–8650. [CrossRef]

32. Samadi, S.; Yousefi, M.; Khalilian, F.; Tabatabaee, A. Synthesis, characterization, and application of Nd, Zr–TiO$_2$/SiO$_2$ nanocomposite thin films as visible light active photocatalyst. *J. Nanostruct. Chem.* **2015**, *5*, 7–15. [CrossRef]

33. Sayagués, M.J.; Avilés, M.A.; Córdoba, J.M.; Gotor, F.J. Self-propagating combustion synthesis via an MSR process: An efficient and simple method to prepare (Ti, Zr, Hf)B$_2$–Al$_2$O$_3$ powder nanocomposites. *Powder Technol.* **2014**, *256*, 244–250. [CrossRef]

34. Teymourian, H.; Salimi, A.; Firoozi, S.; Korani, A.; Soltanian, S. One-pot hydrothermal synthesis of zirconium dioxide nanoparticles decorated reduced graphene oxide composite as high performance electrochemical sensing and biosensing platform. *Electrochim. Acta* **2014**, *143*, 196–206. [CrossRef]

35. Rajabi, M.; Khodai, M.M.; Askari, N. Microwave-assisted sintering of Al–ZrO$_2$ nano-composites. *J. Mater. Sci. Mater. Electron.* **2014**, *25*, 4577–4584. [CrossRef]

36. Sannino, D.; Vaiano, V.; Ciambelli, P.; Isupova, L.A. Mathematical modelling of the heterogeneous photo-Fenton oxidation of acetic acid on structured catalysts. *Chem. Eng. J.* **2013**, *224*, 53–58. [CrossRef]

37. He, J.; Yang, X.; Men, B.; Wang, D. Interfacial mechanisms of heterogeneous Fenton reactions catalyzed by iron-based materials: A review. *J. Environ. Sci.* **2016**, *39*, 97–109. [CrossRef]

38. Kalmakhanova, M.S.; Diaz de Tuesta, J.L.; Massalimova, B.K.; Gomes, H.T. Pillared clays from natural resources as catalysts for catalytic wet peroxide oxidation: Characterization and kinetic insights. *Environ. Eng. Res.* **2020**, *25*, 186–196. [CrossRef]

39. Rache, M.L.; García, A.R.; Zea, H.R.; Silva, A.M.T.; Madeira, L.M.; Ramírez, J.H. Azo-dye orange II degradation by the heterogeneous Fenton-like process using a zeolite Y-Fe catalyst—Kinetics with a model based on the Fermi's equation. *Appl. Catal. B Environ.* **2014**, *146*, 192–200. [CrossRef]

40. Ghime, D.; Ghosh, P. Decolorization of diazo dye trypan blue by electrochemical oxidation: Kinetics with a model based on the Fermi's equation. *J. Environ. Chem. Eng.* **2020**, *8*, 102792. [CrossRef]

41. Herney-Ramirez, J.; Silva, A.M.T.; Vicente, M.A.; Costa, C.A.; Madeira, L.M. Degradation of Acid Orange 7 using a saponite-based catalyst in wet hydrogen peroxide oxidation: Kinetic study with the Fermi's equation. *Appl. Catal. B Environ.* **2011**, *101*, 197–205. [CrossRef]

42. Silva, A.M.T.; Herney-Ramirez, J.; Söylemez, U.; Madeira, L.M. A lumped kinetic model based on the Fermi's equation applied to the catalytic wet hydrogen peroxide oxidation of Acid Orange 7. *Appl. Catal. B Environ.* **2012**, *121–122*, 10–19. [CrossRef]

43. Chapman, H.D. Cation exchange capacity. *Methods Soil Anal. Chem. Microbiol. Prop.* **1965**, *9*, 891–901.

44. APHA. *Standard Methods for the Examination of Water and Wastewater*, 21st ed.; American Public Health Association: Washington, DC, USA, 2005.

45. Lowry, O.H.; Rosebrough, N.J.; Farr, A.L.; Randall, R.J. Protein measurement with the Folin phenol reagent. *J. Biol. Chem* **1951**, *193*, 265–275.

46. Molina, C.B.; Casas, J.A.; Zazo, J.A.; Rodríguez, J.J. A comparison of Al-Fe and Zr-Fe pillared clays for catalytic wet peroxide oxidation. *Chem. Eng. J.* **2006**, *118*, 29–35. [CrossRef]

47. Guimarães, V.; Rodríguez-Castellón, E.; Algarra, M.; Rocha, F.; Bobos, I. Influence of pH, layer charge location and crystal thickness distribution on U(VI) sorption onto heterogeneous dioctahedral smectite. *J. Hazard. Mater.* **2016**, *317*, 246–258. [CrossRef]

48. Wang, Y.; Wang, W.; Wang, A. Efficient adsorption of methylene blue on an alginate-based nanocomposite hydrogel enhanced by organo-illite/smectite clay. *Chem. Eng. J.* **2013**, *228*, 132–139. [CrossRef]

49. Hou, B.; Han, H.; Jia, S.; Zhuang, H.; Xu, P.; Wang, D. Heterogeneous electro-Fenton oxidation of catechol catalyzed by nano-Fe$_3$O$_4$: Kinetics with the Fermi's equation. *J. Taiwan Inst. Chem. Eng.* **2015**, *56*, 138–147. [CrossRef]

50. Zhou, J.; Wu, P.; Dang, Z.; Zhu, N.; Li, P.; Wu, J.; Wang, X. Polymeric Fe/Zr pillared montmorillonite for the removal of Cr(VI) from aqueous solutions. *Chem. Eng. J.* **2010**, *162*, 1035–1044. [CrossRef]
51. Carriazo, J.G.; Guelou, E.; Barrault, J.; Tatibouët, J.M.; Moreno, S. Catalytic wet peroxide oxidation of phenol over Al–Cu or Al–Fe modified clays. *Appl. Clay Sci.* **2003**, *22*, 303–308. [CrossRef]

Permissions

All chapters in this book were first published in MDPI; hereby published with permission under the Creative Commons Attribution License or equivalent. Every chapter published in this book has been scrutinized by our experts. Their significance has been extensively debated. The topics covered herein carry significant findings which will fuel the growth of the discipline. They may even be implemented as practical applications or may be referred to as a beginning point for another development.

The contributors of this book come from diverse backgrounds, making this book a truly international effort. This book will bring forth new frontiers with its revolutionizing research information and detailed analysis of the nascent developments around the world.

We would like to thank all the contributing authors for lending their expertise to make the book truly unique. They have played a crucial role in the development of this book. Without their invaluable contributions this book wouldn't have been possible. They have made vital efforts to compile up to date information on the varied aspects of this subject to make this book a valuable addition to the collection of many professionals and students.

This book was conceptualized with the vision of imparting up-to-date information and advanced data in this field. To ensure the same, a matchless editorial board was set up. Every individual on the board went through rigorous rounds of assessment to prove their worth. After which they invested a large part of their time researching and compiling the most relevant data for our readers.

The editorial board has been involved in producing this book since its inception. They have spent rigorous hours researching and exploring the diverse topics which have resulted in the successful publishing of this book. They have passed on their knowledge of decades through this book. To expedite this challenging task, the publisher supported the team at every step. A small team of assistant editors was also appointed to further simplify the editing procedure and attain best results for the readers.

Apart from the editorial board, the designing team has also invested a significant amount of their time in understanding the subject and creating the most relevant covers. They scrutinized every image to scout for the most suitable representation of the subject and create an appropriate cover for the book.

The publishing team has been an ardent support to the editorial, designing and production team. Their endless efforts to recruit the best for this project, has resulted in the accomplishment of this book. They are a veteran in the field of academics and their pool of knowledge is as vast as their experience in printing. Their expertise and guidance has proved useful at every step. Their uncompromising quality standards have made this book an exceptional effort. Their encouragement from time to time has been an inspiration for everyone.

The publisher and the editorial board hope that this book will prove to be a valuable piece of knowledge for researchers, students, practitioners and scholars across the globe.

List of Contributors

Efraím A. Serna-Galvis and Ricardo A. Torres-Palma
Grupo de Investigación en Remediación Ambiental y Biocatálisis (GIRAB), Instituto de Química, Facultad de Ciencias Exactas y Naturales, Universidad de Antioquia UdeA, Calle 70 No. 52-21, Calle Nueva, 050014 Medellín, Colombia

John F. Guateque-Londoño
Grupo de Investigación en Remediación Ambiental y Biocatálisis (GIRAB), Instituto de Química, Facultad de Ciencias Exactas y Naturales, Universidad de Antioquia UdeA, Calle 70 No. 52-21, Calle Nueva, 050014 Medellín, Colombia
Maestría en Ciencias Químicas, Facultad de Tecnología, Universidad Tecnológica de Pereira, 660001 Pereira, Colombia

Yenny Ávila-Torres
Grupo de Investigación QUIBIO, Facultad de Ciencias Básicas, Universidad Santiago de Cali, Santiago de Cali, 760035 Pampalinda, Colombia

Philipp Otter, Alexander Goldmaier and Florian Benz
AUTARCON GmbH, D-34117 Kassel, Germany

Katharina Mette, Robert Wesch, Tobias Gerhardt and Frank-Marc Krüger
GNF e.V. Volmerstr. 7 B, 12489 Berlin, Germany

Pradyut Malakar
International Centre for Ecological Engineering, University of Kalyani, Kalyani, West Bengal 741235, India

Thomas Grischek
Division of Water Sciences, University of Applied Sciences Dresden, Friedrich-List-Platz 1, 01069 Dresden, Germany

Raúl Acosta-Herazo and María H. Pinzón-Cárdenas
School of Chemical Engineering, Universidad del Valle, Cali A.A. 25360, Colombia

Fiderman Machuca-Martínez
School of Chemical Engineering, Universidad del Valle, Cali A.A. 25360, Colombia
Escuela de Ingeniería Química, Universidad del Valle, Cali 760032, Colombia

Briyith Cañaveral-Velásquez, Katrin Pérez-Giraldo and Miguel A. Mueses
Modeling and Applications of Advanced Oxidation Technologies Research Group, Photocatalysis and Solar Photoreactors Engineering, Department of Chemical Engineering, Universidad de Cartagena, Cartagena A.A. 1382-195, Colombia

Khanh Chau Dao
Department of Health and Applied Sciences, Dong Nai Technology University, Bien Hoa, Dong Nai 810000, Vietnam
Department of Civil Engineering, National Chi Nan University, Nantou Hsien 54561, Taiwan

Chih-Chi Yang, Ku-Fan Chen and Yung-Pin Tsai
Department of Civil Engineering, National Chi Nan University, Nantou Hsien 54561, Taiwan

Fathallah Karimzadeh and Keyvan Raeissi
Department of Materials Engineering, Isfahan University of Technology, Isfahan 84156-83111, Iran

Sami Rtimi, John Kiwi and Cesar Pulgarin
School of Basic Sciences (SB), Institute of Chemical Science and Engineering (ISIC), Group of Advanced Oxidation Processes (GPAO), École Polytechnique Fédérale de Lausanne (EPFL), Station 6, CH-1015 Lausanne, Switzerland

Minoo Karbasi
Department of Materials Engineering, Isfahan University of Technology, Isfahan 84156-83111, Iran
School of Basic Sciences (SB), Institute of Chemical Science and Engineering (ISIC), Group of Advanced Oxidation Processes (GPAO), École Polytechnique Fédérale de Lausanne (EPFL), Station 6, CH-1015 Lausanne, Switzerland

Stefanos Giannakis
Departamento de Ingeniería Civil: Hidráulica, Universidad Politécnica de Madrid (UPM), E.T.S. Ingenieros de Caminos, Canales y Puertos, Energía y Medio Ambiente, Unidad docente Ingeniería Sanitaria, c/ Profesor Aranguren, s/n, ES-28040 Madrid, Spain

Yu-Jung Liu and Shang-Lien Lo
Graduate Institute of Environmental Engineering, National Taiwan University, Taipei 106, Taiwan

Yung-Ling Huang and Ching-Yao Hu
School of Public Health, Taipei Medical University, Taipei 110, Taiwan

Samuel Moles, Rosa Mosteo and María P. Ormad
Water and Environmental Health Research Group, c/ María de Luna 3, 50018 Zaragoza, Spain

Jairo Gómez
Navarra de Infraestructuras Locales SA, av. Barañain 22, 31008 Pamplona, Spain

Joanna Szpunar and Sebastiano Gozzo
Institute of Analytical Sciences and Physico-Chemistry for Environment and Materials (IPREM), Centre National de la Recherche Scientifique (CNRS), Universite de Pau et des Pays de l'Adour, CEDEX 9 Pau, France

Juan R. Castillo
Analytical Spectroscopy and Sensors Group Analytic Chemistry Department, Science Faculty, Environmental Science Institute, University of Zaragoza, 50009 Zaragoza, Spain

Chang-Sheng Guo, Heng Zhang and Jian Xu
Center for Environmental Health Risk Assessment and Research, Chinese Research Academy of Environmental Sciences, Beijing 100012, China

Qi-Yan Feng
School of Environment Science and Spatial Informatics, China University of Mining and Technology, Xuzhou 221116, China

De-Ming Gu
Center for Environmental Health Risk Assessment and Research, Chinese Research Academy of Environmental Sciences, Beijing 100012, China
School of Environment Science and Spatial Informatics, China University of Mining and Technology, Xuzhou 221116, China

Ana L. Camargo-Perea
Grupo de Investigación en Remediación Ambiental y Biocatálisis, Instituto de Química, Universidad de Antioquia UdeA, Calle 70, No. 52-21 Medellín, Colombia

Gustavo A. Peñuela
Grupo GDCON, Facultad de Ingeniería, Sede de Investigaciones Universitarias (SIU), Universidad de Antioquia UdeA, Calle 70, No. 52-21 Medellín, Colombia

Ainhoa Rubio-Clemente
Grupo GDCON, Facultad de Ingeniería, Sede de Investigaciones Universitarias (SIU), Universidad de Antioquia UdeA, Calle 70, No. 52-21 Medellín, Colombia

Facultad de Ingeniería, Tecnológico de Antioquia–Institución Universitaria TdeA, Calle 78b, No. 72A-220 Medellín, Colombia

Jian Wang
State Key Joint Laboratory of Environment Simulation and Pollution Control, School of Environment, Tsinghua University, Beijing 100084, China
State Key Laboratory of Environmental Criteria and Risk Assessment, Chinese Research Academy of Environmental Sciences, Beijing 100012, China

Yonghui Song
State Key Laboratory of Environmental Criteria and Risk Assessment, Chinese Research Academy of Environmental Sciences, Beijing 100012, China

Feng Qian, Cong Du, Huibin Yu and Liancheng Xiang
State Key Laboratory of Environmental Criteria and Risk Assessment, Chinese Research Academy of Environmental Sciences, Beijing 100012, China
Department of Urban Water Environmental Research, Chinese Research Academy of Environmental Sciences, Beijing 100012, China

Cátia A. L. Graça, Ana R. Ribeiro and Adrián M. T. Silva
Laboratory of Separation and Reaction Engineering-Laboratory of Catalysis and Materials (LSRE-LCM), Faculdade de Engenharia, Universidade do Porto, Rua Dr. Roberto Frias, 4200-465 Porto, Portugal

Sara Ribeirinho-Soares and Olga C. Nunes
LEPABE—Laboratory for Process Engineering, Environment, Biotechnology and Energy, Faculdade de Engenharia, Universidade do Porto, Rua Dr. Roberto Frias, 4200-465 Porto, Portugal

Joana Abreu-Silva and Célia M. Manaia
Universidade Católica Portuguesa, CBQF—Centro de Biotecnologia e Química Fina—Laboratório Associado, Escola Superior de Biotecnologia, Rua Diogo Botelho 1327, 4169-005 Porto, Portugal

Inês I. Ramos and Marcela A. Segundo
LAQV, REQUIMTE, Departamento de Ciências Químicas, Faculdade de Farmácia, Universidade do Porto, Rua de Jorge Viterbo Ferreira 228, 4050-313 Porto, Portugal

Sérgio M. Castro-Silva
Adventech-Advanced Environmental Technologies, Centro Empresarial e Tecnológico, Rua de Fundões 151, 3700-121 São João da Madeira, Portugal

Jin Ni, Huimin Shi and Yuansheng Xu
Department of Environmental Engineering, School of Energy and Environmental Engineering, University of Science and Technology Beijing, 30 Xueyuan Road, Haidian District, Beijing 10083, China

Qunhui Wang
Beijing Key Laboratory on Resource-oriented Treatment of Industrial Pollutants, University of Science and Technology Beijing, 30 Xueyuan Road, Beijing 10083, China

Kevin Mauricio Aldana-Villegas and Luis Andres Betancourt-Buitrago
Escuela de Ingeniería Química, Universidad del Valle, Cali 760032, Colombia

Augusto Arce-Sarria
Escuela de Ingeniería Química, Universidad del Valle, Cali 760032, Colombia
Tecnoparque Nodo Cali, GIDEMP Materials and Products Research Group, ASTIN Center, SENA Regional Valle, Cali 760003, Colombia

Jose Ángel Colina-Márquez and Miguel Angel Mueses
Modeling & Application of Advanced Oxidation Processes, Photocatalysis & Solar Photoreactors Engineering, Chemical Engineering Program, Universidad de Cartagena, Cartagena 130001, Colombia

Vanessa Guimarães, Ana R. Teixeira, Marco S. Lucas and José A. Peres
Vila Real Chemistry Center (CQVR), University of Trás-os-Montes and Alto Douro (UTAD), Quinta de Prados, 5000-801 Vila Real, Portugal

Index

Printed in the USA
CPSIA information can be obtained
at www.ICGtesting.com
JSHW062237071123
51533JS00031B/99